Biofouling of Membrane Systems

Biofouling of Membrane Systems

Szilárd Bucs, Joop Kruithof, Mark van Loosdrecht and Johannes Vrouwenvelder

IWA PUBLISHING

Published by **IWA Publishing**
Alliance House
12 Caxton Street
London SW1H 0QS, UK
Telephone: +44 (0)20 7654 5500
Fax: +44 (0)20 7654 5555
Email: publications@iwap.co.uk
Web: www.iwapublishing.com

First published 2018

British Library Cataloguing in Publication Data
A CIP catalogue record for this book is available from the British Library

ISBN: 9781780409580 (Paperback)
ISBN: 9781780409597 (eBook)
ISBN: 9781780409603 (ePUB)

Contents

Preface

Membrane filtration processes such as nanofiltration (NF) and reverse osmosis (RO) can produce high quality water from all types of water sources including seawater and waste water effluent. The global demand for fresh water is growing, therefore the application of these membrane technologies is increasing strongly. A serious operational problem of NF and RO is biofouling – excessive growth of biomass – negatively affecting the membrane systems performance. The IWA book *Biofouling of Spiral Wound Membrane Systems* addresses research on most relevant aspects of biofouling published until 2010 bridging the gap between microbiology, hydraulics and membrane technology.

Compared to the previous 35 years, since 2010, the study of membrane biofouling has increased strongly as reflected by an increase with a factor of more than 3.5 in the number of scientific papers. No single periodic provides an update on recent developments in biofouling research.

This book, *Biofouling of Membrane Systems*, gives a comprehensive overview of key areas of advancements for the period 2010 - 2018 for research on biofouling subdivided in: (i) modelling and numerical analysis, (ii) non-destructive characterization and (iii) feed spacer geometry optimization.

In the long run, biofouling cannot be avoided completely, therefore biofouling control strategies should focus on delaying biofilm formation, reducing its impact on membrane performance and enhancing biofilm removal by advanced cleaning strategies. Proposed research directions are: (i) biofilm structural characterization; (ii) understanding how biofilm properties affect membrane filtration performance, and (iii) developing methods to engineer biofilm properties to enable the removal

Author(s): Szilárd Bucs, Joop Kruithof, Mark van Loosdrecht and Johannes Vrouwenvelder
doi: 10.2166/9781780409597_xiii

of biomass and achieve that biofouling would has a restricted or delayed impact on the filtration process only.

Szilárd Bucs
King Abdullah University of Science and Technology
szilard.bucs@kaust.edu.sa

Joop Kruithof
Wetsus
joop.kruithof@wetsus.nl

Mark van Loosdrecht
Delft University of Technology
m.c.m.vanloosdrecht@tudelft.nl

Johannes Vrouwenvelder
King Abdullah University of Science and Technology
Delft University of Technology
johannes.vrouwenvelder@kaust.edu.sa
20 May 2018

List of authors

Adam Wexler
Wetsus, Centre for Sustainable Water Technology

Amber Siddiqui
King Abdullah University of Science and Technology

Andrea Radu
Delft University of Technology, Department of Biotechnology

Arie Zwijnenburg
Wetsus, Centre for Sustainable Water Technology

James Kidwell
Conwed Plastics

Carsten Schellenberg
LANXESS BU Liquid Purification Technologies, R&D Membranes

Claudia Dreszer
Wetsus, Centre for Sustainable Water Technology

Cristian Picioreanu
Delft University of Technology, Department of Biotechnology

Einar Fridjonsson
University of Western Australia

Hans-Curt Flemming
University of Duisburg-Essen

Jeremy Martson
King Abdullah University of Science and Technology

Johannes Vrouwenvelder
King Abdullah University of Science and Technology

Joop Kruithof
Wetsus, Centre for Sustainable Water Technology

Ludmilla Bereschenko
Delft University of Technology, Department of Biotechnology

Luke Bergwerff
Delft University of Technology, Department of Biotechnology

Marc Staal
King Abdullah University of Science and Technology

Mark van Loosdrecht
Delft University of Technology, Department of Biotechnology

Mike Johns
University of Western Australia

Nadia Farhat
King Abdullah University of Science and Technology

Rodrigo Valladares Linares
King Abdullah University of Science and Technology

Sergey Borisov
Technische Universitat Graz

Sarah Vogt
University of Western Australia

Sandra Drusova
Wetsus, Centre for Sustainable Water Technology

Sarah Creber
University of Cambridge

Szilárd Bucs
King Abdullah University of Science and Technology

Tjesse Overdijk
Wetsus, Centre for Sustainable Water Technology

Mathijs van Steen
Delft University of Technology, Department of Biotechnology

Vasile Lavric
University Politehnica of Bucharest

Victor Haaksman
Delft University of Technology, Department of Biotechnology

Chapter 1

New approaches to characterizing and understanding biofouling of spiral wound membrane systems

INTRODUCTION

Surfaces in contact with water will inevitably be colonized by bacteria forming a biofilm layer. Membrane units are therefore intrinsically excellent systems for bacteria to colonize. Not surprisingly, biofouling is an important problem in the application of membrane filtration processes for water treatment. Biofouling has been, and still is, an important research topic which until now has not been solved. This is largely due to the very complex nature of the problem involving physical, chemical and biological aspects in a complex geometry. Partly because of the complexity of the problem, biofouling studies have a rather anecdotal nature and the information obtained from pilot and full scale installations is rather anecdotal. Biofilm growth occurs on each surface in contact with water, even in ultrapure water systems (Chicote *et al.*, 2004). In principle, for membrane filtration systems, biofilm growth is not equal to biofouling. Only when biofilm growth leads to a significantly decreased membrane flux, a much increased operational pressure of the system or a decrease in retention properties of the membrane is it called biofouling. In other words, while a biofilm has a clear definition, biofouling is related to operational practises. A further problem is caused by the fact that water quality is not feasible for biofouling characterization. In addition, sampling and analysing membrane units is not easy. Membrane processes are typically black box systems that can only be studied effectively by an autopsy at the end of a filter run. Therefore there is a need for well-defined tools allowing the assessment of the complex interactions occurring in such process units. Finally, biofouling in low pressure membranes (microfiltration (MF), ultrafiltration (UF)) and high pressure membranes (nanofiltration (NF), reverse osmosis (RO)) cannot be

Author(s): Szilárd Bucs, Joop Kruithof, Mark van Loosdrecht and Johannes Vrouwenvelder
doi: 10.2166/9781780409597_0001

directly compared. This is due to the fact that different fouling mechanisms play a dominant role in these systems. One example is the concept of critical flux which is applicable for microfiltration and ultrafiltration (Bacchin *et al.*, 1995; Field *et al.*, 1995; Howell, 1995) but is not relevant for NF and RO systems (Vrouwenvelder *et al.*, 2009b, d). In this contribution we focus on NF and RO-membrane fouling.

BIOFOULING STUDIES

Biofouling control is pursued by: (i) biological pre-treatment removing nutrients (e.g. sand filtration), (ii) applying biocide dosage or UV irradiation (metabolic inactivation of bacteria), (iii) membrane and spacer modification; and (iv) chemical cleaning. Development of membranes materials or coatings which prevent adhesion is one general advocated method to prevent biofouling without the need of applying toxic chemicals. In view of the wide range of possibilities of microorganisms to anchor themselves to surfaces, these approaches to avoid this adhesion do not lead to success. Protection against overgrowth by microorganisms in higher organisms (animals and plants), is mainly obtained by continuous renewal of epithelium cells or, e.g. applying strong wax layers or structures (e.g. lotus leaves). The first is difficult in non-living systems, while the latter is usually not a realistic option. Lotus leaves have a very strong hydrophobic nanostructure which makes them effectively dry and not in contact with the water, protecting them against microbial colonization. This characteristic is sometimes suggested for membrane modification (Marmur, 2006), however the hydrophobic nature makes the membranes unsuitable for water permeation (except as a vapour). Surface modification by hydrophilic groups or polymer brushes is also suggested for minimization of microbial adhesion. The effects of these modifications are usually investigated in short term studies. These studies focus on initial adhesion, which is not representative for biofilm formation (Gjaltema *et al.*, 1997). In general, adhesion is influenced by physico-chemical interactions (for instance electrostatic forces, hydrophobicity, etc.) while biofilm formation is dependent on physical interactions (biofilm strength, shear forces, attachment surface roughness). A low adhesion surface might delay biofilm formation, but not prevent it. Moreover in practical systems a surface gets quickly covered by a conditioning layer from organic molecules in the treated water, masking the surface modification effect. In order to study biofouling a 'scaled-down' monitor representative of actual membrane processes is needed in order to be able to do controlled experiments and analyse the system. In essence, for spiral wound NF and RO membrane systems this can be done by a system that represents the geometry of one feed channel with identical geometry and flow rates. The recently developed biofouling monitor (Vrouwenvelder *et al.*, 2006) has such properties. This system makes the biofouling more experimentally accessible. Nevertheless in such experimental systems the complex interaction of processes in, for instance, a biofilm in a NF /RO module are still difficult to assess. It is difficult to change just one parameter at a time; an increased flow rate means more shear, more mass transfer and a different biofilm geometry, all affecting the impact of the biofilm on the filtration processes. In-situ visualization methods such as nuclear magnetic resonance (NMR) imaging (Graf von der Schulenburg *et al.*, 2008; Vrouwenvelder *et al.*, 2009a) or CT-scanning of full membrane modules allows us to assess how processes at the micrometre scale influence the macroscopic behaviour of the membrane processes. Mathematical simulation models can help to quantitatively assess the interaction of the many different processes (flow, diffusion, permeation, microbial growth, etc.), which individually are well known (Picioreanu *et al.*, 2009; Radu *et al.*, 2010). This combination of experimental and modelling approaches has been shown to be a fruitful approach leading to better understanding of biofouling, as discussed below. The main items will be addressed in the following sections

RECENT MICROBIOLOGICAL RESEARCH INTO CHARACTERIZING BIOFOULING

Using flow cells allows regular sampling and analysis of the microbial community that leads to biofouling. (Bereschenko *et al.*, 2010) indicated that in a fresh water RO module, first colonization likely occurs by sphingomonas-type bacteria, a bacterium hitherto not studied in this context. These organisms are well adapted to oligotrophic environments and very rapidly form a gelatinous monolayer on a fresh membrane. These organisms have twitching and swarming motilities, which allow them to rapidly colonize a fresh membrane (Bereschenko *et al.*, 2011; Pang *et al.*, 2005). This thin layer does not directly influence the membrane performance in fresh water systems, but when further colonized by a thicker layer of other microorganisms, fouling occurs. Sphingomonads form in a developed fouling layer only a minor fraction of the microbial population and will not easily be recognized as important, but as they may form the connection between the membrane and the rest of the fouling layer they could form a good target for microbial control. In this respect it is also relevant to realize that these organisms generally make a different extracellular polymeric substances (EPS) than most other bacteria. Herzber *et al.* (2007) concluded that the RO membrane selects for a specific microbial community. Indeed better characterization of the organisms involved might open the way to directed control of the specific bacteria and their EPS involved. Chemical cleaning leads to a reduction of the pressure over the feed spacer in an RO module. However after chemical cleaning still a major fraction of the biofilm is present as dead cell material in a gel matrix (Figure 1.1). Likely pieces of biofilm subjected to large shear have been detached, resulting in less pressure drop, however the majority of the biofilm is still present (Bereschenko *et al.*, 2011). Molecular genomic studies showed that after a chemical cleaning the sphingomonads (living in the deeper layer of the fouling layer and thereby more protected from cleaning) quickly regrow, (Figure 1.2) likely on the cell mass. Within a short time the fouling is established again.

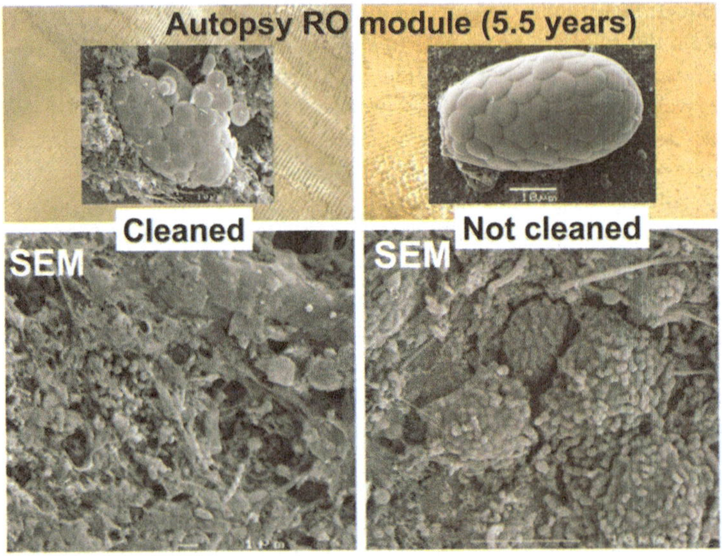

Figure 1.1 EM pictures from full scale RO membranes after 5 years of operation. Left side was cleaned, right membrane module was not cleaned. Both show copious amounts of biomass, although the left side of the biomass shows a more deteriorated image.

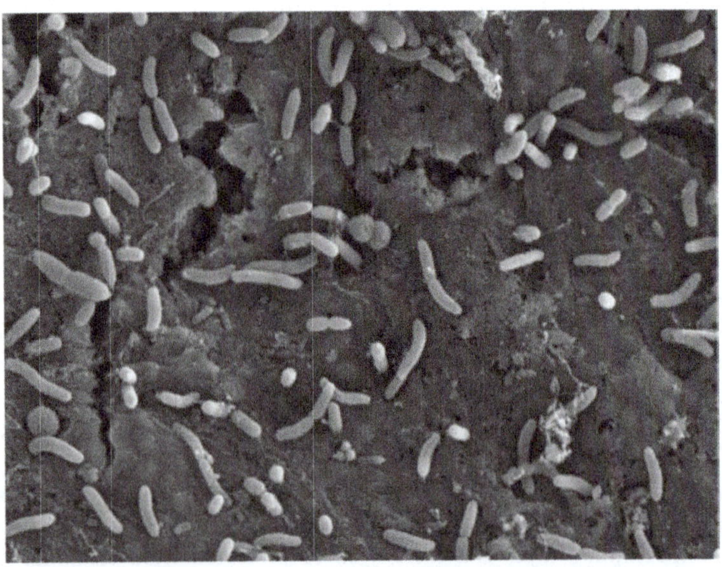

Figure 1.2 Rapid recolonization of a dead 'cleaned' fouling layer by sphingomonads.

FLOW CELL STUDIES

Biofilm morphology (and the related impact on pressure drop) is influenced not only by substrate load but also by linear velocity, substrate concentration and fluid shear on the biofilm (Vrouwenvelder *et al.*, 2009b), showing the complexity in interpreting pressure drop measurements as signal for microbial growth in membrane modules. In practice, applied cleanings usually inactivate but do not remove (most of) the accumulated biomass from the RO/NF membrane module, which is the reason why most cleanings are not effective for biofouling control. The effect of chemical cleaning on biofouled RO systems has been studied with the in-situ MRI technique (Creber *et al.*, 2010b).

Chemical cleaning at an early stage of biofouling was more efficient in removing biomass than cleanings performed at a later stage. Preventive cleaning is probably preferred over curative cleaning. The effect of substrate concentration, linear flow velocity, substrate load and flow direction on pressure drop development and biofilm accumulation was investigated in membrane fouling simulators (Vrouwenvelder *et al.*, 2009b) The pressure drop increase was related to the amount of accumulated biomass and linear flow velocity. Biomass accumulation was related to the substrate load. In RO and NF membrane elements, at linear flow velocities applied in practice, voluminous and filamentous biofilm structures were developed in the feed spacer channel, causing a significant increase in feed channel pressure drop. The amount of accumulated biomass was independent of the applied shear, but was depending on the substrate load. A high shear force resulted in a more compact and less filamentous biofilm structure compared with a low shear force, causing a lower feed channel pressure drop increase. This is in accordance with observation in other biofilm systems. A biofilm grown at low shear is however easier to remove by water flushing compared with a biofilm grown at high shear. The flow regime can be used to manipulate the biofilm morphology, enabling new approaches to control biofouling. Operating membranes at low flow velocity reduces the effect of biomass accumulation on feed channel pressure drop/increase (Vrouwenvelder *et al.*, 2009b). Membrane systems could be operated at an optimal feed channel pressure, tolerating biomass accumulation. In other words: a high biomass concentration causing a low feed channel pressure drop increase only is not a problem. Therefore, from a practical point of view the focus of research should be restricting the feed channel pressure drop and the feed channel pressure drop/increase rather than biomass accumulation. The Opti- flux concept (van Paassen *et al.*, 2005) reduces the operational costs of membrane filtration by applying fewer membrane modules in a pressure vessel than commonly applied (i.e. three modules instead of up to eight). The operational cost reduction is based on lower energy consumption. In addition, a lower flow velocity will impact the effect of the biomass amount on the feed channel pressure drop, so the Optiflux concept may reduce the feed channel pressure drop increase caused by biofouling, reducing the membrane operational costs even further. Biomass accumulation is reduced at the location where impact on the feed channel pressure drop is highest and because of the lower flow, the impact of biomass on pressure drop increase is lower.

Reducing the impact of biofouling by reducing the linear flow velocity implies that the impact biofouling will be balanced with extra concentration polarization. Therefore, studies combining concentration polarization and biofouling are important. Digital Holographic Interferometry has proved to be a valid technique to visualize concentration polarization in-situ in a cross flow RO system. Experimental results showed that the hydrodynamics has great influence on the polarization layer (Fernández-Sempere *et al.*, 2010). From these flow cell studies it becomes clear that the interaction between hydraulics, load, substrate and membrane operational parameters is so complex that only detailed analysis can give sufficient insight into the mechanisms. There is a substantial need for novel measurement techniques that enable non-invasive spatially resolved observation of biofouling in NF and RO membrane modules. Such measurements will enhance our understanding of the key design and operational parameters influencing biofilm fouling. In our research we have demonstrated the first application of NMR microscopy to a spiral wound RO membrane module.

Despite the opaque nature of membrane design, NMR microscopy is shown to be able to provide a non-invasive quantitative measurement of RO membrane biofouling and its impact on hydrodynamics and mass transport. The developed NMR protocols allow the evolution of biofouling to be followed. Minimal biofilm growth is shown to have a substantial impact on: (i) flow field homogeneity, (ii) the spatially resolved velocity field, and (iii) displacement propagators, which are distributions of molecular displacement of a passive tracer (in our case, water) in the membrane. From these measurements, the effective membrane surface area (subjected to flow) is quantified. The observed channelling is in agreement with observation from practice, the above described monitor studies and the below mentioned numerical modelling studies. Biofouling activity quantification in membrane filtration installations addressed assessment of oxygen consumption rates in membrane installations by measuring the oxygen concentration of the water before and after passing through the installation (feed and concentrate), but this measurement was not suitable to assess the biofouling activity during membrane operation (Kappelhof *et al.*, 2003). Studies making a carbon or chemical oxygen demand (COD) balance of biological processes in membrane systems by analysing relevant parameters in feed water, concentrate and permeate have not resulted in a closed balance. Thus, no suitable method has been found yet for non-destructive assessment of biological activity in membrane systems.

MATHEMATICAL MODELLING

Recently, biofilm models have been integrated with computational fluid dynamics (CFD) to investigate biofouling in RO and NF systems (Picioreanu

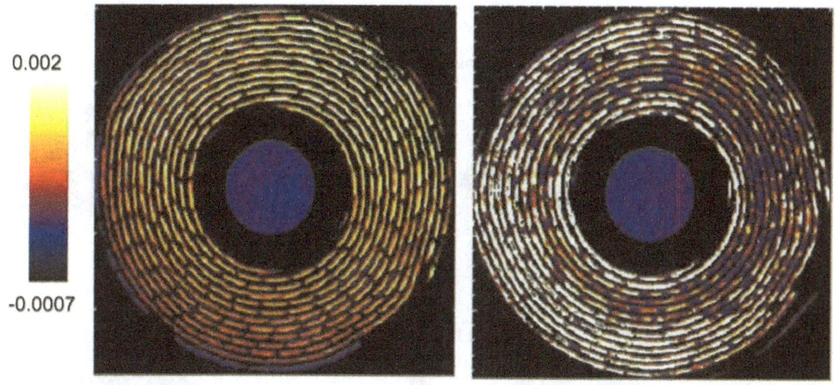

Figure 1.3 NMR velocity imaging of a RO membrane module. Left: Clean module, flow evenly distributed, right: fouled membrane with strongly heterogeneous flow. Colour scale: Liquid velocity in m/s. Adapted from Graf von der Schulenburg *et al.* (2008).

et al., 2009; Pintelon *et al.*, 2010; Radu *et al.*, 2010). These mathematical models are useful tools to analyse complex phenomena occurring when a biofilm is formed in RO or NF feed channels. These models are based on basic physical, chemical and biological processes involved in biofouling. One of the challenges in modelling is to maintain a good compromise between the model complexity and insight gained.

Flow cell studies as well as modelling studies showed that fouling of the spacer very rapidly leads to flow channelling and large areas of the membrane being not accessible any more for fluid flow (Figures 1.3 and 1.4), (Picioreanu *et al.*, 2009; Vrouwenvelder *et al.*, 2009a). These dead zones decrease the effective membrane area and influence solute mass transport. Biofouling is therefore indicated as a spacer problem. For salt water systems however concentration polarization is more important and a thin biofilm will already lead to an increased concentration polarization. Expectedly, the formation of preferential flow channels due to biofilm growth in brackish and seawater systems will lead to a severe degradation of plant performance. Mathematical modelling allows us to assess the relative impact of different fouling mechanisms as a function of process conditions. Biofilm enhanced concentration polarization of nutrients is regularly mentioned to enable even faster biofilm growth. Mathematical modelling makes clear that only in very specific cases is this true, but that for normal operation such a mechanism is quantitatively marginal. Numerical tools have been developed to a stage where the interaction between biofilms and other potential foulants can be studied in a virtual RO system. Given that micro-scale models provide a good understanding of local effects of biofilms (flow channelling, concentration polarization, salt passage, hydraulic resistance), and their relation to global performance parameters of RO

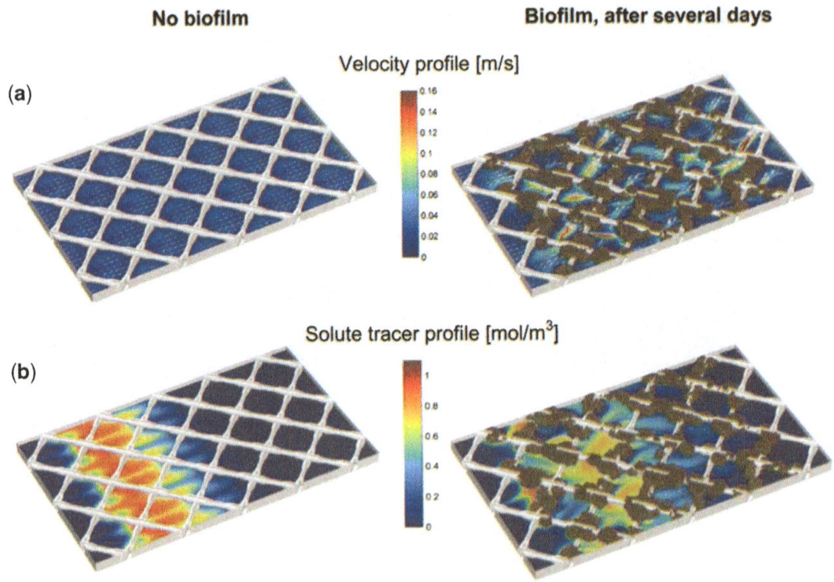

Figure 1.4 Three dimensional simulation of flow and solute mass transport: (a) Simulated flow velocity in a two-dimensional section along the feed channel without biofilm (left) and with a developed biofilm (right). The initial regular flow pattern is changed into a preferential flow with narrow channels. (b) Two dimensional solute tracer distribution. Flow is from left to right.

systems (flux, pressure drop, salt passage), the modelling approach can be used to complement experimental studies.

CONCLUSION

In general, control of biofouling should be based on a good pre-treatment of the water before NF and/or RO to minimize microbial growth. However, as inevitably some microbial growth will occur, systems should be designed in such a way that limited growth will not lead to biofouling and cleaning can be easily carried out (both by better spacer and module design). A good understanding for the design of such systems is only provided by a combination of an experimental scale-down approach and mathematical models supported by critical observation of full scale membrane processes.

Chapter 2

Effect of flow velocity, substrate concentration and hydraulic cleaning on biofouling of reverse osmosis feed channels

INTRODUCTION

Desalination of brackish water and seawater using reverse osmosis (RO) technology has become increasingly important in recent years due to severe water shortages experienced in several regions in the world (Greenlee *et al.*, 2009). Despite recent innovative pre-treatment strategies for the feed water of the RO module, prevention of biofouling remains a problem. Biofilms develop in all RO membrane plants from the start of operation (McDonogh *et al.*, 1994), but what actually has an impact on the system performance are the amount and the place where the biofilm forms. Biofouling, i.e., the substantial decline of membrane performance due to biofilm development, only occurs when a threshold value of biomass is attained in the system (Flemming, 2002). Better understanding of biofilm formation in membrane systems is important so that the operational problems caused by the biofilm can be reduced. Computational approaches can complement experimental studies if one could evaluate by numerical means the effect of different operational conditions on biofouling. In general, models for membrane processes account for the liquid flow pattern and the associated mass transfer of solutes (e.g., salts, organic substrates). Besides these essential membrane-related processes, a numerical study of biofouling must include also a model of biofilm development and of its effect on membrane processes. Many studies reported a complex relationship between liquid flow and biofilm structure (de Beer & Stoodley, 1995; Lewandowski *et al.*, 1993; Manz *et al.*, 2005; Ohl *et al.*, 2004). Mass transport and fluid shear stress are both dependent on the hydrodynamic conditions, thus the flow pattern will significantly influence biofilm processes (as described in models (Duddu *et al.*, 2009; Eberl *et al.*, 2000; Picioreanu *et al.*, 2009, 2001)). One would expect that

Author(s): Szilárd Bucs, Joop Kruithof, Mark van Loosdrecht and Johannes Vrouwenvelder
doi: 10.2166/9781780409597_0009

the higher the fluid velocity, the larger the external mass transfer rate of nutrients to the biofilm and therefore the biofilm growth is faster. On the other hand, the net biofilm accumulation is determined by the balance between biomass growth and detachment rates. High velocity also implies larger shear forces at the biofilm surface. If the biofilm had constant mechanical properties in time, larger shear could lead to more biomass detachment. However, on the long term, denser, stronger and more resilient biofilms develop at large shear rates. Clearly, due to its multiple roles fluid flow has a crucial importance in determining the biofilm amount in a system and its structure (Horn *et al.*, 2003; Kwok *et al.*, 1998; Stoodley *et al.*, 1999a; van Loosdrecht *et al.*, 1995).

Biofilm formation in RO membrane processes has multiple negative effects on process performance: decline in permeate flux, increased salt passage and increased pressure drop over the feed channel. Experiments identified several ways by which biofilm accumulation deteriorates the membrane process performance: additional hydraulic resistance for the trans-membrane flow (McDonogh *et al.*, 1994), biofilm enhanced concentration polarization (Chong *et al.*, 2008a; Herzberg & Elimelech, 2007), increased feed channel pressure drop (Vrouwenvelder *et al.*, 2009a) and the occurrence of stagnant zones and preferential flow channels (Vrouwenvelder *et al.*, 2010b). Recently, biofilm models have been integrated with computational fluid dynamics (CFD) to investigate biofouling in RO and NF systems (Picioreanu *et al.*, 2009; Pintelon *et al.*, 2010; Radu *et al.*, 2010). Picioreanu *et al.* (2009) and Pintelon *et al.* (2010) have developed three-dimensional models for biofilm development in the feed channel of a reverse osmosis membrane device. Their results for a module operated without permeate flux, indicate that biofouling leads to formation of preferential flow paths and feed channel pressure drop increase. Radu *et al.* (2010) included permeate flux in a micro-scale two-dimensional model, and analysed the effect of biofilm development on the main membrane process performance indicators (flux, solute passage and feed channel pressure drop). One has to note that much higher computational requirements appear when considering permeation because correct (i.e., accurate) calculations of the steep concentration polarization layer require fine meshes, thus a directly higher computational burden (both in calculation times and computer memory requirements). The objective of the current study was to investigate the effect of cross-flow velocity and substrate concentration on biofilm formation in feed channels and on the performance of membrane systems. First, simulations at different cross-flow velocities were carried out at a fixed substrate concentration. Second, we evaluated the effect of substrate concentration in the feed on the biofilm structure and amount at a fixed average velocity. Third, different hydraulic cleaning strategies were evaluated.

BIOFOULING MODEL DESCRIPTION

We chose a micro-scale two-dimensional (2-d) model geometry, as a good initial compromise between the calculation time and model realism. The use of a two-dimensional geometry allows a detailed numerical investigation of physical aspects relevant for biofilm formation; however a quantitative comparison with experimental data will require three-dimensional (3-d) models. The feed channels without and with feed spacer were compared in a series of simulations. For the illustration of the spacer effect on biofouling, only the case of spacer transverse filaments situated in the middle of the feed channel was considered in this study. Modelled spacer and channel dimensions are chosen according to practice (Schwinge *et al.*, 2002a): channel width is $L_Y = 10^{-3}$ m, spacer diameter is $d_s = 0.5 \times 10^{-3}$ m, distance between two spacer filaments is $L_s = 4 \times 10^{-3}$ m. Channel length is chosen $L_X = 15 \times 10^{-3}$ m. For numerical convenience the whole computational domain is divided into biofilm and bulk liquid sub-domains. A more detailed description of the model was presented in (Radu *et al.*, 2010), therefore only the main assumptions, equations and further model adaptations are presented here.

Model assumptions and equations

The model used in this work accounts for the following processes: (i) fluid flow; (ii) mass transport of solutes and (iii) biofilm development (including attachment, growth and detachment). Biofilm development strongly affects the flow patterns and solute concentration (salts, substrates) fields in the feed channel. Conversely, the biofilm growth, attachment and detachment (removal) are determined by substrate availability and hydrodynamic conditions. Microbial cells multiply when nutrient is supplied and detach when exposed to high shear stress induced by liquid flow. The fluid flow is calculated from the stationary laminar incompressible Navier–Stokes equations in the liquid sub-domain (Equation 2.1), and from the Brinkman flow equations (Brinkman, 1949) for the porous biofilm sub-domain (Equation 2.2):

$$\rho(\mathbf{u}\cdot\nabla)\mathbf{u} + \nabla p = \nabla\cdot(\eta\nabla\mathbf{u}), \quad \nabla\cdot\mathbf{u} = 0 \tag{2.1}$$

$$\frac{\eta}{\kappa}\mathbf{u} + \nabla p = \frac{1}{\varepsilon}\left[\eta\cdot\nabla^2\mathbf{u} - \left(\frac{2\eta}{3} - \kappa\right)(\nabla\cdot\mathbf{u})\right], \quad \nabla\cdot\mathbf{u} = 0 \tag{2.2}$$

where the state variables are local liquid velocity $u = (u, v)$ and pressure p, and the parameters are liquid density ρ, liquid dynamic viscosity η, biofilm permeability $\kappa = 10^{-16}\ \mathrm{m}^2$ and porosity $\varepsilon = 0.8\ \mathrm{m}^3$ liquid m^{-3} biofilm. The physical parameters (i.e., liquid viscosity and density) depended on the local salt concentration: $\rho(C_1) = (0.999 + 4.099 \times 10^{-5} \times C_1) \times 10^3\ \mathrm{kg\ m}^{-3}$ and $\eta(C_1) = (1.004 + 8.377 \times 10^{-5} \times C) \times 10^{-3}$ Pa s with C_1 salt concentration in mol m^{-3} (Lyster & Cohen, 2007). At the inlet ($x = 0$) a fully developed parabolic velocity profile (laminar, because in all simulations Re < 200) is specified with different average velocities kept constant in time ($U_{in} = 0.033\ \mathrm{m\ s}^{-1}$, $0.066\ \mathrm{m\ s}^{-1}$, $0.1\ \mathrm{m\ s}^{-1}$). At the membrane surface the tangential velocity u is set to zero. The liquid velocity through the membrane (i.e., local permeate flux) is the normal velocity

$v(x) = s L_P(\Delta p - \Delta\pi)$ where $s = -1$ on the lower membrane and $s = 1$ on the upper membrane. Δp is the trans-membrane pressure, $\Delta\pi$ is the osmotic pressure created by the concentration polarization, and $L_P = 9.7 \times 10^{-12}$ m · Pa^{-1} · s^{-1} is the membrane permeability (Lyster & Cohen, 2007). The local trans-membrane pressure is the difference between the local pressure on the feed side on the membrane and an assumed constant permeate pressure ($p_p = 1 \times 10^5$ Pa), so that $\Delta p = p(x) - p_P$. The osmotic pressure difference is proportional with the salt concentration difference on the two sides of the membrane, $\Delta\pi = B(C_1(x) - C_{1P})$ where $B = 0.04572$ N · m · mol^{-1} (Lyster & Cohen, 2007). No-slip conditions ($u = 0$) were imposed on the feed spacer surface. Flow continuity was assumed at the biofilm surface. On the outlet boundary ($x = L_x$) the pressure is set constant to pout $= 16 \times 10^5$ Pa (Greenlee *et al.*, 2009).

The only soluble compounds chosen are salt (because of brackish water desalination conditions) and substrate (that determines biofilm growth rate), with concentrations C_1 and C_2, determined from the mass balances (2.3) and (2.4), respectively:

$$D_1\nabla^2 C_1 - \mathbf{u}\nabla C_1 = 0 \tag{2.3}$$

$$D_2\nabla^2 C_2 - \mathbf{u}\nabla C_2 + r_S = 0 \tag{2.4}$$

We assume the same diffusion coefficients ($D_1 = 1.5 \times 10^{-9}$ m^2 · s^{-1} and $D_2 = 10^{-9}$ m^2 · s^{-1}) in the bulk liquid and in the biofilm sub-domain (justified by the relatively small quantitative effects caused by 20–40% less diffusivity of small molecules on the overall results). For substrate consumption, a single substrate-limitation Monod rate $r_S = \mu_m / Y_{SX} \cdot C_X C_2 / (K_S + C_2)$ was used, with biomass yield $Y_{SX} = 0.5$ C-mol biomass/mol substrate, maximum specific growth rate $\mu_m = 2$ day^{-1} and half saturation coefficient $K_S = 0.025$ mol · m^{-3} set as for heterotrophic microorganisms. On the inlet boundary, the brackish water contains salt and limiting substrate with concentrations $C_{1,\mathrm{in}} = 40$ mol · m^{-3} (Greenlee *et al.*, 2009) and $C_{2,\mathrm{in}} = 0.25$ mol · m^{-3}, respectively. No-diffusion conditions apply on the outlet boundary. The spacer is considered as impermeable wall. The convective flux of solutes ($i = 1, 2$) toward the membrane equals the sum of the diffusive back transport of solute and the convective flux passing through the membrane: $vC_i - D_i \partial C_i / \partial y = vC_i(1 - R)$, at $y = 0$ and $y = L_Y$. The same membrane rejection coefficient, $R = 1$ (i.e., solutes are completely rejected by the membrane), was assumed both for salt and substrate. Flux continuity applies on the biofilm-liquid interface. See for more model details (Radu *et al.*, 2010).

An adapted form of the discrete biofilm model by Picioreanu *et al.* (Picioreanu *et al.*, 2001, 1998b) including biomass accumulation by growth, transport and biofilm/liquid transfer (e.g., attachment and detachment) was used here. The whole computational space is divided into a mesh of square elements, some of these elements containing biomass with concentration C_X. The interface between the bulk liquid and the biofilm is a moving boundary, its position resulting from biomass growth and transport, coupled with attachment and detachment processes. Biomass attachment rate was constant ($r_{att} = 5 \times 10^{-3}$ C-mol · m^{-2} · day^{-1}), at randomly chosen places (on the membranes, spacers and existing biofilm). Biomass growth occurs in each square element containing biofilm with the rate correlated to the substrate consumption rate ($r_X = r_S \cdot Y_{SX}$). It was assumed that the osmotic stress (caused by high salt concentrations) does not affect biomass growth. Biomass transport takes place when the local biomass concentration exceeds a critical value $C_{X,m}$ (8000 C-mol/m^3), using the cellular automata algorithm described in (Picioreanu *et al.*, 1998b). Biomass detachment is based on internal stress created by moving liquid past the biofilm. After solving the liquid flow, the load forces on the biofilm surface are applied in the plane stress equations, solved for the mechanical stress σ. Biofilm breakage occurred when the calculated stress σ in a certain biomass element is higher than a threshold stress $\sigma_{det} = 7$ N · m^2 (Möhle *et al.*, 2007) corresponding to the cohesive strength of the biofilm (Picioreanu *et al.*, 2001). In this stage, detachment means that all biomass elements (squares) where $\sigma > \sigma_{det}$ are removed from the biofilm, their place being taken by liquid. Biomass no longer connected to a solid surface is also removed from the system (resembling the biofilm sloughing).

Model solution

The biofilm structure develops in series of steps in time, determined by the local substrate availability and hydrodynamic conditions. The chosen time step was dependent of the substrate concentration in the feed: $\Delta t = 1$ h, 2 h and 6 h for substrate concentrations $C_{2,in} = 2.5$ mol · m^{-3}, 0.25 mol · m^{-3} and 0.025 mol · m^{-3}, respectively. A sequential approach was used based on the assumption that hydrodynamics, mass transport and biomass growth occur at very different time scales. The characteristic time for biofilm development is very long (~days) compared with characteristic times for achieving steady flow (~s) and mass transport (~min) in the biofilm (Picioreanu *et al.*, 2001). Therefore, a reasonable approximation is to solve only stationary flow and mass balances in between each biomass development steps. A detailed model solution strategy is presented in (Radu *et al.*, 2010). The main algorithm steps performed in the time loop are:

(1) Solving liquid flow equations and the mass balance for salt simultaneously in stationary conditions to get **u**, p, and C_1 for a given geometry of the biofilm matrix.
(2) Solving the stationary substrate mass balance for concentration C_2.
(3) Computing new biomass distribution C_X. First, biomass growth in each volume element is calculated based on the substrate concentration resulted in step (2), secondly biomass is redistributed with a spreading algorithm (cellular automaton). The biomass is then detached due to shear stress induced in the biofilm structure by liquid flow. Detachment events are

performed in a loop, until a mechanical equilibrium (i.e., no detachment) is reached for the biofilm in the channel. After detachment, new biomass is attached and with the newly obtained biomass distribution the calculations for the next time step start.

The main computer code used for model solution is implemented in Matlab (MATLAB 2008b, Mathworks, Natick, MA, www.mathworks.com) and makes use of finite element methods from COMSOL Multiphysics (COMSOL 3.5a, Comsol Inc, Burlington, MA, www.comsol.com) to solve the fluid dynamics, solute mass transport and plane stress equations and of own Java code for the biomass spreading algorithm. The finite element mesh is re-created at each time step because the sub-domains and their boundaries change in time. Tests on the numerical accuracy showed the need of using a very fine mesh next to the membrane surface, because of steep concentration gradients in the boundary layer. The mesh has a maximum size of 5 μm on the membrane surface, the size increasing towards the bulk liquid up to a maximum of 30 μm.

RESULTS AND DISCUSSION

Effect of cross-flow velocity on biofilm development in the feed channel

Local biofilm accumulation on the membranes results from a balance between the rates of biomass attachment and growth (as processes leading to increased biofilm volume) and the rate of detachment (which decreases the biofilm volume). An example of typical computed biofilm development in the feed channel of a membrane device is presented in Figure 2.1. The biofilm starts growing in small colonies from randomly attached cells, both on the membranes and spacer filaments. These early semi-circular colonies will merge and form a continuous biofilm on the membrane and streamer-like structures on the spacer (e.g., Figure 2.1, 192 h). Streamer formation on the feed spacer has been observed in experimental studies using membrane fouling simulators (Vrouwenvelder *et al.*, 2009a). Although short streamers begin to grow downstream spacer filaments, they will eventually break because the model assumes a brittle mechanical behaviour of the biofilm. More realistically, viscoelastic biofilm mechanics should be considered as shown in experiments by Stoodley *et al.* (1999b) and Klapper *et al.* (Klapper *et al.*, 2002). The biofilm detachment plays an important role deciding whether an initially attached cell will form a stable colony or will be washed-out with the fluid. When the spacer filaments are in the middle of the feed channel, the highest shear rate regions are: (i) on the upstream side of the spacer, and (ii) on the membrane next to the spacer. In these areas, the biofilm hardly develops.

The liquid shear continuously erodes the biofilm surface, creating on the long term hydrodynamic (streamlined) biofilm shapes. As expected, at high liquid flow

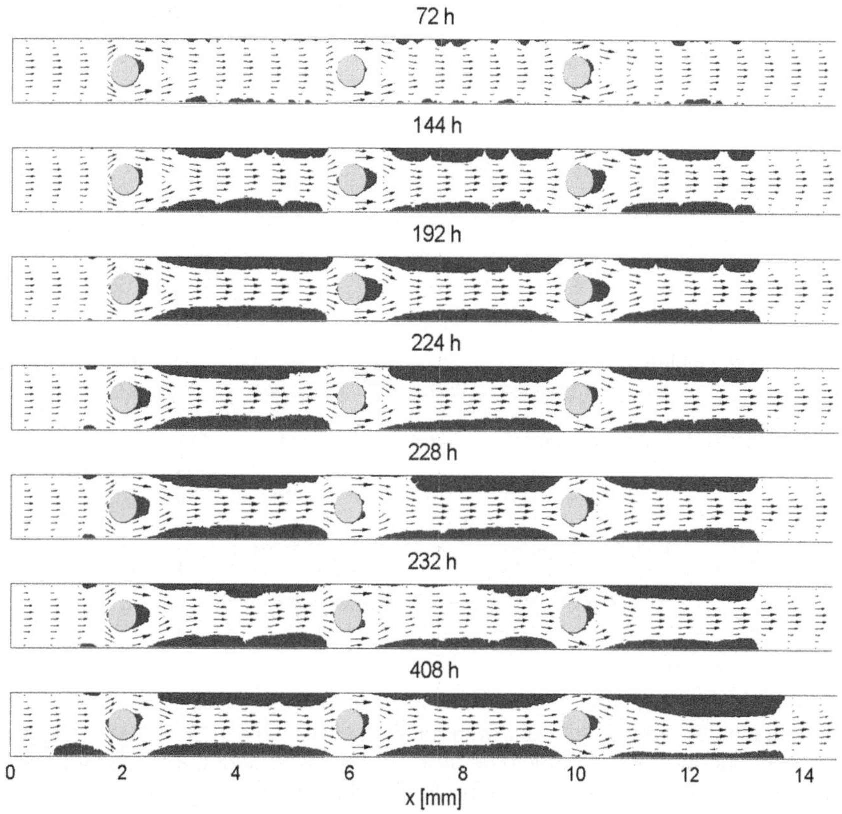

Figure 2.1 Illustration of biofilm development in time in the feed channel with spacer ($U_{in} = 0.066$ m · s^{-1}). The arrows indicate the velocity vectors. 72 hours: early biofilm colonies spread over the membrane surface; 144–224 hours: small biofilm streamers form on the spacers in the middle of the channel, then break. Biofilm colonies on the membrane expand laterally; 224–232 hours: a sloughing event – in a relatively short period, large patches of biomass are removed from the feed channel; 192–408 hours: a quasi-steady state was achieved for biomass in the channel.

rates, the shear is also higher so that more biomass is detached. In consequence, thinner biofilms (and less overall biomass amount) are observed in the simulations at average inlet liquid velocity $U_{in} = 0.1$ m · s^{-1} compared with the biofilms at $U_{in} = 0.033$ and 0.066 m · s^{-1} (Figure 2.2), irrespective of feed spacer presence. At low shear rates, the biofilm may develop in such extent that it blocks completely the flow between membrane and spacer. However, total clogging of the feed channel is not expected at constant liquid inlet flow rate. The narrowing of flow channels

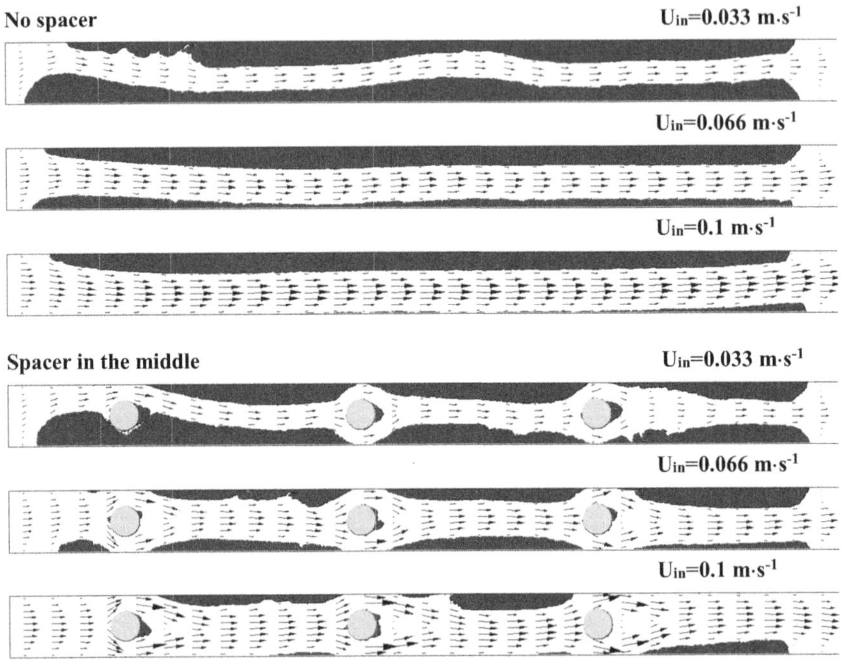

Figure 2.2 Effect of cross-flow velocity on the steady state biofilm structures, in the feed channel without spacer and with middle spacer (shown as light gray circles). All figures show biomass structures after 16 days. The dark gray areas represent the biofilm. The arrows lengths are proportional to the liquid velocity.

leads to increased velocity and shear stress, thus promoting more biofilm erosion, which keeps a certain flow section always open.

In time, the detachment rate is able to balance the growth and attachment rates, so that a quasi-steady state is achieved in the biofilm amount (Kwok *et al.*, 1998; van Loosdrecht *et al.*, 1995). Interestingly, sloughing events also take place occasionally. An example is apparent in the simulation shown by the image sequences from 224 to 232 h in Figure 2.1. On the upper membrane between 6 and 10 mm from the inlet a large patch of biomass is removed in a relatively short time interval. After large biofilm removal events, re-growth will occur and bring the biomass layer back to a "maximum" average thickness. It can be observed in Figure 2.1 that after the sloughing event starting at 192 h, the biofilm will re-grow so that at 408 h approximately the same biomass amount is reached as before the major detachment episode. At large flow velocities (e.g., 0.1 m · s^{-1}, in Figure 2.2), the biomass is non-uniformly distributed on the two membranes. While the biofilm thickness increases on one membrane, the biofilm on the other membrane is highly eroded.

Biofilm influence on the permeate flux and pressure drop

The influence of overall biomass accumulation on the permeate flux and pressure drop in the feed channel is presented in Figure 2.3 for six simulated cases (three different velocities for each channel with and without spacer). Biomass accumulation follows the experimentally observed trend with three phases: exponential (non-limited) growth, (mass-transfer) limited linear growth and, finally, the quasi-stationary state (Herzberg & Elimelech, 2007). The higher the liquid velocity, the less biomass accumulates in the channel, both with or without spacer (Figure 2.3a and 2.3b). At all studied velocities the spacer presence in the middle of the channel leads to less biomass formation. This is obviously an effect of the larger shear rates produced when the flow section is narrowing. Increasing the liquid flow rate has actually several contradictory effects. First, the thickness of the mass transfer boundary layer for substrate is reduced, which could lead to an accelerated biofilm growth due to larger substrate transfer rate. Second, higher cross-flow velocity reduces the substrate concentration polarization, thus the substrate availability next to the membrane. Finally, higher flow rates apply more shear on the biofilm and, consequently, determine more detachment. The detachment, however, seems to be the decisive effect that limits the overall biomass accumulation in all the studied cases. It should however be mentioned that several studies reported an increased biofilm density and strength when grown in high shear conditions (Kwok *et al.*, 1998). Although it is still unclear what are the mechanisms for this process, further model improvements could consider variable biofilm strength leading to differential detachment rates (for example, as a function of cell content, EPS fraction or biomass density, like in Xavier *et al.* (De Bivar Xavier *et al.*, 2005).

In reverse osmosis processes, the permeate flux is one of the main performance indicators. The permeate flux obtained appears to be in direct correlation with the biofouling extent (Figure 2.3c and 2.3d). The flux is initially almost the same in all cases in the clean channel, but declines less in the particular case of feed channel with spacer in the middle (less biomass formed due to more shear).

In practice, the feed channel pressure drop increase is the main decision factor for stopping the plant to perform cleaning operations. Figure 2.3e and 2.3f present the feed channel pressure drop in the small simulated channel length (1.5 cm). The same general trend for feed channel pressure drop increase due to biofilm formation has been reported in several experimental studies (Pintelon *et al.*, 2010; Vrouwenvelder *et al.*, 2009a). Obviously, the pressure drop is much larger in full-scale applications, where eight modules of 1 meter in series are used. Without spacer the pressure drop is initially significantly lower than with spacers, and this difference is more marked at higher flow rates. Interestingly, model results indicate that the relative increase in pressure drop due to biofilm formation is the least pronounced at the highest flow velocities. This is again correlated to the higher shear and less biomass formation at high flow rates. These results suggest that increasing the flow rate could represent a strategy to alleviate biofouling if energy costs allow it.

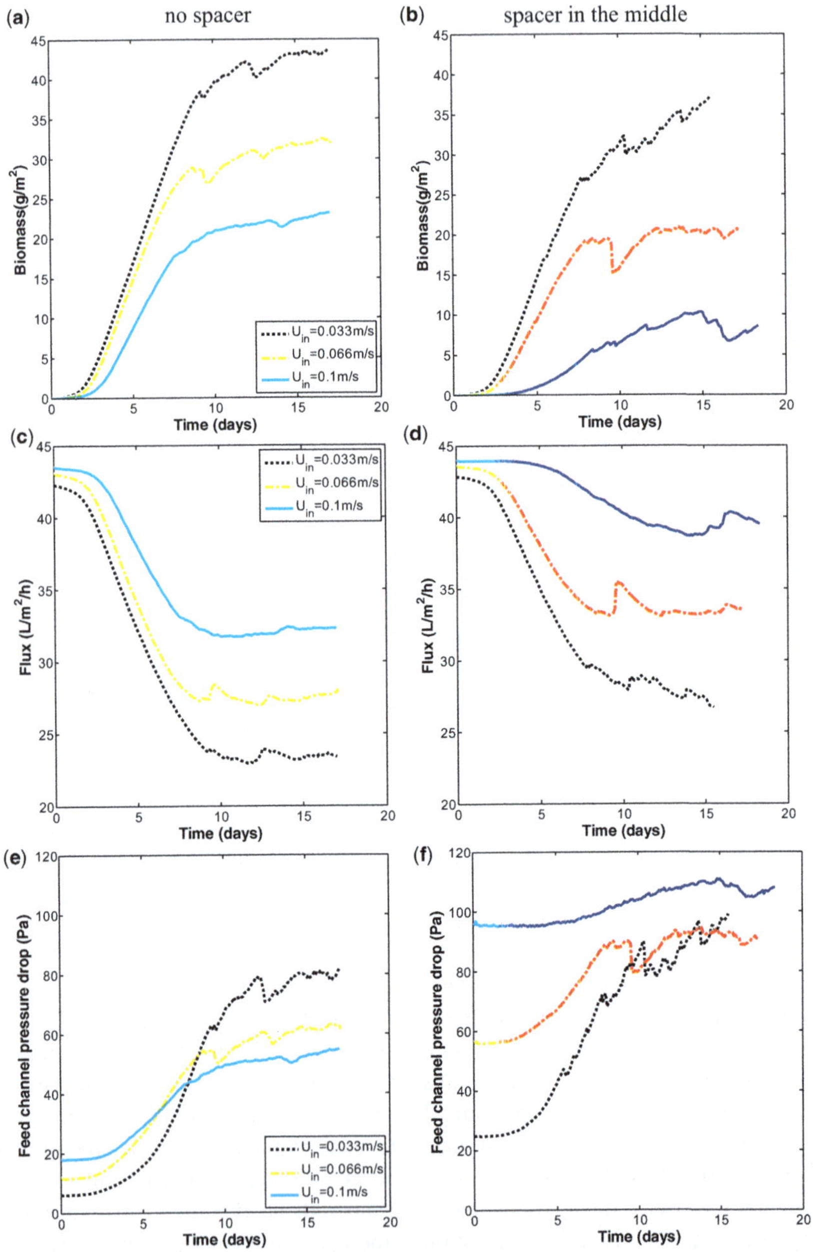

Figure 2.3 Compared biomass amount (a, b), permeate flux (c, d) and feed channel pressure drop (e, f) developed in time at different inlet flow velocities (U_{in}). Left: feed channel without spacer; Right: spacer in the middle of the feed channel.

Hydraulic membrane cleaning

Changes in hydraulic regime might be used to remove biofilms (Manz *et al.*, 2005; Ohl *et al.*, 2004; Vrouwenvelder *et al.*, 2010c). Results of various simulated cleaning strategies are presented in Figure 2.4. First, for the operation at $U_{in} = 0.033\ m \cdot s^{-1}$, we increased the average inlet velocity in a step-up to $U_{in} = 0.1\ m \cdot s^{-1}$ at day 8, when the biofouling was very significant. Large sloughing occurred immediately and the biomass amount dropped shortly. Eventually, when keeping this high velocity after day 8, steady biomass amount established again after approx. 2 days but at slightly higher levels than those reached when the biomass developed from the beginning at $U_{in} = 0.1\ m \cdot s^{-1}$. Clearly, this operation would not be economically very efficient due to the high energy costs, lower water recovery and investment in more powerful pumping equipment.

Second, the flow velocity was increased every 48 h for 6 h to $U_{in} = 0.1\ m \cdot s^{-1}$, (thus a three-fold increase), leading to alternating periods of biomass sloughing and re-growth. An increased biomass detachment when higher flow rates are suddenly applied has been also shown experimentally in test tubes (Manz *et al.*, 2005). This operation mode could be less expensive than the previous case. However, this alternating flow regime would still be difficult to be realized in practice due to pumping limitations.

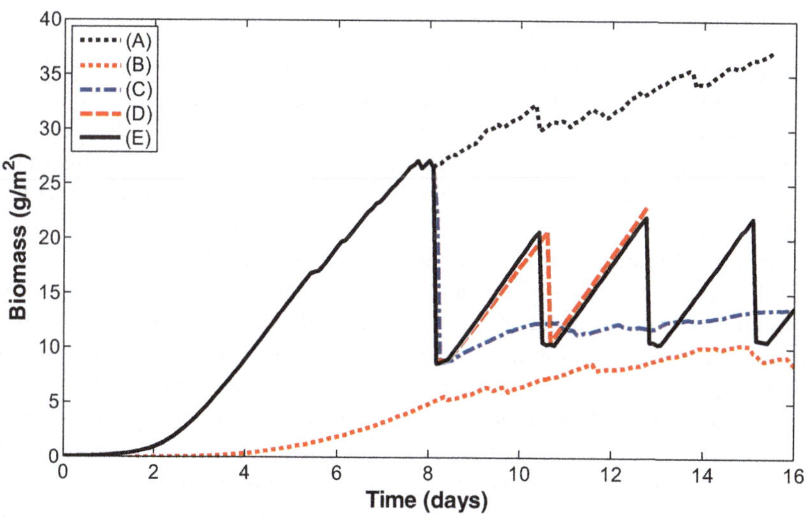

Figure 2.4 Effect of different hydraulic cleaning strategies on biofilm development in the feed spacer channel: (A) constant $U_{in} = 0.033\ m \cdot s^{-1}$, (B) constant $U_{in} = 0.1\ m \cdot s^{-1}$, (C) sudden increase of U_{in} from 0.033 to 0.1 $m \cdot s^{-1}$ at day 8, (D) periodic reversal and increased flow (0.1 $m \cdot s^{-1}$) after day 8, (E) periodic increased forward velocity after day 8.

Third, we investigated the effects of periodically reversing the flow direction. It was assumed that no permeation occurs during the cleaning, that the cleaning is performed with water containing 100 times less substrate than the feed water and that the reversed flow has 3-fold increased velocity ($0.1\ m \cdot s^{-1}$) instead of $0.033\ m \cdot s^{-1}$, under which the biomass has developed. The results were similar to those obtained at periodically increased forward flow. However, the computational model should be improved to take into account the fate (e.g., re-attachment and filtration) of the removed biomass, which transported by the liquid could eventually block other parts of the installation.

In conclusion, having less biofouling at high flow rates does not necessarily mean a more economically efficient way of operation due to the larger energy input needed and the lower water recovery.

Nevertheless, there are some important practical implications related to the use of an increased cross-flow. The benefits of a higher cross-flow in reducing concentration polarization, salt precipitation and colloidal fouling have been emphasized in several studies (Chong *et al.*, 2008b; Fernández-Sempere *et al.*, 2010). Since often different types of foulants influence each other, such a strategy may reduce the overall performance degradation. Defining optimal operation conditions would certainly involve a more elaborated analysis of process-associated costs.

Effect of inlet substrate concentration on biofouling

Since biofilm development is determined by substrate availability and detachment rate, we studied the effect of different levels of substrate ($C_{2,in} = 0.025$, 0.25 and $2.5\ mol \cdot m^{-3}$) present in the feed water on biofouling. Model results indicate that, at steady state, comparable amounts of biomass are formed in the channel for high as well as for low substrate concentrations in the feed (Figure 2.5). This suggests again that the limitative factor for biofilm development, within the current model parameters, is the biomass detachment. Low substrate levels may only help to delay the undesired effects of biofilm developed in a RO membrane device, assuming that the biofilm is a problem ("biofouling") only when a threshold biomass amount accumulates in the feed channels. The quasi-steady state is reached within 50 days for the most substrate limited case ($0.025\ mol \cdot m^{-3}$), while at higher substrate availability ($2.5\ mol \cdot m^{-3}$) the steady biofilm thickness is achieved within ~12 days.

An interesting observation is that in reverse osmosis systems substrate accumulates near the membrane surface, a phenomenon called "concentration polarization". For inert solutes (e.g., salt) the biofilm strongly increases the concentration polarization. Enhanced concentration polarization is recognized as one of the main causes by which biofilms contribute to flux decline, because a higher salt concentration at the membrane means enhanced osmotic pressure opposing the transmembrane pressure, thus lower effective driving force for permeation. A more detailed theoretical analysis can be found in (Radu *et al.*, 2010) and experimental evidence in (Herzberg & Elimelech, 2007).

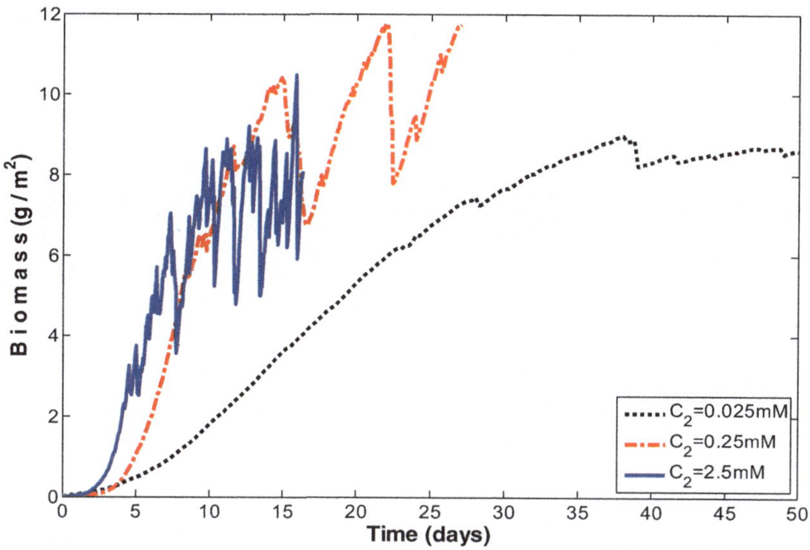

Figure 2.5 Biomass amount formed in the feed channel with spacer for different substrate concentrations at an average inlet fluid velocity of 0.1 m · s^{-1}. For high substrate concentration (2.5 mM) faster biomass growth can be observed, together with more biomass sloughing, resulting in a highly fluctuating steady state biomass amount.

For the substrates utilized by the biofilm, the concentration polarization effects can be more complex: the substrate accumulates due to water permeation and membrane rejection, but it is also consumed if a biofilm is present. Figure 2.6 shows the substrate distribution corresponding to biofilm structures at different moments and inlet substrate concentrations. In a first stage, high concentrations of substrate are present in the colonies situated on the membrane (Figure 2.6a – 4 days). On the contrary, the substrate level is rather low in the colonies developed on the spacer. This clearly demonstrates that the rejected substrate accumulates at the membrane surface (i.e., forming a substrate polarization layer) and that the biofilm enhances the concentration polarization. Interestingly, as the biofilm thickness increases, the overall substrate consumption rate will dominate over the accumulation due to polarization, so that a switch takes place from increasing to decreasing substrate concentrations in the biofilm (Figure 2.7). In this second stage, while the smaller colonies still benefit from increased substrate levels, substrate depletion occurs in the larger colonies, especially near the membrane. Therefore the biofilm developed on the membrane will contribute to a decreased passage of substrates to permeate. Note that in Figure 2.7 the substrate peak at around $y = 0.15$ mm is due to substrate brought by the liquid flow (convection) from regions of higher substrate concentration polarization situated near the membrane upstream this colony.

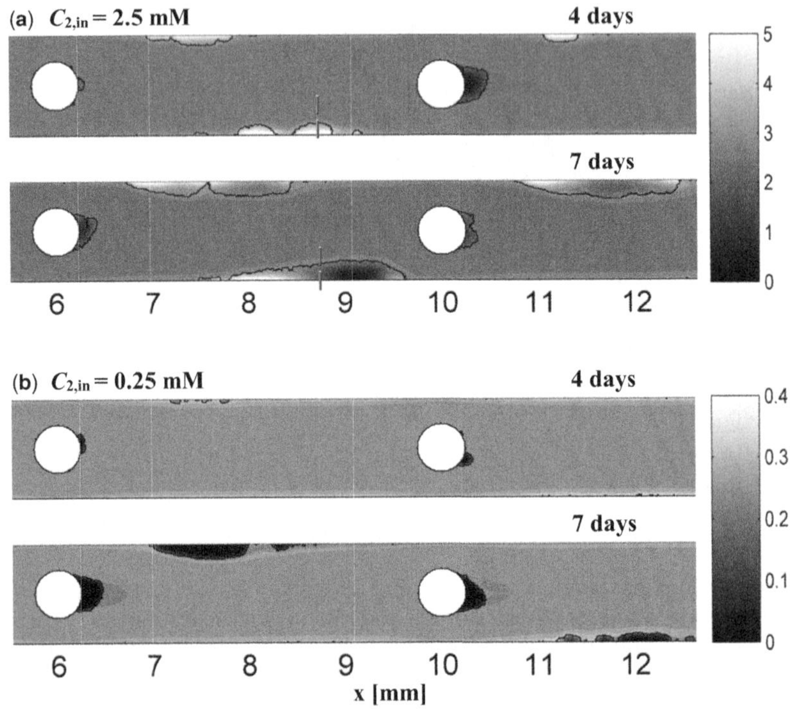

Figure 2.6 Details with 2-d substrate concentration distribution and biofilm structures at different substrate concentrations ($C_{2,in}$) in the feed water. Grayscale: substrate concentration, C_2 [mol · m^{-3}]. The average inlet fluid velocity was 0.1 m · s^{-1}.

This effect could be observed only in a truly two-dimensional model including liquid flow.

At low inlet substrate concentrations (0.25 and 0.025 mol · m^{-3}), most of the biofilm is substrate-limited from early stages (Figure 2.6b). Biomass growth occurs only in a thin outer layer of the biofilm (the "active layer" or "penetration depth").

It has been argued that the substrate concentration polarization may contribute to enhanced biofilm growth (Herzberg & Elimelech, 2007), but a more detailed model-based analysis is needed to confirm or reject this hypothesis. It can also be assumed that different compositions of biofouling layers (i.e., various microbial types and microorganism/gel ratios) develop in distinct areas of the feed channel due to variable substrate availability. One more factor that may contribute to different biofilm accumulation in systems with liquid permeation is the flow though the biofilm matrix (see streamlines in Figure 2.8), which could lead to enhanced convective substrate flux. Further experimental studies in flow cells operated with permeation (thus with concentration polarization) and without permeation (no

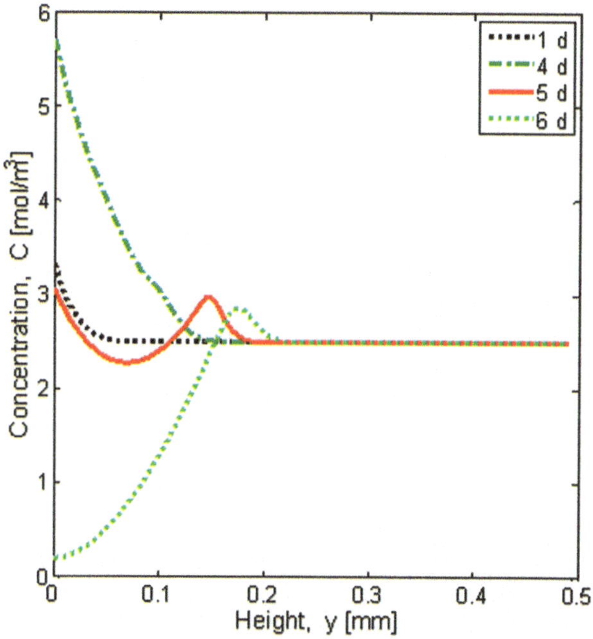

Figure 2.7 One-dimensional substrate profiles at 1, 4, 5 and 6 days in a biofilm colony situated at $x = 8.75$ mm on the lower membrane (see red line in Figure 2.6). The switch from substrate accumulation due to concentration polarization (day 1 to 4) to depletion (day 6) is governed by biofilm thickness. Substrate concentration was 2.5 mol · m^{-3} and average inlet fluid velocity 0.1 m · s^{-1}. Legend: time in days. Note that the substrate peak around $y = 0.15$ mm is due to substrate brought by the liquid from regions of substrate concentration polarization situated near the membrane upstream this colony.

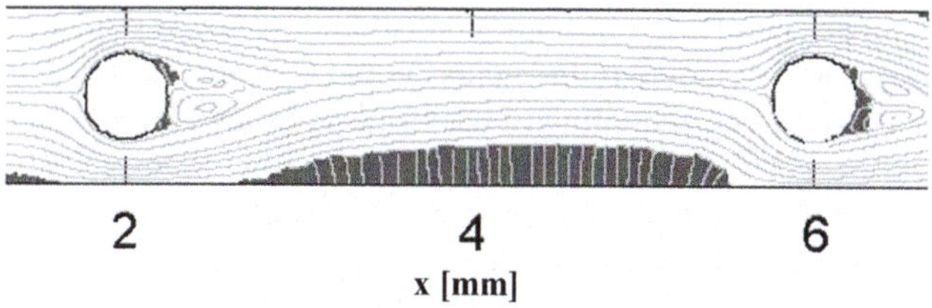

Figure 2.8 Detail with liquid flow through the biofilm. The light gray curves are liquid streamlines. The dark gray areas represent the biofilm colonies. Simulation conditions: substrate concentration 0.25 mol · m^{-3} and average inlet fluid velocity 0.1 m · s^{-1}.

polarization) could provide more insight, as well as applying multispecies biofilm models (Picioreanu *et al.*, 2004) in the RO biofouling process.

Reproducibility of biofouling simulations

The main stochastic biofilm processes are attachment and detachment; this aspect is being recognized an important factor that may affect the reproducibility of certain biofilm processes (Chambless & Stewart, 2007; Lewandowski *et al.*, 2004). In the current model, attachment is assumed to be occurring with a constant rate, but at any randomly chosen place, both on membranes and spacer.

In this context, the reproducibility of biofouling simulations was tested in a series of five runs starting from different seeds of the random number generator used in the computed code. In addition, the effect of attachment rate and detachment (via cross-flow velocity) on the reproducibility of the numerical results was evaluated. Three scenarios were studied: (A) "low" attachment rate (5×10^{-3} C-mol · m^{-2} · day^{-1}) and "high" detachment ($U_{in} = 0.1$ m · s^{-1}), (B) "high" attachment rate (15×10^{-3} C-mol · m^{-2} · day^{-1}) and "high" detachment ($U_{in} = 0.1$ m · s^{-1}), (C) "low" attachment rate (5×10^{-3} C-mol · m^{-2} · day^{-1}) and "low" detachment ($U_{in} = 0.033$ m · s^{-1}).

It can be assumed that the attachment rate may play a role in the biomass development especially in the early stages: an increased attachment rate increases the probability of attaching a cell in the right place, i.e., where it will not be immediately detached due to shear stress caused by the fluid flow. Indeed, numerical results show better reproducibility for the total biomass amount at higher attachments rates (Figure 2.9 series (A) compared to series (B)). At the same time, as expected, biomass accumulation is significantly faster for higher attachment rate. However, it can be noticed that the amount of biomass in the system at steady state is independent of the attachment rate. A relatively constant amount of biomass obtained at steady state for different attachment rates shows that growth and detachment are important in the model and the biofilm formation in the model is not based on a filtration effect (i.e., the simple accumulation of cells due to deposition). However, under severe nutrient limitations, it is possible that initial attachment of cells has a major contribution to the overall development of the biofilm. Additional model refinements should consider selective attachment of bacteria, as a result of i.e., coupling of bacterial motion with the flow field.

We would expect that for the same attachment rate, a lower detachment rate increases the probability that cells once attached will remain in place. For "low" attachment (5 cells/h), biomass development was studied under two cross-flow velocities: $U_{in} = 0.033$ m · s^{-1} ("low" detachment) and $U_{in} = 0.1$ m · s^{-1} ("high" detachment). As expected, for a lower detachment better reproducibility of results was obtained (compare biomass accumulation in Figure 2.9 series (A) with that in series (C)). Until ~ day 8 (figure series (C)) the same amount of biomass exists in the channel for all simulation runs at low cross-flow velocity. A good agreement is

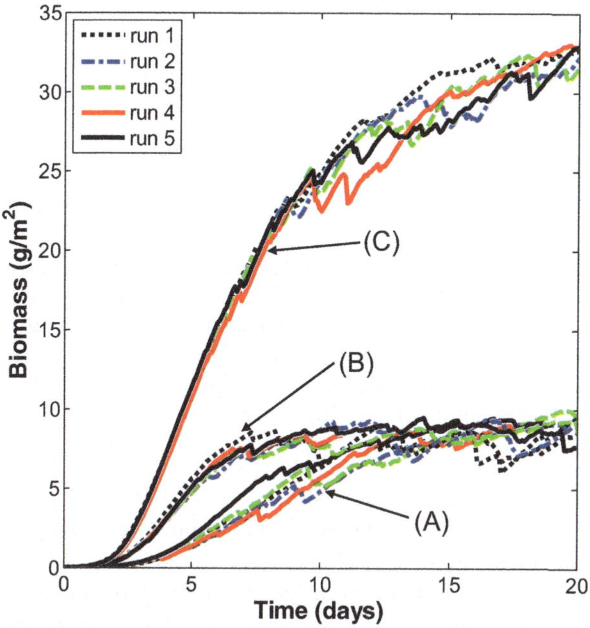

Figure 2.9 Evolution of biomass in the feed spacer channel obtained from simulation runs using various seeds for the random number generator and different conditions: (A) $U_{in} = 0.1\ m \cdot s^{-1}$; attachment rate of 5×10^{-3} C-mol $\cdot\ m^{-2} \cdot day^{-1}$; (B) $U_{in} = 0.1\ m \cdot s^{-1}$; attachment rate of 15×10^{-3} C-mol $\cdot\ m^{-2} \cdot day^{-1}$ h; (C) $U_{in} = 0.033\ m \cdot s^{-1}$; attachment rate of 5×10^{-3} C-mol $\cdot\ m^{-2} \cdot day^{-1}$. All other model parameters were identical.

observed between the replicates of each simulation during the exponential growth phase and the mass-transfer limited phase. After the first major sloughing event (approx. day 9), the amount of biomass in the system changes unpredictably, but still the same "noisy" steady state is reached (~32 g biomass $\cdot\ m^{-2}$). At a high cross-flow velocity ("high" detachment) differences in the amount of biomass in the channel can already be observed earlier, from day 4, right after the exponential growth phase (Figure 2.9 series (A)). However, replicates indicate that the general conclusion reached remains valid: although the attachment is modelled as a stochastic event its effect on the biomass accumulation remains limited, i.e., a higher cross-flow velocity leads to less biomass in the channel.

Overall, less reproducibility was obtained for the results at highest detachment and lowest attachment rate (Figure 2.9 series (A)). The cause of these slightly different total biomass accumulation curves is the biomass location: more biomass attached to the membrane in the exponential growth phase seems to lead to more biomass formed later, during the substrate limitation phase. Figure 2.10 shows comparatively the biomass structures in the channel at day 3 obtained in runs 4 and

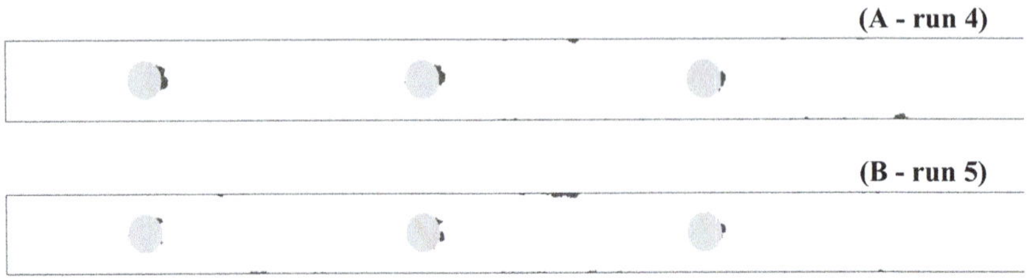

Figure 2.10 Biomass distribution in the channel at t = 3 days, for runs 4 (A) and 5 (B) at $U_{in} = 0.1\ m \cdot s^{-1}$; attachment rate of 5×10^{-3} C-mol $\cdot\ m^{-2} \cdot day^{-1}$. Although the total amount of biomass is almost identical, for (A) 37% of the total biomass is on the membranes, while for (B) 74% is on the membranes. This may lead to faster biomass growth in case (B) in the mass-transfer limited phase, but eventually the same steady state is obtained (see Figure 2.9).

5. Although the biomass amount for these two simulations is very similar (0.352 and 0.362 $g \cdot m^{-2}$), the biomass distribution in the channel can be different. For run 4, 63% of the total biomass developed on the spacer and only 36% on the membranes (Figure 2.10(A)); for run 5 (Figure 2.10(B)), there was more biomass present on the membranes (74%) compared to the spacer (26%). In this particular 2-d case with spacer filaments in the middle of the channel, due to the concentration polarization effect the substrate availability next to the membrane is higher than next to the spacer and this may explain a slightly higher growth rate on the membranes. On the other hand, also the biomass detachment rate is different at various places (e.g., on the membranes and spacer).

CONCLUSIONS

The proposed numerical model can provide significant insight on biofilm formation in relation to different operational conditions of reverse osmosis membrane systems. Simulations suggested places in the feed channel where biofilms are more likely to form, usually in regions of low liquid shear stress. In all simulations at constant inlet flow rate a steady state was reached in biomass amount, as a consequence of an increasing detachment rate in time, which eventually could balance the biomass growth rate. At high liquid flow rates, the higher shear leads to more detachment and in consequence to thinner biofilms. The amount of biomass at steady state is fluctuating in time because of intermittent sloughing events. The permeate flux decline is directly correlated with the amount of biomass in the feed channel. The relative increase in pressure drop is less pronounced at higher flow velocities, due to less biomass formation. Simulations clearly illustrate how the rejected substrate accumulates at the membrane surface due to concentration polarization. With an increased biofilm thickness, however, the overall substrate consumption rate dominates the accumulation due to polarization. Therefore, the accumulation of rejected substrate could only accelerate in a limited extent the biofilm growth. Finally, the dynamics of biomass accumulation in the channel for several replicate model runs proves not to be strongly influenced by the stochasticity of the microbial attachment mechanism considered in the model.

A two-dimensional model is limited in interpreting experimental data for spiral-wound reverse osmosis process on a quantitative basis. However, this is a necessary step towards much more demanding three-dimensional calculations. We believe that the numerical approach combining computational fluid dynamics with biofilm models will properly complement experimental observations. The model can theoretically explain results from particle image velocimetry (Willems *et al.*, 2010) and magnetic resonance measurements (Vrouwenvelder *et al.*, 2010b) of flow fields or the salt concentration polarization (Chong *et al.*, 2008a; Herzberg & Elimelech, 2007). Moreover, the model suggests that new measurement techniques need to be developed, e.g., for accurate substrate measurements and for biofilm mechanical properties. In this way, the modelling approach will improve the understanding of biofouling in spiral-wound reverse osmosis and nanofiltration processes.

Chapter 3

Modelling the effect of biofilm formation on reverse osmosis performance: Flux, feed channel pressure drop and solute passage

INTRODUCTION

The scarce availability of clean water in certain areas has led to an increased interest for fresh water production via desalination of brackish and seawater or, e.g., treated wastewater. In recent years, reverse osmosis (RO) has become the leading global technology to desalinate water because of the significant progress in membrane technology and the advantages RO offers compared to thermal desalination techniques (Kaghazchi *et al.*, 2010; Pankratz, 2000).

From brackish water, with salinity between fresh and seawater, drinking water can be produced at lower energy costs than from seawater. For this reason, most of the desalination plants use brackish water as feed (Greenlee *et al.*, 2009). One of the limitations for RO use in desalination is a strong tendency to membrane fouling (Tran *et al.*, 2007). Several forms of fouling were recognized: (i) mineral scaling (salt precipitation from supersaturated solutions), (ii) colloidal fouling (cake formation by suspended particles), (iii) organic fouling (due to the presence of natural organic matter) and (iv) biofouling (accumulation of biomass over a tolerable amount due to growth of microbial films – called biofilms). Most often these forms of fouling influence each other and do not occur independently. While scaling and colloidal fouling can be prevented with adequate water pre-treatment, biofouling remains a major concern in RO membrane plant operation (Baker & Dudley, 1998a). Biofouling affects not only the quantity (declined permeate flux) and quality (increased salt passage) of produced water, but also the associated operational costs by increased pressure drop over the feed channel.

The way biofouling deteriorates the membrane process performance has been studied mainly experimentally. Multiple causes were identified: additional

Author(s): Szilárd Bucs, Joop Kruithof, Mark van Loosdrecht and Johannes Vrouwenvelder
doi: 10.2166/9781780409597_0029

hydraulic resistance for the trans-membrane flow (McDonogh *et al.*, 1994), biofilm enhanced concentration polarization (Chong *et al.*, 2008a; Herzberg & Elimelech, 2007) and increased feed channel pressure drop (Vrouwenvelder *et al.*, 2009a).

As the membrane is permeable for water but it rejects most of the solute, a thin layer of concentrated solutes termed concentration polarization (CP) layer forms near the membrane surface. The CP layer substantially decreases the permeation driving force by increasing the osmotic pressure at the membrane surface. Direct measurements of solute gradients near the membrane are limited, in spite of several techniques available for in-situ monitoring (Ghidossi *et al.*, 2006). Due to difficulties in measuring the CP layers for reverse osmosis membranes, this issue has been mainly addressed by modelling studies (Lyster & Cohen, 2007; Subramani *et al.*, 2006). With the exception of Chong *et al.* (2008a), none of the proposed models explicitly included biofilm presence. The classical film theory can provide a quick estimate for concentration polarization in open channels (no spacers, no biofilms), but for more complex geometries (i.e., containing spacers or geometrically irregular biofilms) a numerical approach is needed (Geraldes *et al.*, 2000).

Herzberg and Elimelech (2007) suggested that the biofilm could contribute to an enhanced concentration polarization by hindering back diffusion of salts to the bulk liquid. Kim *et al.* (2006) presented an analytical mode for solute transport within the biofilm. Their results indicate that the EPS may be important in hindering the back diffusion of solutes from the membrane surface. The work of Chong *et al.* (2008a) gave a measurement of the average CP value, using a tracer technique. In addition, they attempted to create a simple model to predict the membrane system performance in the presence of homogeneous and constant thickness biofilms covering the membrane. This situation may only occur in the absence of a feed channel spacer, whereas in field applications where the spacer is needed to separate the membrane leaves, a more "patchy" biofilm forms (Vrouwenvelder *et al.*, 2009a; Vrouwenvelder *et al.*, 2010c). Therefore, the development of a more elaborate model in two or three spatial dimensions is needed for a better description of experimental observations.

The recent development of sophisticated CFD (computational fluid dynamics) packages enabled three-dimensional (3-d) studies of the influence of spacer geometry on the flow and mass transfer (Fimbres-Weihs & Wiley, 2007; Koutsou *et al.*, 2009; Lau *et al.*, 2009; Li *et al.*, 2005). However, the effect of biofilm growth on the hydrodynamics has only recently been incorporated in a three-dimensional model of biofouling in RO systems (Picioreanu *et al.*, 2009) and qualitatively compared with experimental data (Pintelon *et al.*, 2010; Vrouwenvelder *et al.*, 2010b). In these studies, the membrane fouling simulator was operated without permeation, so the CP phenomenon could not be addressed. Moreover, the influence of biofilm growth on flux decline and solute rejection in an RO system has not been systematically evaluated by numerical approaches.

The objective of this work was to develop, based on first principles, a spatially two-dimensional (2-d) numerical model to study the dynamics of global and local impact of biofouling in a reverse osmosis membrane device. The model describes:

(i) the flux loss and (ii) pressure drop increase in time, and (iii) the salt rejection decline, as a result of a growing biofilm with different feed spacer geometries, for brackish water desalination. We also investigated the local consequences of biofilm development on the membrane performance. In particular, the relative contributions of biofilm hydraulic resistance to trans-membrane flow, biofilm enhanced concentration polarization and increased feed channel pressure drop were evaluated. The numerical approach can be useful, as it allows uncoupling of several phenomena and contributes thus to a better understanding of the mechanisms through which the biofilm can affect membrane system performance.

MODEL DEVELOPMENT

In this study we build up a numerical model and investigate by computer simulations the role of biofilms in the decrease of performance of spiral-wound reverse osmosis membrane devices. Hereby we focus on application to brackish water desalination. However, the model is also relevant for other reverse osmosis and nanofiltration applications. The model includes two-dimensional liquid flow through the feed channel and permeable membranes, coupled with the mass transport of solutes, together with the biofilm development.

Model geometry

Previous studies have shown that a fully three-dimensional (3-d) description of flow and mass transport is needed to have a correct quantitative representation of the process (Picioreanu *et al.*, 2009; Vrouwenvelder *et al.*, 2010b). However, when accounting for mass transfer through the membrane the concentration polarization (CP) plays an important role. Because steep concentration gradients near the membrane surface must be calculated on a very fine mesh, large number of variables, large memory and relatively long computing times are needed. Therefore, 3-d models including CP in complex geometry with biofilm growth in time would currently be computationally very demanding. We believe that a two-dimensional (2-d) model is a good initial compromise between the calculation time and model realism. It is usually accepted that 2-d calculations can be a screening device for future evaluations of more complex geometries and, at the same time, allow the effect of numerical parameters and modelling assumptions to be studied (Schwinge *et al.*, 2002a).

Although not completely realistic, the two-dimensional geometry can still provide useful approximations in some particular cases. For instance, slices cut through a woven spacer in ladder configuration (Figure 3.1a) lead to 2-d geometries known as "zig-zag spacer" (Figure 3.1c) (Schwinge *et al.*, 2002a; Song & Ma, 2005). Another studied configuration (Schwinge *et al.*, 2002a) is the "cavity spacer" (Figure 3.1b) resulted from a cut through the non-woven ladder spacer (Figure 3.1c). A slice cut through a spacer having the axial filaments thicker than the transverse filaments (Figure 3.1e) can lead to the 2-d geometry known as

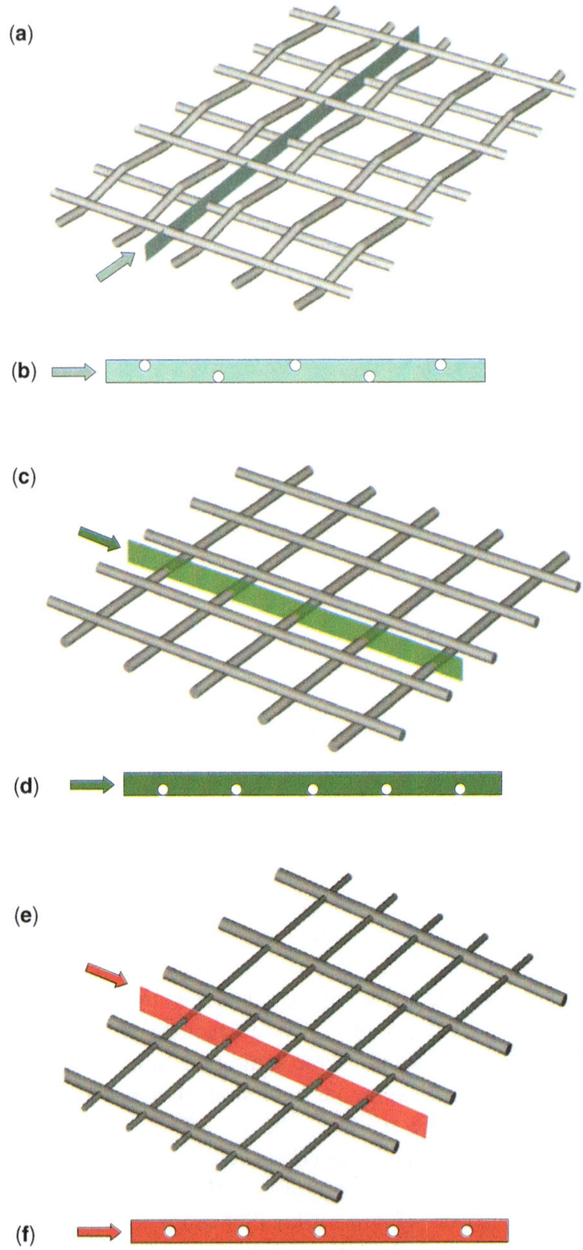

Figure 3.1 Schematic view of spacer configurations. (a) 3-d woven-like spacer; (b) 2-d cross-section: zigzag spacer; (c) 3-d spaghetti-like spacer; (d) 2-d cross-section: cavity spacer; (e) 3-d middle-layer spacer; (f) 2-d cross-section: submerged spacer.

"submerged spacer" (Figure 3.1f). These are just a few examples, as other spacers used in practice or reported in the literature could also be approximated by one of these simplified 2-d geometries. The validity of the 2-d approximation for some cases was discussed in (Shakaib *et al.*, 2007). In addition to these configurations with spacer, the empty channel (i.e., channel without spacer) was also evaluated in simulations presented in this paper.

An example of model geometry is shown in Figure 3.2, for the submerged spacer case. The geometrical characteristics chosen for this study are representative in practice in the case of spiral wound reverse osmosis processes (Schwinge *et al.*, 2002a; Vrouwenvelder *et al.*, 2009a; Vrouwenvelder *et al.*, 2010b). The channel height is $L_Y = 1$ mm. The length of the modelled channel ($L_X = 15$ mm) is small compared to the membrane module size used in practice. The effect of the curvature of the module is neglected, as some studies have emphasized that fluid dynamics is not significantly different in flat and curved channels (Ranade & Kumar, 2006; Schock & Miquel, 1987), while others suggest possible influences (Wardeh & Morvan, 2009). The spacer filament diameter d_s is chosen so that $d_s/L_Y = 0.5$; the distance between two consecutive

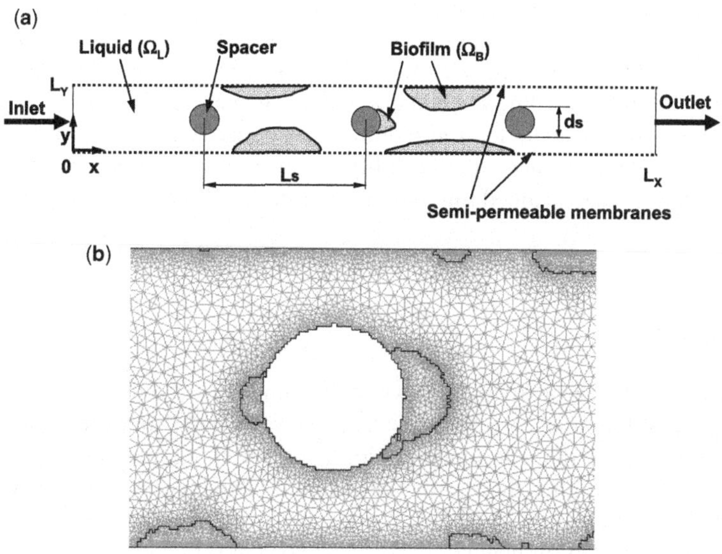

Figure 3.2 *Computational domain and finite element mesh.* (a) Illustration of the computational domain of size $L_X \times L_Y$. For numerical reasons, liquid (Ω_L) and biofilm (Ω_B) sub-domains are explicitly defined. (b) Detail of the finite element mesh used to solve the hydrodynamics and mass transport equations. The mesh size of 5 μm on the boundaries was increased towards the bulk liquid up to a maximum of 30 μm. Geometry dimensions are in Table 3.1.

filaments L_s equals 4 times the channel height. In the computational domain, no spacers were placed near the liquid inlet or outlet to ensure a fully developed flow upstream the spacer array and to avoid the outlet interference with the flow recirculation regions after the last spacer filament. We consider for computational convenience the whole domain Ω divided into two sub-domains: biofilm Ω_B and bulk liquid Ω_L.

Model equations

Fluid flow

Due to the geometrical complexity when including spacers in the channel and especially when irregularly shaped biofilms develop, the flow pattern in the liquid sub-domain Ω_L must be calculated directly from the incompressible Navier–Stokes equations, for laminar regime:

$$\begin{aligned} &\rho(\mathbf{u}\cdot\nabla)\mathbf{u} + \nabla p = \nabla\cdot(\eta\nabla\mathbf{u}) \\ &\nabla\cdot\mathbf{u} = 0 \end{aligned} \tag{3.1}$$

where $\mathbf{u}$ is the vector of local liquid velocity (with u and v the velocity components tangential and normal on the membrane, respectively), p is the pressure, ρ is the liquid density and η is the liquid dynamic viscosity. The assumption of a steady-state laminar flow is justified at the Re numbers used in the simulations (Re $\approx$ 160). At higher Re, unsteady flows would develop, as shown in (Schwinge *et al.*, 2003).

For the flow calculations, the biofilm sub-domain Ω_B was assimilated with a porous medium, having a permeability $\kappa = 10^{-16}\ \mathrm{m^2\,s^{-1}}$ and a porosity $\varepsilon = 0.8$. Brinkman flow was assumed through the biofilm (Brinkman, 1949):

$$\begin{aligned} &\frac{\eta}{\kappa}\mathbf{u} + \nabla p = \frac{1}{\varepsilon}\left[\eta\cdot\nabla^2\mathbf{u} - \left(\frac{2\eta}{3} - \kappa\right)(\nabla\cdot\mathbf{u})\right] \\ &\nabla\cdot\mathbf{u} = 0 \end{aligned} \tag{3.2}$$

The dependence of different physical parameters (i.e., liquid viscosity and density) of solute concentrations needs to be considered in both sub-domains (see Table 3.1), as the salt concentration near the membrane can reach values up to 2–5 times higher than in the bulk liquid. This necessity was underlined by the work of Lyster and Cohen (Lyster & Cohen, 2007). Consequently, the flow velocity is also influenced by the local concentration of salt, which in turn is altered by the hydrodynamics (because the flow determines the convective transport of salt).

At the inlet ($x = 0$) a fully developed laminar velocity profile (parabolic) is specified with an average velocity, $u_{\mathrm{in}} = 0.1\ \mathrm{m\,s^{-1}}$, value kept constant in time. The liquid velocity through the membrane is the velocity component v (normal to the membrane), which takes variable values at different positions x along the

Table 3.1 Biofouling model parameters.

Parameter	Symbol	Value	Units
Geometrical parameters			
Channel length	L_X	15×10^{-3}	m
Channel width	L_Y	1×10^{-3}	m
Limits for biomass attachment on the channel length	L_i	1.35×10^{-3}	m
	L_f	12.9×10^{-3}	m
Spacer diameter	d_s	0.5×10^{-3}	m
Distance between spacers	L_s	4×10^{-3}	m
Physical parameters			
Liquid density[A]	ρ	$\rho(C)^a$	kg m^{-3}
Liquid dynamic viscosity[B]	η	$\eta(C)^b$	Pa s^{-1}
Osmotic pressure[C]	Π	$\Pi(C)^c$	bar
Diffusion coefficient for salt[D]	D	$D(C)^d$	m^2 s^{-1}
Diffusion coefficient for substrate	D_S	10^{-9}	m^2 s^{-1}
Biofilm properties			
Biofilm hydraulic permeability	κ	10^{-16}	m^2 s^{-1}
Biofilm porosity	ε	0.8	m^3 liquid m^{-3} biomass
Maximum specific biomass growth rate	μ_{max}	2	day^{-1}
Biomass yield on substrate	Y_{SX}	0.5	mol C-mol^{-1}
Monod half-saturation constant	K_S	0.025	mol m^{-3}

(continued)

Table 3.1 Biofouling model parameters (*Continued*).

Parameter	Symbol	Value	Units
Biomass attachment rate	$r_{X,\mathrm{att}}$	5×10^{-3}	C-mol m^{-2} day^{-1}
Maximum biomass concentration in the biofilm	$C_{X,m}$	8000	C-mol m^{-3}
Elastic modulus	E	64	N m^{-2}
Poisson ratio	ν_P	0.3	–
Threshold stress for biomass detachment	σ_{det}	7	N m^{-2}
Operation conditions			
Mean inlet velocity	u_{in}	0.1	m s^{-1}
Inlet concentration of salt	C_{in}	40	mol m^{-3}
Inlet concentration of substrate	$C_{S,\mathrm{in}}$	0.25	mol m^{-3}
Membrane permeability	L_p	9.7×10^{-12}	m Pa^{-1} s^{-1}
Rejection coefficient	R	0.98	
Outlet pressure	p_{out}	15	bar

[a] $\rho(C) = (0.999 + 4.099 \times 10^{-5} \times C) \times 10^{3}$.
[b] $\eta(C) = (1.004 + 8.377 \times 10^{-5} \times C) \times 10^{-3}$.
[c] $D(C) = (1.418 + 0.1159 \times C)/(1 + 0.085 \times C + 1.514 \times 10^{-5} \times C^{2}) \times 10^{-9}$.
[d] $\Pi(C) = 0.04572 \times C$; where C is the salt concentration in mM NaCl.

membrane length. At the membrane surface, the tangential velocity u is set to zero. This results in the following boundary conditions on the lower membrane:

$$\begin{aligned} u(x,0) &= 0 \\ v(x,0) &= -L_P(\Delta p - \Delta\pi) \end{aligned} \tag{3.3}$$

and on the upper membrane:

$$\begin{aligned} u(x,L_Y) &= 0 \\ v(x,L_Y) &= L_P(\Delta p - \Delta\pi) \end{aligned} \tag{3.4}$$

The local permeate flux, v, is proportional with the trans-membrane pressure Δp (or TMP), counteracted by the osmotic pressure $\Delta\pi$ created by the concentration polarization. The proportionality coefficient is the membrane permeability L_P. The local TMP is defined as the difference between the local pressure on the feed side on the membrane and an assumed constant permeate pressure $p_P = 1$ bar (100 kPa), so that $\Delta p = p(x,\ 0) - p_P$ and $p(x,\ L_Y) - p_P$. The osmotic pressure difference is proportional (osmotic coefficient B) with the salt concentration difference on the two sides of the membrane: $\Delta\pi = B\ (C(x,\ 0) - C_P)$ and $B\ (C(x,\ L_Y) - C_P)$. For numerical stability reasons, transition zones were created near inlet ($x = [0,\ 0.02 \cdot L_X]$) and outlet ($x = [0.98 \cdot L_X,\ L_X]$), where the membrane permeability was smoothly allowed to change from zero to L_P. No slip conditions were imposed on the spacer surface, i.e., $\mathbf{u} = 0$. Flow continuity was assumed at the biofilm surface. On the outlet boundary ($x = L_x$) the pressure is set constant to p_{out}.

Mass balances

The salt and substrate are assumed to be the only soluble compounds relevant for the scope of this study. Salt is chosen because it is the main contributor to concentration polarization. The substrate is needed because changes in its availability determine the development of different biofilm structures. The mass balance for salt (concentration C) in the channel includes transport by convection and diffusion (Eq. 3.5). For the substrate, besides transport, there is also consumption due to microbial growth (Eq. 3.6). In this first approach only one substrate (e.g., organic compounds) of concentration C_S limits the biofilm growth.

$$D\nabla^2 C - \mathbf{u}\nabla C = 0 \tag{3.5}$$

$$D_S\nabla^2 C_S - \mathbf{u}\nabla C_S + r_S = 0 \tag{3.6}$$

For the salt, the diffusion coefficient D depends on the local concentration C (Table 3.1). However, due to the very low concentrations of substrate the diffusion coefficient D_S is considered constant. For simplicity, we assume the same diffusion coefficients in the bulk liquid and in the biofilm sub-domain. For substrate consumption, a simple Monod kinetics was used:

$$r_S = Y_{SX}\mu_m \frac{C_S}{K_S + C_S} C_X \tag{3.7}$$

The reaction rate parameters (the yield Y_{SX}, the maximum specific growth rate μ_m and the half saturation coefficient K_S) are set as for heterotrophic microorganisms according to activated sludge models (Table 3.1). The local biomass concentration in the biofilm C_X results from the biofilm sub-model. Possible substrate consumption by suspended biomass in the liquid is neglected.

On the inlet boundary ($x = 0$) the brackish water contains salt and the limiting substrate with concentrations C_{in} and $C_{S,\text{in}}$, respectively, and these concentrations are constant in time. No-diffusion conditions ($D(\partial C/\partial x) = 0$, $D_S(\partial C_S/\partial x) = 0$) apply on the outlet boundary ($x = L_X$). The whole spacer surface is considered as impermeable wall (zero-flux condition). The convective flux of solute toward the membrane (at $y = 0$ and $y = L_Y$) equals the sum of the diffusive back transport of solute and the convective flux passing through the membrane.

$$\left(vC - D\frac{\partial C}{\partial y} \right)_{y=0 \text{ or } L_Y} = vC(1 - R) \tag{3.8}$$

$$\left(vC_S - D_S\frac{\partial C_S}{\partial y} \right)_{y=0 \text{ or } L_Y} = vC_S(1 - R) \tag{3.9}$$

The same membrane rejection coefficient, $R = 0.98$, was assumed both for salt and substrate. Flux continuity applies on the biofilm-liquid interface.

Biofilm sub-model

As a biofilm develops on the membrane and spacer surface, the concentration and velocity fields in the channel will change. On the other hand, biofilm development is influenced by substrate availability and hydrodynamic conditions. Microbial cells grow and divide if nutrient is supplied and are detached if they are exposed to high shear stress induced by liquid flow.

We adapted here the discrete biofilm model by Picioreanu *et al.* (Picioreanu *et al.*, 2001, 1998b). The new model includes biomass accumulation by growth, transport and biofilm/liquid transfer (e.g., attachment and detachment). In this approach, the whole computational space is divided on a grid of square elements. In the biofilm domain the squares contain biomass with a concentration C_X, whereas in the liquid space $C_X = 0$. The interface between the bulk liquid and the biofilm is a moving boundary, resulting from biomass growth and transport, coupled with attachment and detachment processes.

Biomass attachment is included here with a constant rate of $r_{X,\text{att}} = 6 \times 10^{-4}$ (C-mol biomass) (m^{-2} module) day^{-1}. For the model solution this is equivalent to 5 grid elements containing biomass newly added each hour. The attachment place is chosen randomly: on the membranes, on the spacers and on existing biofilm colonies.

Biomass growth occurs in each square element containing biofilm with the rate r_X correlated to the substrate consumption rate (Eq. 3.7):

$$\frac{dC_X}{dt} = r_X = \mu_m \frac{C_S}{K_S + C_S} C_X \tag{3.10}$$

It was assumed that the osmotic stress (caused by high salt concentrations) does not inhibit biomass growth. If natural selection takes place, only microbial strains resistant to high salt concentration will grow in the channel. In addition, it has been proved by Chong *et al.* (2008a) that the viability of certain bacterial strains is not affected by high concentrations of sodium chloride (up to 10000 ppm).

Biomass transport takes place when the local biomass concentration exceeds a critical value $C_{X,m}$. The redistribution of biomass in space is performed using the cellular automata algorithm described in (Picioreanu *et al.*, 1998b).

Biomass detachment is based on internal stress created by moving liquid past the biofilm. The model was described in detail in (Schwinge *et al.*, 2003), thus only the main assumptions will be presented here. The biofilm is considered a solid elastic material in a state of plane stress (because of the 2-D geometry). After solving the liquid flow (Eqs 3.1 and 3.2), the load forces on the biofilm surface are known. The biofilm-membrane and biofilm-spacer interfaces are fixed (no movement). The plane stress equations are solved for the mechanical stress σ.

Biofilm breakage was considered to occur when the calculated Von Mises stress σ in a certain biomass element is higher than a threshold stress σ_{det} corresponding to the cohesive strength of the biofilm. In this stage, detachment means that all biomass elements (squares) where $\sigma > \sigma_{\text{det}}$ are removed from the biofilm, their place being taken by liquid. We assume that detached biofilm patches are quickly taken away by the flow and washed out from the system (i.e., no filtration effect taken into account). This detachment stage resembles biofilm erosion. In a second stage, all suspended biomass (i.e., no longer connected to a solid surface – either biofilm or membrane or spacer) is also removed in a process resembling the biofilm sloughing.

Model parameters

For the case study reported in this work conditions applicable to brackish water desalination were chosen: inlet salt concentration 40 mM NaCl (Greenlee *et al.*, 2009), operating pressure set to 15 bar in the outlet (Greenlee *et al.*, 2009), average inlet flow velocity of 0.1 m s^{-1} (Chong *et al.*, 2008a; Lyster & Cohen, 2007;

Vrouwenvelder *et al.*, 2009a), and a membrane permeability of $9.7 \cdot 10^{-12}$ m Pa^{-1} s^{-1} (Lyster & Cohen, 2007). The salt and substrate membrane rejection coefficients were set to 0.98, which is in the range of values reported in the literature for RO processes (Greenlee *et al.*, 2009). The effect of different rejection coefficients, different salt concentrations and operation pressures on biofouling will be investigated in future studies. A relatively high value for the concentration of substrate in the inlet was deliberately chosen (0.25 mM), in order to achieve substantial biomass accumulation over a short period of time (strategy used also in several experimental studies on membrane biofouling (Herzberg & Elimelech, 2007; Vrouwenvelder *et al.*, 2009b). Physical properties (ρ, η, Π and D) were a function of salt concentration, according to the expressions proposed in (Lyster & Cohen, 2007). A constant value for the substrate diffusivity D_S was assumed, similar to the diffusion coefficient for acetate (Wanner & Gujer, 1986).

While operational conditions and the liquid and solute physical parameters are rather well characterized, the biofilm-related parameters are in general less known and reliable. Typical biokinetic parameters (μ_{max}, K_S, Y_{SX}) and biofilm biomass density ($C_{X,m}$) were selected as for heterotrophic microorganisms (cf. activated sludge models, (Wanner & Gujer, 1986)). Biofilm permeability is an important parameter, not very well characterized experimentally yet. Values in the order of 10^{-15} m^2 are currently accepted for bacterial cell cakes, while the extracellular polymeric substances (EPS-like gels) are less permeable, e.g., 10^{-16} to 10^{-17} m^2 (McDonogh *et al.*, 1994; Schock & Miquel, 1987). The value assumed here (10^{-16}) corresponds to a biofilm matrix containing bacterial cells and EPS (McDonogh *et al.*, 1994). Biofilm mechanical properties (E and ν_P) cannot be easily measured so we chose values according to Picioreanu *et al.* (Picioreanu *et al.*, 2001). Biofilm detachment is still a subject of ongoing research (Kapellos *et al.*, 2007). The value selected for the threshold stress for detachment is in the range of values reported in (Möhle *et al.*, 2007). Finally, a constant biomass attachment rate was chosen for inoculation, sufficiently high to ensure that the biomass is not completely removed in the initial biofilm formation stage. However, in our simulations, in later biofilm phases the chosen attachment rate has no significant influence on the overall biomass accumulation in the system.

Model solution

The model was solved by combining MATLAB (MATLAB 2008b, Mathworks, Natick, MA, www.mathworks.com) code with the finite element method implemented in COMSOL Multiphysics (COMSOL 3.5a, Comsol Inc, Burlington, MA, www.comsol.com). Own Java code was used for the cellular automata biofilm model. The main steps of the algorithm are presented in Figure 3.3.

The first section of the MATLAB framework defines the model input parameters and the feed channel geometry with or without spacers. In the second section, the model equations with corresponding boundary conditions are defined and solved at different times, corresponding to different biofilm structures. The biofilm structure develops in a time loop, determined by the local substrate availability and hydrodynamic conditions (the chosen time step was $\Delta t = 2$ h). A sequential approach was used (Figure 3.3) based on the assumption that a time scale separation of hydrodynamics, mass transport and biomass growth can be considered when characterizing biofilm development (Picioreanu *et al.*, 2000).

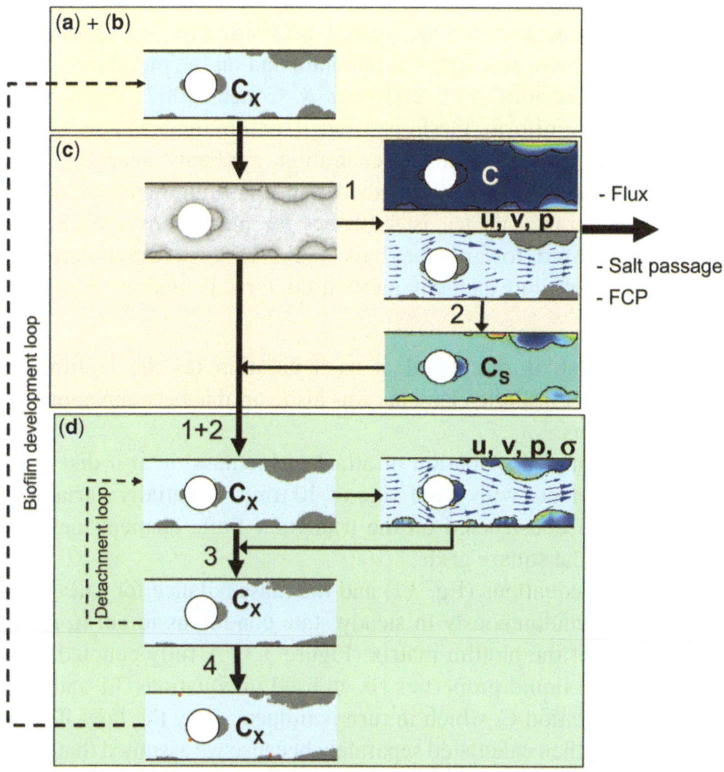

Figure 3.3 *Main steps of the numerical algorithm.* For a given biofilm geometry (either following initial biomass attachment or at any other intermediate state), (a) first the subdomains, equations, boundary conditions are defined, then (b) the finite-element mesh is created. Next, (c1) the hydrodynamics and mass transport for salt are simultaneously solved, followed by (c2) the solution of substrate transport and reaction. Based on the resulted substrate distribution, (d1 + d2) the biomass grows and spreads according to the cellular automaton mechanism. (d3) Mechanical stress in the biofilm is calculated and detachment steps are repeatedly performed (re-meshing after each step), followed by (d4) biomass attachment (newly attached biomass shown in red). Finally, time is updated and with the newly obtained biofilm geometry the whole cycle is repeated. The flux, salt passage and passage and feed channel pressure (FCP) drop are calculated from results of step c1.

At any time t the following sequence of calculations was used:

- The liquid and biofilm sub-domains are defined, together with the model equations and the boundary conditions for the current geometry. For this, the following COMSOL application modes were constructed: (i) incompressible Navier-Stokes for momentum transport, (ii) convection and diffusion for mass transport of salt and substrate and (iii) plane stress for detachment.
- The finite element mesh is created according to the specifications for maximum element size in the subdomain and on the membrane boundaries. As stated in previous studies (Lyster & Cohen, 2007; Wiley & Fletcher, 2003), a non-uniform mesh is needed to obtain accurate solutions for mass transfer due to high concentration gradients near the membrane. Preliminary simulations established that a minimum mesh size of 5 μm on the membrane surface is sufficient for mesh convergence. The mesh size is increased towards the bulk liquid up to a maximum of 30 μm. An example of finite element mesh used for calculations is illustrated in Figure 3.2b.

Because the mesh is re-created at each iteration (as the biofilm geometry changes), the number of finite elements was also variable between approx. 200000 and 270000 elements.

The biofilm (spatial distribution of attached biomass) is also discretized on a grid of square elements with fixed size of 10 μm. A spatially variable biomass function $C_X(x, y)$ is constructed on the triangular finite element mesh by linear interpolation from the square grid.

The liquid flow equations (Eq. 3.1) and the mass balance for salt (Eqs 3.5 and 3.8) were solved simultaneously in steady state conditions to get $\mathbf{u}$, p, and C for a given geometry of the biofilm matrix (Figure 3.3). A fully coupled approach is needed because the liquid properties (ρ, η) used in equations 3.1 and 3.2 depend on the salt concentration C, which in turn is influenced by the flow. The substrate concentration C_S is then calculated separately because we assumed that the physical properties of the brackish water do not depend on C_S.

Biofilm development after initial biomass attachment was calculated by a sequence of steps:

(1) biofilm *growth* to obtain the new biomass distribution $C_X(t + \Delta t, x, y)$ from $C_X(t, x, y)$ and the previously calculated substrate concentration $C_S(t, x, y)$.
(2) *spreading* (transport) of biomass by cellular automata algorithms implemented in own Java code, as described in (Wardeh & Morvan, 2009).
(3) biomass *detachment* as a result of mechanical stress induced by the liquid flow. First the flow pattern is recalculated for the new biomass distribution resulted after spreading. The biofilm surface is a sub-domain boundary loaded with forces generated by the flowing liquid. The Von Mises stress σ calculated on the triangular finite element mesh is then mapped to the square

elements discretizing the biofilm. In this stage, detachment means that all biomass elements (squares) where $\sigma > \sigma_{det}$ are removed from the biofilm, their place being taken by liquid. After a detachment event a new biomass distribution C_X results, the biofilm sub-domain boundaries change and the flow field and mechanical stress must be recalculated. Several detachment steps are performed until the mechanical equilibrium is reached (i.e., no detachment takes place anymore).

(4) *attachment* of new biomass was performed by placing new biomass in a fixed number of grid elements occupied by liquid, at places randomly chosen on the membranes, spacers and on existing biofilm colonies.

The algorithm steps (d1–d4) update the biomass distribution on the square mesh, re-shaping the system geometry. With the new biofilm geometry obtained, a new time step starts at step (a). Simulations were carried out until the biofilm attained a quasi-steady state, i.e., when the biofilm reached an almost constant average thickness. All model calculations reported here were performed on a HP workstation with two Intel Xeon E5430 quad-core processors and 32 Gb RAM. Typical runtime for one simulated case is ~3 days.

RESULTS

Two-dimensional simulations with biofilm formation using different feed channel geometries (without spacer and with three spacer configurations) were performed. We studied the impact of the biofilm developed in the membrane device on the flux, permeate quality and feed channel pressure drop under different conditions. Various mechanisms by which the biofilm contributes to performance decline were identified and discussed.

Biofilm development in the feed channel

Biofilm formation

It has been shown by several studies that there is a mutual influence between biofilm development and liquid flow (Ohl *et al.*, 2004; Picioreanu *et al.*, 2009; Stoodley *et al.*, 1999a; van Loosdrecht *et al.*, 1995). In all cases simulated here, small colonies appear initially in isolated places on the membrane and on the spacer (Figure 3.4). After a few days, for the cavity spacer configuration the biofilm colonizes the spacer surface and expands sideways. As a result of detachment due to high shear, there is no biomass on the upper membrane in places opposite to the spacer (Figure 3.4, 9 days). Finally, a quasi-steady state is achieved in the biomass developed in the channel when the biomass growth rate is balanced by the detachment rate (Figure 3.4, 13 and 15 days).

The quasi-steady state in biofilm thickness has been previously reported in many other biofilm applications, both by experimental (Flemming, 2002; Ohl *et al.*, 2004) and modelling studies (De Bivar Xavier *et al.*, 2005; Picioreanu *et al.*,

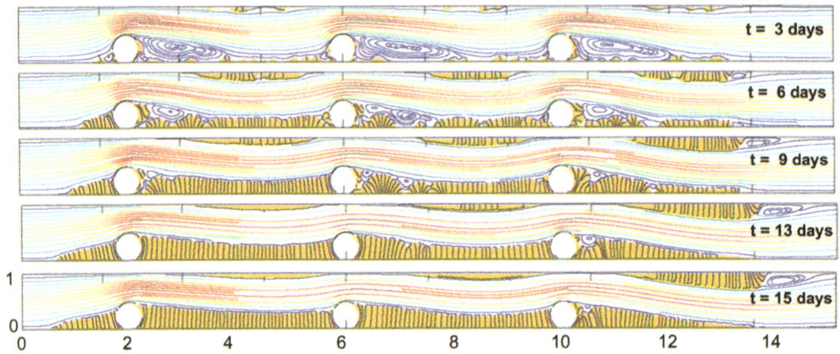

Figure 3.4 *Simulated biofilm development in time and streamlines in the feed channel with cavity spacers.* The biofilm (yellow areas) develops mainly on the lower membrane, supported by spacers (white circles). Streamlines colours change with the liquid velocity from dark blue (minimum, close to 0) to red (maximum, 0.28 m/s). Axes labels in mm.

2001). Due to the few stochastic elements in the model, differences between runs started with different seeds of the random number generator can occur. However, it has been shown that the only differences occur after the first biofilm sloughing events and eventually, the same "noisy steady state" is reached.

Model results demonstrate that spacer presence and arrangement has an important influence on the steady state biofilm distribution pattern in the feed channel (Figure 3.5). In the channel without spacer the colonies merge into a continuous biofilm (Figure 3.5a). In the spacer filled channel, the biofilm pattern depends on the spacer configuration. For the submerged configuration, no

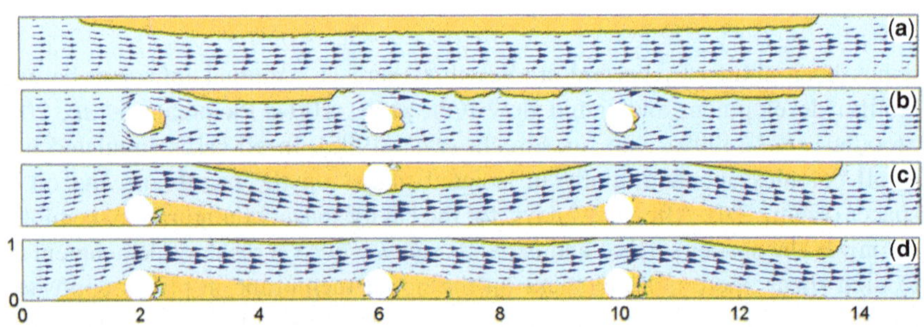

Figure 3.5 *Compared biofilm distributions and velocity vector plots for different feed channel geometries after 15 days.* (a) channel without spacers; (b) channel with submerged spacers; (c) channel with zigzag spacer; (d) channel with cavity spacers. The yellow areas represent the biofilm and the white circles the spacer. Axes labels in mm.

biofilm develops on the membrane at places above and below the spacer. Rather, a quasi-uniform layer is formed on the membrane between the spacer filaments (Figure 3.5b). Interestingly, in this configuration, streamer-like biofilms start developing behind the filaments, resembling structures observed experimentally (Vrouwenvelder *et al.*, 2010b). However, in the simulations these streamers easily break off, because in the model the biofilm is considered a brittle material. In reality, the biofilm behaves closely to a viscoelastic material allowing vibrations without breaking ((Stoodley *et al.*, 1998a) experimental work; (Taherzadeh *et al.*, 2010) theoretical model). The biofilm formation in zigzag and cavity configurations shows similar tendencies (Figure 3.5c and 3.5d): the biofilm seems to be held together by the spacer filaments. This shielding effect leads to significantly more biomass in zigzag and cavity configurations compared to the submerged spacer.

Substrate distribution

Biofilm development is mainly determined by two factors: substrate availability which influences biomass growth and hydrodynamic conditions which affect biomass detachment (in addition to substrate transport). Figure 3.6a

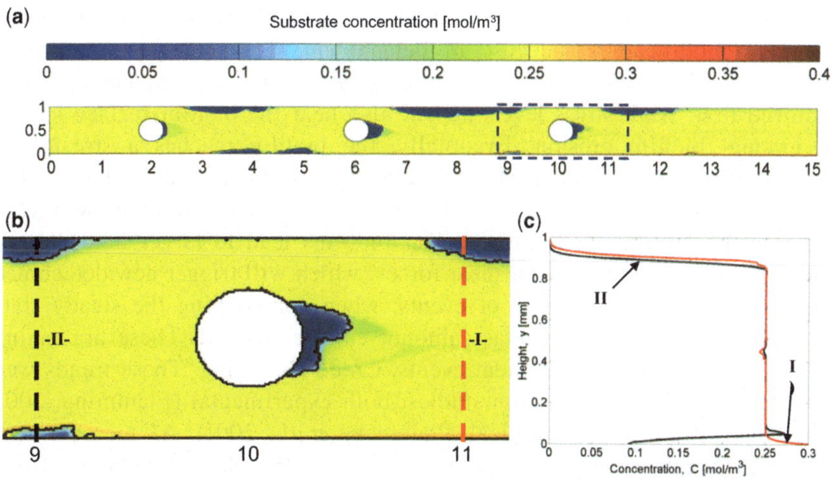

Figure 3.6 *Concentration polarization and consumption of substrate.* (a) 2-d substrate concentration distribution in the channel at $t = 8$ days. Flow direction is from left to right. (b) Detail of the substrate distribution in the channel. The same color scale as in (a) was used, from dark blue (minimum) to red (maximum concentration). (c) Cross-sections on the y-direction showing: I (red line) – substrate consumption in a biofilm colony on the upper membrane and substrate concentration polarization next to the lower membrane; II (black line) – substrate consumption in two biofilm colonies of different heights. The cross-sections are indicated by dashed lines in (b), at $x = 9$ mm (black) and $x = 11$ mm (red).

shows the substrate distribution corresponding to a biofilm structure at $t = 8$ days. In this stage, substrate depletion occurs in all the colonies situated on the membrane or on the spacer. On the contrary, the substrate concentration is higher on the membrane than in the bulk liquid in places without biofilm (Figure 3.6b – zoom – see concentration profile I next to the lower membrane, $y = 0$). This clearly demonstrates that the rejected substrate accumulates at the membrane surface (i.e., forming a substrate polarization layer). It should be noticed that the smaller microbial colonies benefit from a high substrate concentration in their vicinity: compare upper and lower colonies along line I in Figure 3.6b and the substrate profiles along line I in Figure 3.6c. As biofilm thickness increases, the overall substrate consumption rate will dominate over the substrate accumulation due to polarization, so that the substrate levels next to the biofilm-liquid interface will be lower than in the bulk. Most of the biofilm is substrate-limited, so that growth occurs only in a thin outer layer (the "active layer").

Biomass detachment

The biofilm detachment is based on shear stress calculations. Figure 3.7 illustrates the evolution of stress in a colony and the subsequent biofilm structure modifications because of detachment. The colony side frontally exposed to flow suffers from the highest stress and consequently will be removed first. High stress levels appear also near the biofilm surface leading to gradual biofilm erosion. Eventually, the biofilm reaches a streamlined shape, with overall lower levels of internal stress. This situation lasts only a limited period of time because re-growth in the detachment areas will again increase the biofilm thickness. Thicker biofilms lead to larger fluid velocity and consequently to increased shear forces, which will trigger new detachment events. Due to this sequence of events, when approaching the steady state, fluctuations in the total biomass amount can be noticed. These are mainly the result of massive detachment events, called sloughing. These trends were also described by other biofilm studies, both experimental (Flemming, 2002; Ohl *et al.*, 2004) and theoretical (Picioreanu *et al.*, 2001). As measurements of biofilm mechanical properties (i.e., Poisson ratio, elastic modulus, critical strength for detachment) are difficult, scarce and controversial in literature, this semi-empirical detachment model can be regarded here only as an adequate method for balancing biofilm growth.

Effect of biofouling on global membrane performance

Two main indicators for a membrane process performance are the permeate flux and salt passage in permeate. We calculated the average flux J ($L\ m^{-2}\ h^{-1}$) by integrating the local permeate velocity (v) over the length of which microbial

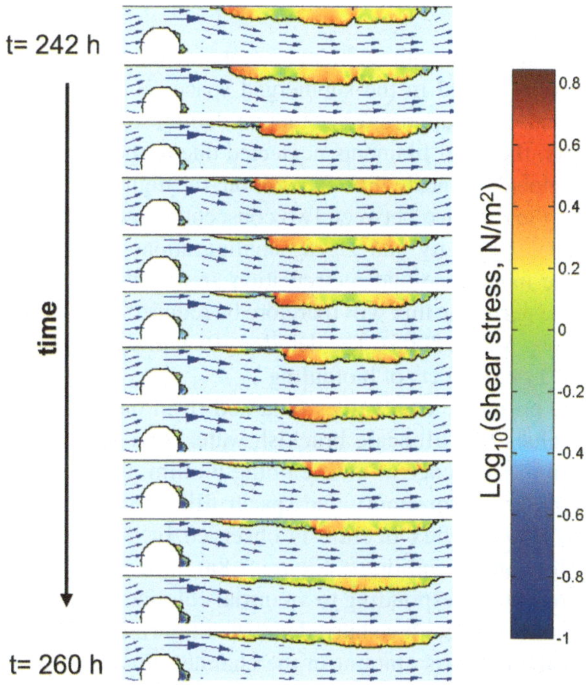

Figure 3.7 *Illustration of the mechanical stress (Von Misses) in a biofilm colony and the subsequent successive detachment events at times between* 242 *and* 260 h. Most stress is concentrated near the biofilm surface, leading to erosion: detachment of small biofilm surface pieces. Areas facing frontally the flow experience also high stress, leading to sloughing: fast detachment of larger biofilm chunks. Biofilm re-growth after detachment is also visible. In this example, the biofilm at 260 h reached a hydrodynamic shape that is rather stable mechanically. Note the logarithmic colour scale.

attachment was allowed (between 9% and 86% of L_X, i.e., L_i and L_f, respectively) for both lower and upper membranes:

$$J = \frac{1}{2(L_f - L_i)} \int_{L_i}^{L_f} [-\nu(x,0) + \nu(x,L_Y)]\mathrm{d}x \tag{3.11}$$

The salt passage is expressed by the average salt concentration (mol m^{-3}), calculated also by integration of salt flux across the membranes, divided by the total permeate flux:

$$\bar{C}_p = \frac{\int_{L_i}^{L_f} [-\nu(x,0)C(x,0) + \nu(x,L_Y)C(x,L_Y)](1-R)\mathrm{d}x}{\int_{L_i}^{L_f} [-\nu(x,0) + \nu(x,L_Y)]\mathrm{d}x} \tag{3.12}$$

Permeate flux

For spacer-filled channels without biomass (e.g., at the initial operation stage), the highest flux was obtained in the submerged spacer configuration (Figure 3.8a). This observation is also supported by other modelling studies (Wardeh & Morvan, 2009). In reality, the axial filaments neglected due to the 2-d approximation will also affect the global permeate flux. In addition, the highest flux performance may also depend on the distance between two consecutive spacer filaments (Schwinge *et al.*, 2002a). In the cavity and zigzag configurations, the spacer is in contact with part of the membrane and recirculation areas with low velocity will form (Figure 3.4). The permeate flux will therefore be reduced by a higher concentration polarization (Song & Ma, 2005). The detailed effect of polarization is explained in paragraph 3.3, "Local effects of biofouling".

The average permeate flux values obtained from our simulations fall in the range of the values for full-scale brackish water desalination plants (Greenlee *et al.*, 2009). In the presence of a developing biofilm, the permeate flux decline follows the same trend for empty and spacer filled channels. In average, in all studied cases, a minor flux decline is observed in the first period (up to ~3 days in the simulated conditions; Figure 3.8a), when only small amounts of biomass are present corresponding to an initial colonization phase (Figure 3.8b). A second phase follows with a sharp decrease of permeate flux, correlated with an increase of the biomass amount in the channel. Finally, a quasi-steady state is reached for the flux value, as well as for the accumulated biomass. When approaching the steady state, fluctuations of the biomass amount in the channel will also affect the flux. These fluctuations are the result of major detachment events (sloughing).

A similar tendency in the flux decline (lag, sharp decline and plateau) was reported by Herzberg and Elimelech (Herzberg & Elimelech, 2007) in an experimental biofouling study in a flow cell without spacers, operated at constant feed pressure. The time scale of their observation was shorter (i.e., the steady state for flux decline was reached in 24 h) due to several reasons: the use of a fast-growing microorganism (*Pseudomonas*), an enhanced concentration of suspended biomass in the feed and possibly attachment and higher substrate levels. Although the trend in flux decline is similar for all simulated cases, the spacer presence and configuration affect the permeate flux (Figure 3.8a). Due to less biomass formation (higher shear stress) (Figure 3.8b), the submerged spacer leads to the highest fluxes compared with the other configurations.

The permeate flux obtained on the upper and lower membranes can be quite different (Figure 3.8c) because of the spatially heterogeneous biofilm development (Figure 3.8d). For a cavity configuration most of the biomass is situated on the lower membrane (where the spacer is located) and very little on the upper membrane, due to the shear stress. The simulations clearly show that more biomass on the membrane leads to lower permeate flux.

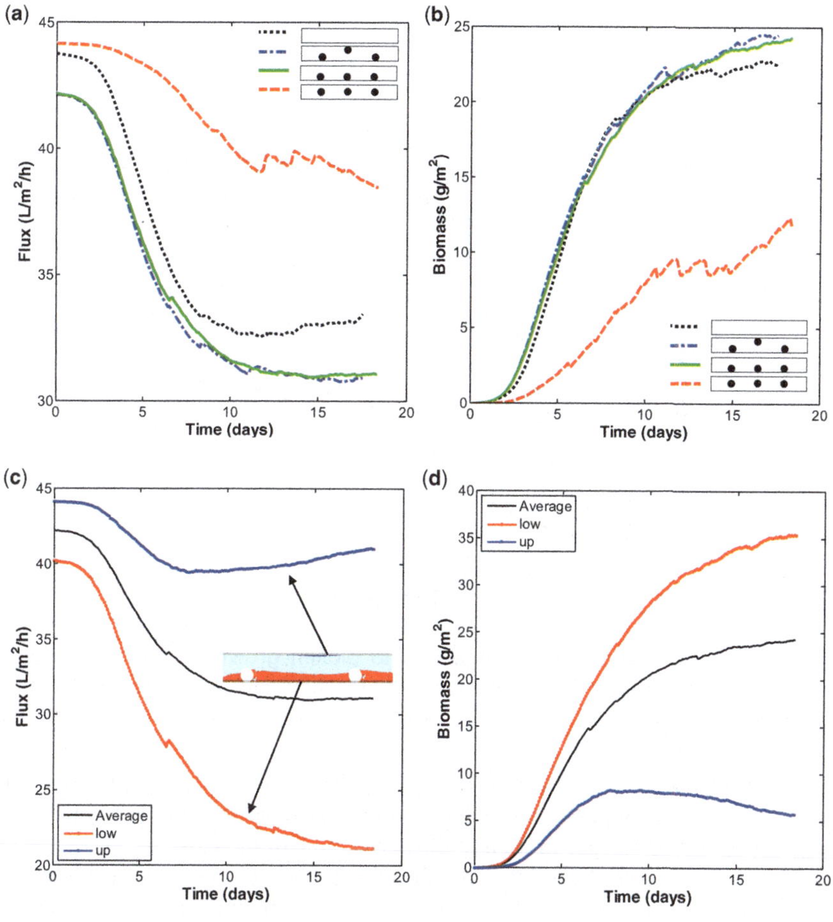

Figure 3.8 *Simulated flux decline and biomass growth in time.* (a) Comparative average flux decline for different channel geometries; (b) Comparative biomass development for different channel geometries (corresponding to (a); (c) Average flux decline on the upper and lower membranes in a channel with cavity spacers due to biofouling; (d) Biomass developed on the two membranes in a channel with cavity spacers (corresponding to (c)).

Salt passage

It has been observed in practice that the salt rejection of the reverse osmosis membrane is altered during operation. Fouling may have an important influence in this sense. The observed increase of salt concentration in the permeate could be attributed to an enhanced salt passage through the membrane, but also to a

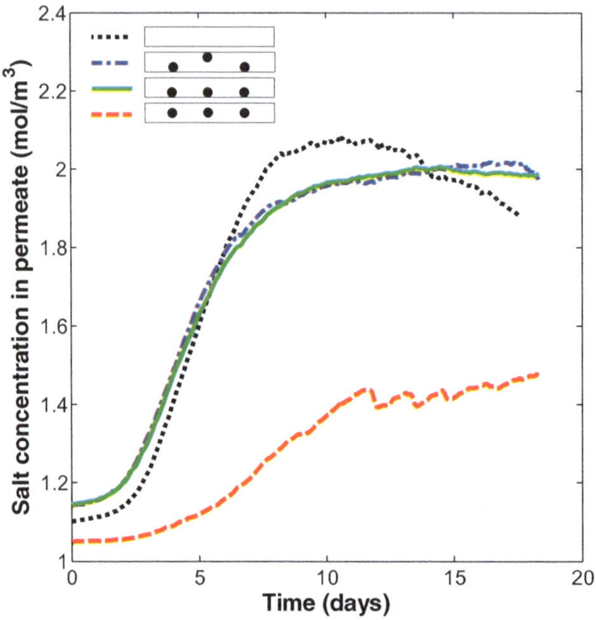

Figure 3.9 Increased salt passage due to biofilm growth in the feed channel, compared for different channel geometries. Results correspond to the simulation from Figure 3.8a, b.

"dilution effect" (Herzberg & Elimelech, 2007). This effect means that if the amount of water that passes through the membrane declines and the amount of salt is constant, the salt concentration in the permeate will increase.

Simulations results show that the development of the biofilm in the channel increases the salt concentration in permeate (Figure 3.9). It is expected that a higher salt concentration near the membrane leads to a higher salt passage (see eq. 3.8). The lowest salt concentration is achieved for the submerged spacer configuration, for which also the least biomass developed in the channel. However, for the channel without spacers, although the amount of the biomass is slightly less than in the zigzag and cavity configurations, the salt passage appears to be more pronounced.

Because the overall permeate flux for the empty channel is less than for the other two configurations, it may be possible to partly attribute the higher salt concentration in permeate to the "dilution effect".

Local effects of biofouling

Concentration polarization

One of the main concerns in reverse osmosis processes for brackish or seawater is concentration polarization, as this is recognized as an essential cause for flux decline. Concentration polarization (CP) is formation of a thin layer of concentrated solutes next to the membrane surface, which substantially decreases the permeation driving force by increasing the osmotic pressure.

Recently, two experimental studies (Chong *et al.*, 2008a; Herzberg & Elimelech, 2007) highlighted possible biofilm effects on enhancing concentration polarization. However, none of the previous modelling studies accounted for the influence of biofilm development on concentration polarization. We numerically evaluated the influence of biofilm formation on the concentration polarization for channels with and without various feed spacers.

A time-series of salt concentration profiles made in a biofilm colony shows that the thicker the biofilm the more salt accumulates (Figure 3.10, channel without spacer, at $x = 1300$ μm in the direction y normal to the membrane). In the operation at constant feed flowrate, as the biofilm grows, the channel cross-section available for flow is reduced so that the fluid flow is accelerated and the mass transfer boundary layer thickness is expected to decrease. However, the enhancement of external mass transfer is minor compared with the increase of internal diffusion resistance given by the biofilm (Figure 3.10).

In the biofilm absence and in a channel without spacers, it is well known that the CP layer continuously increases in the main flow direction (Fletcher & Wiley, 2004; Wardeh & Morvan, 2009). If the channel contains spacers, our model results predict that a very high salt concentration develops at places where the spacer is in contact with the membrane (Figure 3.11a, b). Experimental studies (Pervov, 1991) provided photographic evidence that the spots where the spacer is in contact with the membrane represent "deadlocks", i.e., very high salt concentrations are reached in those areas, which easily lead to scale formation. Our model results are consistent with conclusions of previous modelling studies indicating that the spacer has a dual effect. On one hand, it contributes to a disruption of the polarization layer, e.g., on a membrane opposite to the spacer, the CP layer is thinner (e.g., Figure 3.11b, $x = 6$ to 7 mm). On the other hand, an aggravation of the concentration polarization (Figure 3.11b, $x = 2$ and 10 mm) occurs in the deadlocks (Schwinge *et al.*, 2002a; Song & Ma, 2005).

The salt concentrations are increasing with increasing biofilm thickness in time (Figure 3.11b). This intensification of concentration polarization is immediately correlated to a decrease of local permeate flux (Figure 3.11c). The local flux decline, however, can be caused not only by the biofilm enhanced concentration polarization but also by the extra hydraulic resistance of the biofilm layer.

Hydraulic resistance

A biofilm layer adds an extra hydraulic resistance to trans-membrane flow, to an extent depending on biofilm permeability. In a system with permeate production, a small part of the flow in the feed channel is directed through the biofilm towards the

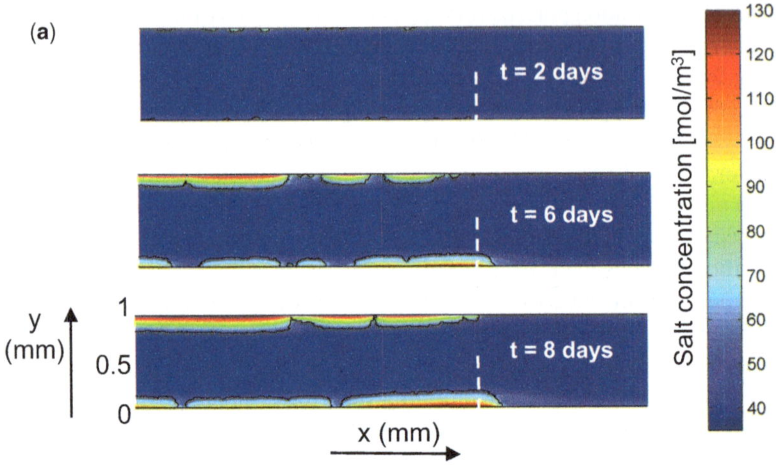

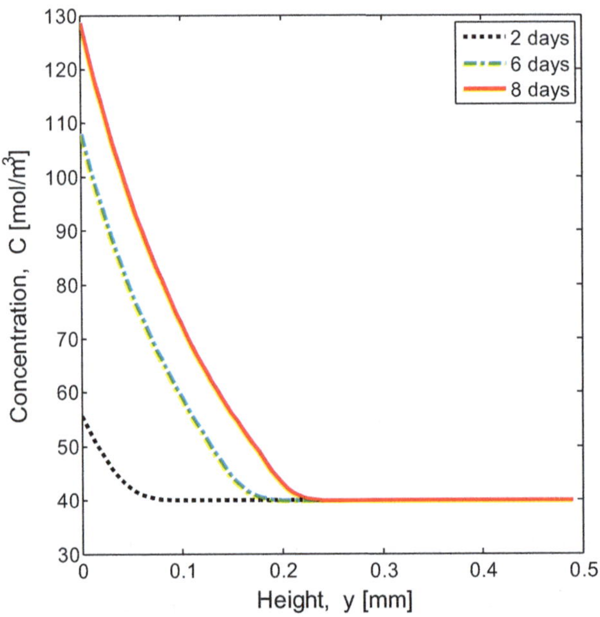

Figure 3.10 *Salt concentration distribution, C(x, y), in a channel without spacers.* (a) Detail with 2-d salt concentration after 2, 6 and 8 days. Color scale: concentration in mol/m^3. (b) Cross-section of salt concentration distribution from the lower membrane ($y = 0$) towards the bulk liquid, at $x = 13$ mm as indicated by the white dashed line from figure (a). As the biofilm grows in thickness (from ~50 μm at day 2 to ~150 μm at day 8), the salt concentration at the membrane surface strongly increases from 55 to 130 mol/m^3. The biofilm constitutes a barrier for the back-diffusion of salt.

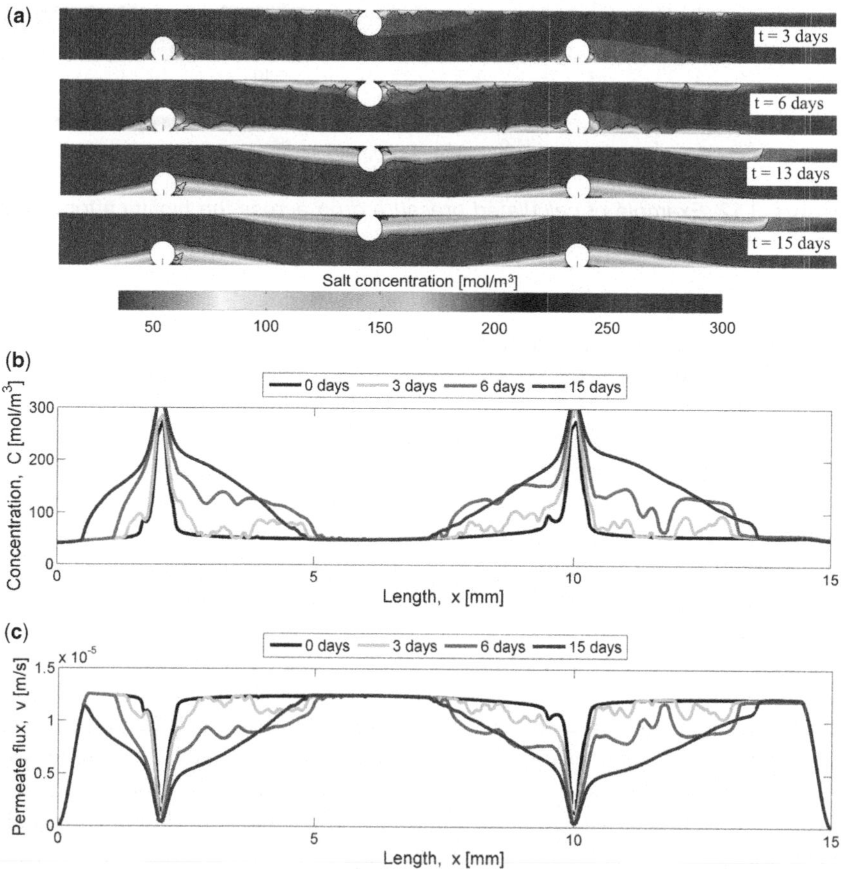

Figure 3.11 *Effect of biofilm development on salt concentration and permeate flux.* (a) 2-d salt concentration distribution in the channel; Color scale: concentration in mol/m^3; (b) salt concentration on the lower membrane; (c) the local permeate flux through the lower membrane. Peaks in salt concentration near the zigzag spacers and at places with biofilm colonies (where the salt transport is exclusively by slow diffusion) correspond to very low permeate flux.

membrane surface. The flow through the biofilm and its magnitude are visualized by the streamlines presented in Figure 3.4. Although the typical velocity inside the biofilm (~10^{-5} ms^{-1}) is small relative to that in the bulk liquid (~0.1 ms^{-1}), it is important to consider flow through the biofilm to calculate the pressure drop across the biofilm and to evaluate the importance of convective salt transport. The resulting pressure gradient in the biofilm layer, $\partial p/\partial y$, will be much higher than along the feed channel, $\partial p/\partial x$ (Figure 3.12). However, it is worth noting that the

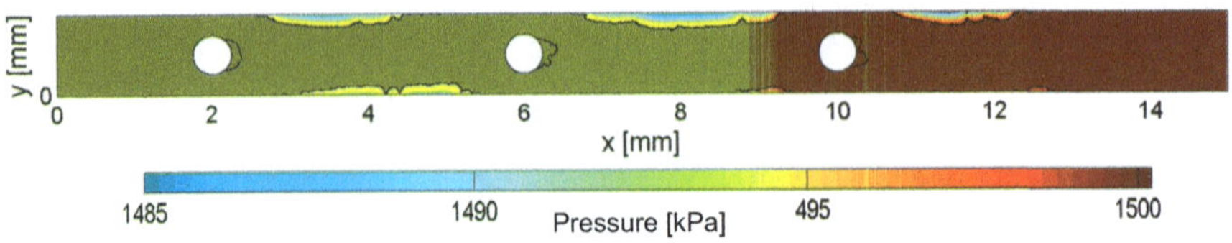

Figure 3.12 *Example of calculated pressure drop across the biofilm* after 9 days for a channel with submerged spacers.

pressure drop across the biofilm (~0.15 MPa) represents only around 1% of the initial net driving force for the process (15 MPa), given the biofilm permeability used in this study.

To study the relative contribution of biofilm enhanced concentration polarization and biofilm permeability on the permeate flux decline, we performed also simulations assuming that demineralized water is fed to the RO membrane system. If the feed contains no salt, the biofilm can only act on the local permeate flux through its permeability, as there is no more osmotic effect. Three biofilm structures grown in the channel without spacer after 3, 6 and 13 days were selected for comparison in Figure 3.13.

Even with little biofouling (day 3), the flux is lower in the presence of salts due to the strong effects of concentration polarization. For a considerable biofilm thickness (max. ~300 μm, day 13), the flux in the presence of salts drops to only ~50% from that obtained with demineralized water (Figure 3.13). Simulation results therefore suggest that with the current parameters, the biofilm enhanced osmotic pressure has a far more important effect than the biofilm hydraulic resistance.

Feed channel pressure drop

The feed channel pressure drop may also contribute to the driving force decline along the channel. The largest pressure drop along the feed channel occurs in the spacer region (Figure 3.14a), where the hydraulic area is the narrowest, as expected and shown also in previous modelling studies (e.g., Picioreanu *et al.*, 2009).

The change in time of total feed pressure drop over the 1.5 cm computational domain due to biofilm growth for different spacer geometries is shown in Figure 3.14b. A constant flow rate was maintained in the inlet and constant pressure in the outlet. The feed pressure drop increases with an increasing biomass amount and it reaches a plateau value when the biomass attains the steady state. The spacer configurations used lead to marked differences: the submerged spacer promotes the least biofilm growth due to the high shear, but it leads to the highest feed channel pressure drop. Less biofouling thus may come at the price of more energy

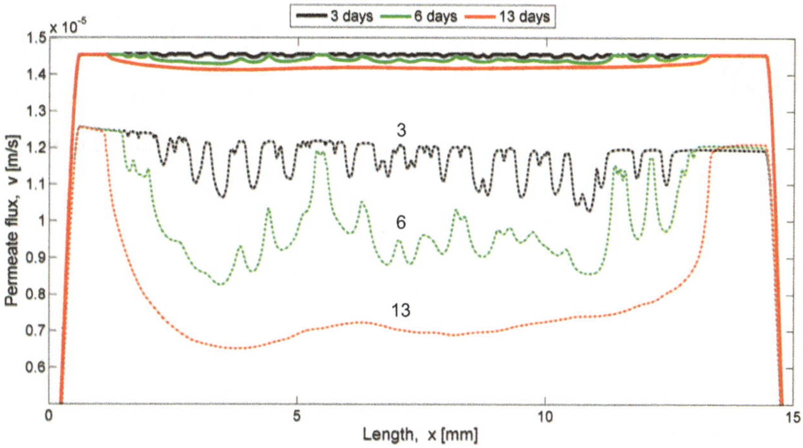

Figure 3.13 Relative contributions of biofilm enhanced concentration polarization and biofilm hydraulic resistance on local permeate velocity. (a) Demineralized water in the feed (solid lines); (b) Salty water (40 mM NaCl) in the feed (dashed lines). Even with little biofouling (day 3), the flux is lower in the presence of salts due to the effects of concentration polarization on the effective driving force for the process by increasing osmotic pressure. For a considerable biofilm thickness (day 13), the flux in the presence of salts is only ~50% from that obtained from demineralized water. The effects of biofilm-enhanced concentration polarization appear to be more significant than those of biofilm hydraulic resistance in the desalination of brackish water.

input in the membrane device. As expected, for the channel without spacers the pressure drop is the lowest.

The channel length simulated in this study is only 1.5 cm, so the feed pressure drop is relatively low. However, because (at least for uniformly fouled channels) the overall pressure drop should be proportional to the length of the channel this effect can become significant for long channels. In practice it has been noticed that membrane devices are not uniformly fouled. The most severe biofouling is initiated in the first membrane sections, spreading then gradually along the membrane length as the biofilm detaches and colonizes further downstream sections (Tran *et al.*, 2007).

Applying therefore this model on several connected membrane segments situated at different positions along the membrane module could provide useful insight on the spatial biofilm distribution along the module, and on overall effects of biofouling on module performance.

DISCUSSION

The model presented in this article is the first to describe numerically the micro-scale effects of biofilm growth in a reverse osmosis desalination module

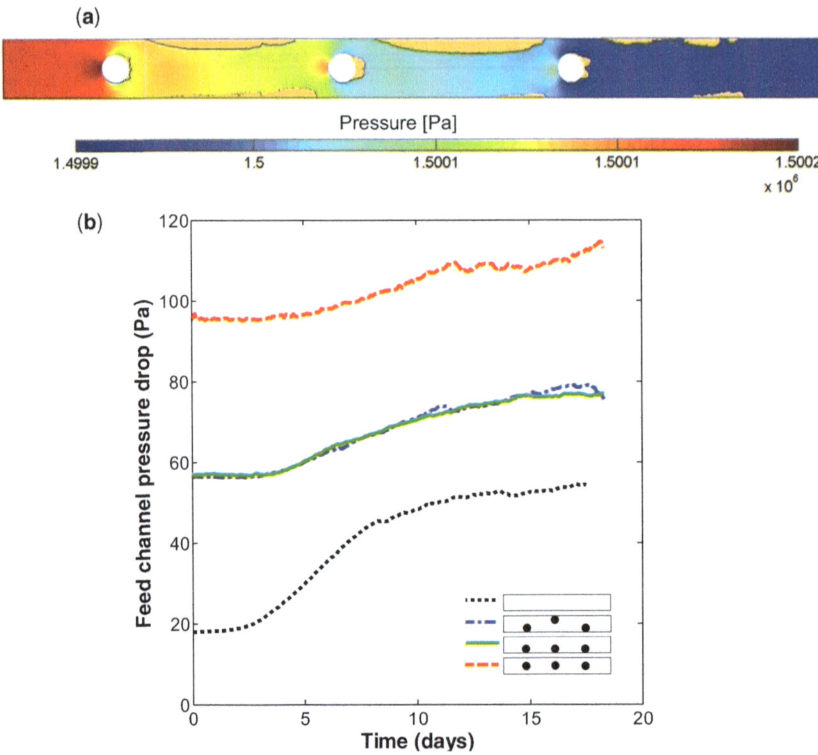

Figure 3.14 *Feed channel pressure drop in different channel geometries.* (a) 2-d distribution of pressure in a channel with submerged spacers at 12 days. Note that the pressure drop per length along the channel is much smaller than that in the biofilm across the channel (Figure 3.12). (b) Comparison of feed channel pressure drop evolution due to biomass growth in channels without spacer and with zigzag, cavity or submerged spacers. The pressure drop is calculated over the whole domain length of 15 mm. The pressure drop for zigzag and cavity spacers develops almost identically.

on: trans-membrane (permeate) flux, salt passage and feed channel pressure drop. The maximum driving force for obtaining permeate flux is the pressure difference between the feed and permeate channels, which has to overcome the membrane hydraulic resistance. For salty water the driving force is diminished by the osmotic pressure created by concentration polarization. Furthermore, the biofilm may cause a triple detrimental effect: reducing the driving force by (i) decreasing the feed pressure along the channel and (ii) enhancing the osmotic pressure by concentration polarization, and increasing the overall hydraulic resistance by (iii) introducing an additional hydraulic resistance to that of the membrane.

Concentration polarization and biofilm permeability

The model results suggested that, for brackish or seawater desalination, the contribution of biofilm enhanced concentration polarization on the permeate flux decline may be significantly larger than the contribution of biofilm-reduced permeability.

In principle, diffusion of salt and substrate is not very much affected by the biofilm, compared with the diffusion in pure water.

Experimental studies have shown that the diffusion coefficients for small molecules (e.g., O_2, NaCl, acetate) in the biofilm are usually not smaller than 70% of those in water (Horn & Morgenroth, 2006; Stewart, 1998; Wanner & Gujer, 1986). Therefore, the main role of the biofilm is to increase the thickness of the diffusion-dominated layer near the membrane. By hindering the liquid convection in the direction parallel to the membrane and increasing the diffusion length towards the membrane, the biofilm very significantly increases the concentration polarization layer. A question remains whether the biofilm growth actually benefits from the higher substrate concentration resulting from the polarization phenomenon. The effect of substrate accumulation near the membrane on biofilm growth will be presented in a forthcoming study.

Although for the concentration polarization the physical assumptions are quite well established, the biofilm hydraulic permeability remains a controversial issue in spite of a few experimental reports. The biofilm hydraulic permeability depends on the relative content of bacterial cells and gel-like polymeric matrix (EPS) embedding the bacterial cells (McDonogh *et al.*, 1994). Our model simulations used biofilm permeability values currently available in the literature for biofilms containing both cells and EPS (Fowler & Robertson, 1991; Johnson & Deen, 1996; McDonogh *et al.*, 1994). Therefore, we consider that, qualitatively, the modeling results from this study (for brackish water) are valid as they were obtained within a reasonable range of hydraulic permeability values. For a fresh surface water treatment plant, the feed does not contain so much salt. In that case, less biofilm enhanced concentration polarization effects are expected and the biofilm hydraulic resistance will have a higher relative importance for the flux decline.

It is not clear yet what is more harmful in terms of flux decline: a more impermeable or a permeable biofilm? A more permeable biofilm could also lead to less flux because of an increased concentration polarization. Whether this flux decrease is significant relative to the decrease due to concentration polarization needs to be evaluated from a future parametric study. Nevertheless, the necessity for reliable measurements of biofilm/cells/EPS permeability remains. Even though the effect of biofilm composition on permeability is not completely elucidated, it is clear that an increased biofilm thickness affects the flux decline. Because the biofilm thickness is determined by the balance between growth and detachment rates, the model also underlines the necessity of better measurements of other important parameters, such as the biofilm mechanical properties (e.g., cohesion strength or threshold stress for detachment, viscoelasticity and elasticity modulus).

Spacer importance

Often, experimental studies on biofouling are performed in flow cells without spacers (Chong *et al.*, 2008a; Herzberg & Elimelech, 2007; Herzberg *et al.*, 2009), emphasizing the role of biofilm-enhanced concentration polarization, but neglecting the potential role of spacers in biofilm formation and reverse osmosis processes. Similarly, many numerical studies modelled concentration polarization effects and other aspects in reverse osmosis feed channel in the absence of a spacer (Kim & Hoek, 2005; Lyster & Cohen, 2007). Recent experimental studies pointed however to the importance of spacers for biofilm development (Picioreanu *et al.*, 2009; Vrouwenvelder *et al.*, 2009a; Vrouwenvelder *et al.*, 2010). Several numerical studies included the presence of spacers (Fimbres-Weihs & Wiley, 2007; Koutsou *et al.*, 2009; Li *et al.*, 2005; Ranade & Kumar, 2006; Schwinge *et al.*, 2003; Song & Ma, 2005) with an emphasis on studying the flow pattern and mass transfer in the feed channel. Although these studies provide a good basis for a mathematical model of the reverse osmosis process, there was a need to develop a numerical model that could describe biofilm growth and its effects in a spacer-filled channel.

Our results indicate that the spacer presence and configuration have an important influence on the RO performance when a biofilm develops in the feed channel. Because the spacer effect is significant, a proper scale-down of the RO system needs to include spacers in order to correctly estimate the effects of biofilm development on the membrane module performance (permeate fluxes, salt passage and feed channel pressure drop).

Interestingly, we found that among the feed channel geometries used in this study, the submerged spacers were the most favourable in terms of both permeate flux and salt concentration in the permeate. As a trade-off, compared with other configurations, submerged spacers lead to larger feed channel pressure drop and thus to more energy consumption. The spacer fibres placed half way between the two membranes leaves clearly reduce biofilm development by promoting high shear and, consequently, more biomass detachment. However, by using 2-d geometry the potential impact of the spacer filaments parallel to the flow ("axial filaments") was not considered. For certain geometries, it has been shown by (Shakaib *et al.*, 2007) that axial filaments do not significantly influence the flow pattern in a parallel 2-d plane if that plane is situated "far enough" from the filaments. Obviously, the flow cannot be well approximated by using 2-d geometry in the region near the axial filaments or when the flow attack angle is at 45° (e.g., diamond spacer configuration).

The strong effect of the spacer on the system behaviour urges to use realistic spacer geometries for numerical studies (see also Picioreanu *et al.*, 2009). With the modelling approach developed in this study different spacer and feed channel geometries could be evaluated in respect to their potential biofilm formation. The model could be used as a tool to optimize the spacer design and membrane device construction and operation with reduced impact of biofilm accumulation on plant performance.

Further model development and use

The model presented here can be easily adapted for other feed quality and operational parameters, allowing thus an estimation of biofilm effects on process performance under different conditions. In addition, the effectiveness of biofilm backwash could also be studied using the model.

Even though the simulations performed in this work were in two spatial dimensions, they could contribute to the qualitative understanding of interactions between different processes occurring during biofouling of a reverse osmosis channel. When considering 2-d geometry, the changes in flow velocity, pressure, solute concentration and biomass in the z-direction are fully neglected. This simplification may affect not only local permeation rate, but also global performance parameters. In addition, the 2-d geometry does not allow the description of preferential flow channels and its consequences (Picioreanu *et al.*, 2009; Pintelon *et al.*, 2010). Numerical results from this 2-d study should therefore be translated with care to real 3-d applications. It is obvious that an increase of geometric complexity (e.g., 3-d simulations) is needed to obtain a better quantitative comparison with experimental measurements. Although simple in principle, the model extension to three dimensions will require significant adaptations of the numerical approach, so that a good compromise can be reached between accuracy of the results and the massive computing requirements.

CONCLUSIONS

A two-dimensional numerical model based on fundamental transport equations was created to study the local and macroscopic effects of biofouling in a reverse osmosis membrane device. The mechanistic model explains several experimental observations, most importantly the loss of permeate flux and the increase of salt passage in time due to biofilm formation.

Due to balanced biofilm growth and detachment a maximum attached biomass amount can be reached, leading to a quasi-steady state for the flux, as observed experimentally. The interplay of biomass growth with detachment also shows why the spatial biofilm distribution in the feed channel is irregular and suggests the places where most biomass accumulation is expected and the mechanisms by which the biofilm is streamlined by the erosive action of the flow.

It was found that the spacer configuration plays a significant role on the amount of biomass formed in the channel and, consequently, on the average permeate flux and salt passage. The submerged spacer gave the best performance among the studied geometries.

Three mechanisms by which the development of a biofilm in a RO membrane device contributes to the flux decline and increased salt passage into permeate were identified: (i) biofilm enhanced concentration polarization; (ii) increased hydraulic resistance to trans-membrane flow; and (iii) increased feed channel pressure drop. Among these, for brackish water and within the range of parameter values reported in the literature, the biofilm enhanced concentration polarization seems to have the most influence on the permeate flux decline.

The developed model could be used as a tool to further investigate the effect of process conditions on biofilm formation and to optimize the membrane device construction and operation.

Chapter 4

Combined biofouling and scaling in membrane feed channels: A new modelling approach

INTRODUCTION

Reverse osmosis (RO) and nanofiltration (NF) are high pressure membrane technologies that allow the production of good quality water from a variety of sources (i.e. groundwater, brackish water, and seawater and wastewater effluent). Despite progress in developing fouling-resistant membranes (Kang & Cao, 2012) and new cleaning protocols (Chesters, 2009; Cornelissen *et al.*, 2007a), fouling due to biofilms, mineral precipitates, colloidal and organic matter still remains a challenge for these systems. Due to the complex composition of feed water for many membrane applications, it can be expected that the various types of fouling do not occur independently. While colloidal and organic fouling may be reduced when using an extensive pre-treatment system, biofilms and precipitates seem to be resilient in many plants (Darton *et al.*, 2004; Vrouwenvelder *et al.*, 2008b; Xu *et al.*, 2010). During membrane autopsy for a seawater plant, calcium sulphate precipitate was found and attributed to flow channelling and stagnant zones caused by the biofilms present in the system (Baker & Dudley, 1998a). A systematic chemical and microbiological analysis coupled with scanning electron microscopy (SEM) for a conventional two-stage RO train revealed the coexistence of biofouling and inorganic foulants within all membrane elements (Schneider *et al.*, 2005). While membrane autopsy studies are useful diagnostic tools, these cannot provide details of the sequence of events that take place in fouling.

On the other hand, laboratory-scale studies carried out under well-defined conditions have focused primarily on a single type of foulant: biofouling (Herzberg & Elimelech, 2007; Vrouwenvelder *et al.*, 2010b), scaling (Lyster *et al.*, 2010; Rahardianto *et al.*, 2006) or organics (Ang *et al.*, 2011). Only relatively recently

Author(s): Szilárd Bucs, Joop Kruithof, Mark van Loosdrecht and Johannes Vrouwenvelder
doi: 10.2166/9781780409597_0061

have combined fouling and the potential interactive effects between foulants started to gain more attention (Li & Elimelech, 2006). Mineral precipitation associated with the existence of a biofilm was reported for a variety of processes, ranging from the genesis of sediments (Riding, 2000) and dental caries formation (Fejerskov, 2004) to corrosion of metallic surfaces (Beech & Sunner, 2004) and wastewater treatment processes (Arvin & Kristensen, 1983; Lee & Rittmann, 2003). Several authors have suggested that the presence of a biofilm can affect mineral scale formation in membrane systems. Ultrasonic time domain reflectometry monitoring of $CaSO_4 \cdot 2H_2O$ scaling with and without a biofilm present on a NF membrane under cross-flow conditions, has revealed more precipitated gypsum when a biofilm developed on the membrane (Hou *et al.*, 2010). Thompson *et al.* (2012) observed via real-time imaging enhanced gypsum nucleation in the presence of biofilms for RO membranes with tertiary wastewater effluents. Moreover, SEM imaging showed crystals embedded in the biofilm. Liu and Mi (2012) used a microscope observation system for studying combined gypsum and alginate fouling (potentially similar to bacterial exopolymers) for forward osmosis (FO) membranes. Their results indicate a synergistic effect for the two foulants: a more pronounced flux decline was observed for combined fouling than by simply considering the algebraic sum of the individual fouling types. The reason for considering the effects of scaling and biofouling together is that conventional measures applied to prevent scaling can impact the formation of biofilms: anti-scalants have been suggested to serve as a nutrient and accelerate biofilm formation (Sweity *et al.*, 2013; Vrouwenvelder *et al.*, 1998). Mathematical models can be useful tools for testing various hypotheses and assessing the contribution of different mechanisms in the context of complex processes, such as those affecting fouling. Micro-scale models for scaling (Lyster *et al.*, 2010; Radu *et al.*, 2014a) and biofouling (Picioreanu *et al.*, 2009; Radu *et al.*, 2010) in membrane processes have been developed to evaluate the formation of foulant layers in relation to macro-scale performance indicators. However, little work is available on the mechanistic modelling of coexisting precipitate and biofilm phases. Cooke *et al.* (1999) developed a one-dimensional model for biochemically driven mineral precipitation in anaerobic biofilms occurring in landfill leachate collection systems. Growth and loss of biofilm are coupled with mineral precipitation of calcite, in order to illustrate change in the porosity due to clogging in column test experiments. Zhang and Klapper (2010) proposed a numerical model with the focus on microbial-induced calcite precipitation. While accounting for complex solution chemistry, biofilm and precipitate formation are represented in a continuum phase with different volume fractions. One of the difficulties when applying this approach is the consideration of the discrete events of cell attachment and crystal nucleation. On the other hand, Olivera-Nappa *et al.* (2010) proposed a particle-based approach for combining the biofilm and precipitate phases within a model applied to bioleaching. In this model, hard spheres representing both the biomass and sulphur deposits can attach, grow and move to create porous solid phases in which contact-area dependent reactions take place. The goal of the present study was to develop a two dimensional numerical model for investigating how bio- films influence salt precipitation in RO/NF membrane systems. Previously developed models (Picioreanu *et al.*, 2004; Radu *et al.*, 2014a, 2010) are integrated and extended to create a new tool for analysing some of the interactions between gypsum scaling and biofouling in spiral wound membrane feed channels.

MODEL DESCRIPTION

Continuum-based fluid dynamics and mass transport calculations are coupled with a particle-based approach for representation of foulants in the feed spacer channel. Based on previous sub-models for biofilm development (Picioreanu *et al.*, 2004) and mineral precipitation (Radu *et al.*, 2014a), a new model was proposed for combined fouling, including specific interactions between the two solid phases.

Geometry, phases, computational domains

Ideally, the computational representation of the feed channel for RO/NF systems should include a 3-D net shaped spacer that keeps two membrane sheets apart (Figure 4.1a). Given the high computational requirements typically associated with 3-D simulations of flow and solute transport (Fimbres-Weihs & Wiley, 2010), a 2-D simplification was chosen for the current study (Figure 4.1a and 4.1b). The use of a 2-D geometry can still capture heterogeneity in transport processes associated with the presence of feed spacers (i.e. local variations in cross-flow and solute concentration at the membrane surface), while needing reasonable calculation time and computer memory. The geometrical characteristics of the feed spacer are identical to those described in previous modelling work (Radu *et al.*, 2014a; Schwinge *et al.*, 2002a). The foulants in the feed channel are represented by a collection of non-overlapping solid particles (Figure 4.1c) characterized by type (s_i), mass (m_i), diameter (d_i) and position (x_i). The biofilm comprises a single type of particle ('generic' biomass), while the precipitate (gypsum) includes inert particles (inner crystal) and active particles (outer crystal). The particles should only be regarded as discretization elements for the foulant phases. That is, a particle is not equivalent to a cell or a crystal, but rather to an aggregate of microbial cells or collection of crystals (Picioreanu *et al.*, 2004; Radu *et al.*, 2014a). In order to couple the particle-based representation of foulant with the continuum description of fluid dynamics and solute distribution, sub-domains must be constructed based on the collection of particles. An agglomeration of particles of the same kind is treated as a continuum with a certain void fraction, forming a sub-domain. The computational domain is first discretized on a rectangular grid ($\Delta x \times \Delta y$) and each grid element is associated with a phase: liquid, spacer, biofilm or precipitate. All neighbouring grid elements containing the same phase will then form a cluster, called a sub-domain (Figure 4.1d). Initially, only spacer and liquid phases are present, but in time the biofilm and precipitate develop, altering the channel geometry. Further, in each grid element occupied by foulant, the concentrations of

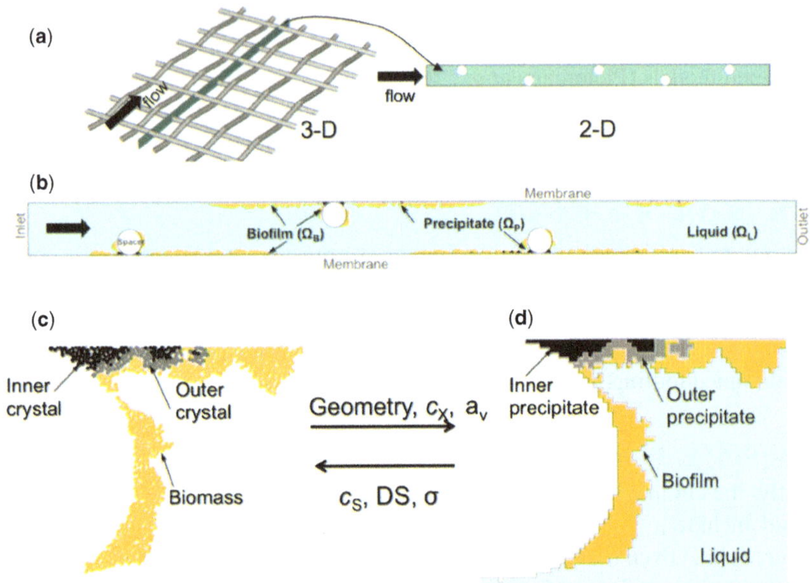

Figure 4.1 (a) Computational domain representing membrane feed spacer channel and foulants. (b) Particle-based representation of foulants: biomass and crystal. (c) Sub-domains in the fouled feed channel: biofilm, precipitate and liquid. The particle distribution determines the geometry of sub-domains and boundaries, the local biomass concentration c_X and the specific surface area for crystals, a_v. In turn, the calculated substrate and salt concentrations together with the mechanical stress, σ, affect the particle formation, growth and removal.

biomass or crystal phase are calculated based on the mass of particles situated in this grid element. Finally, specific equations (i.e. reactions and transport processes) and physical properties (i.e. porosity, permeability) are defined for each sub-domain, function of the phase present. The spacer is an inert and impermeable material (i.e. polypropylene) inside which no flow or mass transfer takes place.

Fluid flow and solute transport

Fluid flow

Cross-flow velocities typically encountered in RO/NF installations (i.e. 0.07–0.2 m s^{-1}, Vrouwenvelder *et al.*, 2009b) correspond to a laminar regime (Fimbres-Weihs & Wiley, 2010). The development of fouling may obstruct the flow in some parts of the channel and result in higher velocities in fouling-free areas when the system is operated at constant inlet flow rate.

However, given that a significant channel cross-section remains open even when fouling occurs, laminar flow is a reasonable assumption. Steady state

Navier-Stokes equations for incompressible fluid $\nabla u = 0, p(u\nabla)u + \nabla p = \nabla(\eta \nabla u)$ are used to calculate the flow in the liquid sub-domain. Given that both a biofilm and an agglomeration of crystals can be permeable to liquid flow (Dreszer *et al.*, 2013; Rahardianto *et al.*, 2006), Brinkman equations $((\eta/\kappa_f)u + \nabla p = (\eta/\varepsilon_f)\nabla^2 u, \nabla u = 0)$ describe flow in the biofilm and precipitate sub-domains. Liquid density ρ and viscosity η were constant (Table 4.1). The biofilm and precipitate have, however, different properties in terms of porosity (ε_f) and permeability (κ_f). Although mostly not the case, for the biofilm, the porosity (ε_{bio}) and permeability (κ_{bio}) were assumed to be invariant and values are set according to the literature (McDonogh *et al.*, 1994; Zhang & Bishop, 1994). For the precipitate, very limited quantitative information on the properties of the deposits occurring in RO systems is available. Intuitively, as crystals grow, pore connectivity is expected to decrease significantly, reducing the liquid flow through the precipitate layer. To mimic such a behaviour, ε_{ppt} was calculated as a function of the local mass of solids and varied between 0.99 (for a low amount of precipitate, ie the initial nuclei formed) and 0.01 (i.e. for a very dense mature inner precipitate layer). In addition, a decrease in the precipitate permeability (κ_{ppt}) with decreasing porosity was considered (Radu *et al.*, 2014a). At the channel inlet, a fully developed laminar flow profile was assumed, with an average velocity u_{in}, while a fixed pressure was defined for the outlet (p_{out}). The spacer is impermeable, thus a no-slip condition was applied at its surface. The membranes are permeable, with a local permeate velocity (v) determined by the hydraulic (Δp) and osmotic ($\Delta \Pi$) pressure difference between the feed and permeate sides (Lyster & Cohen, 2007; Radu *et al.*, 2010). The osmotic pressure is a calculated function of salt concentrations (Radu *et al.*, 2014a). A continuity boundary condition is used for all remaining interfaces within the channel (i.e. biofilm–liquid, precipitate–liquid, and biofilm–precipitate).

Solute transport

Feed water for RO/NF membrane systems can contain a variety of solutes in different ratios. In this numerical study, three different solutes were considered: a substrate (concentration c_S) which serves as a growth-limiting nutrient for biofilm development, a potentially precipitating salt ($CaSO_4$, c_1) which can form gypsum and an inert (i.e. not forming crystals in the current conditions) salt (NaCl, c_2), which is the main contributor to total dissolved solids in many water sources. The two ions of a salt were treated as a single solute for computational efficiency (i.e. both Ca^{2+} and SO_4^{2-} were represented as c_1), although this may not be an accurate representation at very high ionic strengths or more complex solution compositions. In the general solute mass balance $D_i\nabla^2 c_i - u\nabla c_i + r_i = 0$, the transport processes (diffusion and convection) are similar for all solutes, but the reactions differ for each solute i in each sub-domain. Diffusion coefficients (D_i) were assumed independently of concentration and had identical values in the biofilm and the liquid. Within the precipitate layers, when more crystal mass

Table 4.1 Variables and parameters of the model.

Variable Name	Description	Value	Units	Source
ρ	Liquid density	1000	$kg \cdot m^{-3}$	–
η	Liquid dynamic viscosity	10^{-3}	$Pa \cdot s^{-1}$	–
D_S	Diffusion coefficient substrate	10^{-9}	$m^2 \cdot s^{-1}$	Wanner (2007)
D_2	Diffusion coefficient NaCl	1.3×10^{-9}	$m^2 \cdot s^{-1}$	Lyster (2009)
D_1	Diffusion coefficient $CaSO_4$	10^{-9}	$m^2 \cdot s^{-1}$	Lyster (2009)
c_1	$CaSO_4$ concentration	State variable	$mol \cdot m^{-3}$	
c_2	NaCl concentration	State variable	$mol \cdot m^{-3}$	
c_S	Substrate concentration	State variable	$mol \cdot m^{-3}$	
U	Velocity vector	State variable	$m \cdot s^{-1}$	
σ	Von Misses stress	State variable	$N \cdot m^{-2}$	
p	Pressure	State variable	Pa	
$\gamma_{Ca^{2+}}$	Activity coefficient Ca^{2+}	Variable	–	Calculated Pitzer model
$\gamma_{SO_4^{2-}}$	Activity coefficient SO_4^{2-}	Variable	–	Pitzer model
a_{H_2O}	Water activity	Variable	–	Pitzer model
DS	Degree of saturation	Variable	–	
R_n	Nucleation rate	Variable	$\# \, m^{-2} \cdot h^{-1}$	Calculated Mullin, Lyster
R_{att}	Attachment rate	Look up	$C\text{-}mol_X \cdot h^{-1} \cdot m^{-1}$	Assumed
$\mathbf{x}_i$	Position vector of particle *i*	Variable	(m, m, m)	–
d_i	Diameter of particle *i*	Variable	m	–
m_i	Mass of particle *i*	Variable	kg	–
Aci	Surface area for growth particle *i*	Variable	m^{-2}	Calculated

$\sigma_{det,bio}$	Threshold stress detachment biomass	7	$N \cdot m^{-2}$	
$\sigma_{det,ppt}$	Threshold stress detachment precipitate	100	$N \cdot m^{-2}$	Assumed
$K_{CaSO_4 \cdot 2H_2O}$	Equilibrium constant gypsum	$10^{-4.58}$	$mol^2 \cdot m^{-6}$	Sheikholeslami and Ong
K_S	Half saturation coefficient	0.025	$mol \cdot m^{-3}$	Henze
μ_m	Maximum growth rate	2	day^{-1}	Henze
Y_{SX}	Yield of biomass on substrate	0.5	$mol \cdot C\text{-}mol_X^{-1}$	Henze
c_X	Biomass concentration	Variable	$C\text{-}mol_X \cdot m^{-3}$	Calculated
p_{out}	Operating pressure	Variable	Pa	(Table 2)
Π	Osmotic pressure	Variable	Pa	Calculated Pitzer
ε_{bio}	Biofilm porosity	0.8		Zhang, Bishop
ε_{ppt}	Precipitate porosity	Variable		Calculated
κ_{bio}	Biofilm permeability	10^{-16}	m^2	Derived [Rah]
κ_{ppt}	Precipitate maximum permeability	10^{-19}	m^2	Derived [Rah]
ρ_{bio}	Biomass density	290	$kg \cdot m^{-3}$	Wanner
ρ_{ppt}	Gypsum density	2700	$kg \cdot m^{-3}$	An
Mw_G	Molecular weight	172.17	$g \cdot mol^{-1}$	Calculated
A_N	Nucleation rate parameter	1.75×10^6	$m^{-2} \cdot s^{-1}$	Lyster
a_N	Nucleation rate parameter	2.36	–	Lyster
k_0	Generalized gypsum growth rate constant	5×10^{-8}	$mol \cdot m^{-2} \cdot s^{-1}$	Chosen
a_v	Specific surface precipitate growth	Variable		Calculated
N	Fractal number crystal growth	1.36	–	Nielsen
u_{in}	Mean inlet velocity	0.1	$m \cdot s^{-1}$	Vrouwenvelder
$c_{1,in}$	Inlet concentration of $CaSO_4$	19.7–26.8	$mol \cdot m^{-3}$	Table 2

(*Continued*)

Table 4.1 Variables and parameters of the model (*Continued*).

$c_{2,in}$	Inlet concentration of NaCl	0–100	$mol \cdot m^{-3}$	Table 2
$c_{S,in}$	Inlet concentration of substrate	0.017	$mol \cdot m^{-3}$	Table 2
T	Temperature	298	K	Chosen
L_P	Membrane permeability	9.7×10^{-12}	$m \cdot Pa^{-1} \cdot s^{-1}$	Radu
R_l	Rejection coefficient	0.98	–	Practice
U	Axial velocity component	State variable	$m \cdot s^{-1}$	–
V	Transverse velocity component	State variable	$m \cdot s^{-1}$	–
Δt	Time step	2	h	Chosen
O_{max}	Maximum particle overlap	0.05	–	Chosen
δ_m	Mesh size at membrane surface	5×10^{-6}	m	Chosen
δ_L	Mesh size in the bulk liquid	20×10^{-6}	m	Chosen
Δx	Discretization of domain	10^{-5}	m	Chosen
L_z	Virtual 3d dimension	10^{-2}	m	Chosen
r_S	Substrate consumption rate in biofilm	Variable	$mol \cdot m^{-3} \cdot s^{-1}$	Calculated
r_1	$CaSO_4$ consumption rate in outer precipitate	Variable	$mol \cdot m^{-3} \cdot s^{-1}$	Calculated
L_x	Channel length	15×10^{-3}	m	Chosen
L_y	Channel width	1×10^{-3}	m	Schwinge
d_s	Spacer diameter	0.5×10^{-3}	m	Schwinge
L_s	Distance between spacers	4×10^{-3}	m	Schwinge

Table 4.2 Simulation conditions.

Case	Fouling Scenario	$c_{1,in}$ (mM) $CaSO_4$	$c_{2,in}$ (mM) NaCl	$c_{s,in}$ (mM) Substrate	DS_{in}	Pressure (bar)	Initial Flux ($L \cdot m^2 \cdot h^{-1}$)
1	Combined	19.7	0	0.017	1.3	5	14.1
2	Biofouling			0.017			
3	Scaling			0			
4	Combined	26.8	100	0.017	1.3	10	13.4
5	Biofouling			0.017			
6	Scaling			0			

accumulates per volume, the diffusion of solutes was reduced by a factor ε_{ppt} compared to diffusion in liquid. In the liquid and inner precipitate sub-domains no reactions take place. Substrate is only consumed within the biofilm (rate r_s) and $CaSO_4$ is only consumed within the outer precipitate sub-domain (rate r_1). NaCl (inert salt) does not take part in any chemical reactions. Substrate consumption followed Monod kinetics, $r_S = Y_{SX}\mu_{max}c_S/(K_S + c_S)c_X$ with rate parameters (the yield Y_{SX}, the maximum specific growth rate μ_{max} and the half saturation coefficient K_S) chosen as for heterotrophic microorganisms (Radu *et al.*, 2010). The local biomass concentration in the biofilm, $c_X(x,y)$, resulted from the particle-based biofilm model. Calcium sulphate was consumed for crystal growth ($r_1 = k_S a_v(\sqrt{DS} - 1)^2$) only if the solution was supersaturated locally (DS > 1). Ionic activity coefficients $\gamma_{Ca^{2+}}$ and $\gamma_{SO_4^{2-}}$ calculated with the Pitzer model were used within the degree of saturation, $DS = \gamma_{Ca^{2+}} + \gamma_{SO_4^{2-}} - c_1^2 a_{H_2O}^2 / K_{CaSO_4 \cdot 2H_2O}$. Details of thee calculation of specific surfaces for precipitate growth a_υ and choice of growth rate constant k_s are given in Radu *et al.* (Radu *et al.*, 2014a). At the channel inlet, fixed concentrations of the three solutes were assumed ($c_i = c_i$, in, Table 4.2). In the outlet, no-diffusion conditions were assumed for each solute, $D_i\,\partial c_i/\partial x = 0$. No flux conditions were set for all solutes at the spacer surface.

Solute passage through the membranes resulted from a mass balance at the membrane surface, $vc_i - D_i\partial c_i/\partial y = vc_i(1 - R_i)$. For all remaining interior boundaries, flux continuity was applied.

Foulant development

Coupled biofouling and scaling are here for the first time integrated in a numerical model describing fouling of the RO/NF membrane feed spacer channel. The model with several types of particles (biomass and inner/outer crystal) has distinct features from previously developed multi-species biofilm models (Picioreanu *et al.*, 2004) and mineral formation (Radu *et al.*, 2014a). First, the particles appear in the channel as a result of different processes: attachment for biomass and nucleation for crystals. Secondly, the increase in mass is a function of particle type: volume-based growth for biomass and surface based growth for the outer precipitate. Thirdly, the particles used here have different mechanical properties, reflected in the procedure applied to transport particles (i.e. the shoving algorithm).

Initiation of fouling, attachment and nucleation

The development of a biofilm in the feed channel starts with biomass attachment, while precipitate formation is initiated by crystal nucleation. In the model, biomass particles can attach randomly to any solid surface in contact with the liquid (i.e. spacer, membrane, biofilm or precipitate). A constant rate of attachment (R_{att}) was considered everywhere. The formation of new crystal particles (i.e. due to nucleation) is governed by the local degree of saturation. Gypsum crystals can appear at the surface of the membrane, spacer or within the biofilm, if the solution is supersaturated locally (DS > 1). Bulk nucleation is not considered in the current model, as it is a much slower process compared to surface nucleation (Dydo *et al.*, 2004). The number of nuclei that can form per unit time and unit area is defined in accordance with the traditional nucleation theory ($R_N = A_N e^{-a_N/\ln(\mathrm{DS})^2}$), with nucleation rate constants A_N and a_N taken for gypsum formation on RO membranes from Lyster *et al.* (2010). Since quantitative information regarding the effect of biofilm or extracellular polymers on nucleation is rather controversial (Dupraz *et al.*, 2009), it was assumed here that the biofilm does not alter the intrinsic nucleation kinetics of gypsum. Therefore, no changes in interfacial tension, crystal morphology or chemical potential of ions due to the presence of biofilm are incorporated in the model. A comprehensive description of how nucleation events are evaluated within the model for each time step is given in Radu *et al.* (Radu *et al.*, 2014a).

Growth of foulant layer

The increase in the amount of foulant is based on different mechanisms: biofilms grow in volume, while precipitates grow through the surface. Within the model, the particle mass increase is governed by the local solute concentrations: biomass particles grow as a function of substrate concentration, whereas crystal particles grow as a function of salt concentrations, which define the degree of saturation. During one time step Δt, each biomass particle increases following a Monod growth rate, $m_i(t+\Delta t) = m_i(t) + \mu_{\max} c_S/(K_S + c_S) m_i(t)\Delta t$. Subsequently, the particle diameter is recalculated, considering the density of biomass, $d_i(t+\Delta t) = \sqrt{4m_i(t+\Delta t)/(\rho_b \pi L_z)}$ (Picioreanu *et al.*, 2004). In order to mimic surface based growth for mineral scale formation, not all the crystal particles can contribute to an increase in the precipitate volume. Particles situated further than a chosen distance $2\Delta x$ from an interface with the liquid or biofilm sub-domains are considered inert (inner crystal). This distance was set in order to obtain a continuous outer layer at the surface of each precipitate domain, also accounting for possible diagonal vicinities with liquid or biofilm sub-domains (Radu *et al.*, 2014a). Crystal particles in the vicinity of the interface are active (outer crystal) and contribute to precipitate growth. The inner particles are immobile and should be imagined as a compact cake of closely packed crystals. The outer particles undergo three processes: growth, division and redistribution in a sequence that conveniently simulates the precipitate expansion. During one time step, each outer particle increases in mass $m_i(t+\Delta t) = m_i(t) + k_S A_{c,i}(\sqrt{\mathrm{DS}} - 1)^2 M_{\mathrm{CaSO_4 \cdot 2H_2O}} \Delta t$. Calculations for the surface area $A_{c,i}$ are presented in Radu *et al.* (Radu *et al.*, 2014a). The particle size after growth is calculated based on the gypsum density, $d_i(t+\Delta t) = \sqrt{4m_i(t+\Delta t)/\rho\pi L_z}$. When any biomass or outer crystal particle becomes larger than the grid size, it divides asymmetrically into two smaller particles varying between 45–55% of the mass of the original particle (Picioreanu *et al.*, 2004). After division (during each time step) overlaps between the particles can occur. A shoving algorithm is applied to rearrange particles so that these do not superpose beyond a chosen threshold ($0.05\Delta x$) and to ensure that a single type of particle is present in each grid element. The conventional implementation of the shoving framework (Kreft *et al.*, 2001) used to redistribute biomass particles was modified in this work to account for the biomass–crystal interactions. Particles are assigned a weight in their displacement distance, a function of their type. The largest weight is assigned to the inner crystal particles, which makes them practically immobile. The outer crystal has a weight larger than the biomass particles, which means that biomass can be pushed by a growing crystal, but not the other way around. By this method, crystal particles are relocated first and their position is known when displacing the biomass particles.

Detachment

The foulants can be removed from the channel by high mechanical stress due to fluid flow. A detailed description of the biofilm detachment procedure is available in Picioreanu *et al.* (Picioreanu *et al.*, 2001) and Radu *et al.* (Radu *et al.*, 2010). First, the plane stress equations in the biofilm and precipitate subdomain are solved with shear stress imposed on the interfaces with liquid. Then, particles are removed in places where the Von Mises stress (σ) exceeds the foulant cohesive strength (σ_{det}). For the precipitate, much higher cohesive strength ($\sigma_{\mathrm{det}} = 100$ Pa) was assumed compared to the biofilm ($\sigma_{\mathrm{det}} = 7$ Pa), resulting in negligible precipitate removal. If a precipitate sub-domain is situated in the middle of the biofilm and the surrounding biofilm is removed, the precipitate will become an 'island' (i.e. suspended, not connected to the membrane/spacer/other biofilm) and will also be removed.

Model solution

The fouling model is based on MATLAB (MATLAB, 2010b, MathWorks, Natick, MA, USA; www.mathworks.com) and Java codes for the particle-based representation of foulants, coupled with finite-element methods from a commercial solver (COMSOL 3.5a, Comsol, Inc., Burlington, MA, USA; www.comsol.com) for flow, solute transport and mechanical stress calculations. After defining the generic feed spacer channel geometry, the model parameters (Table 4.1) and equations, the

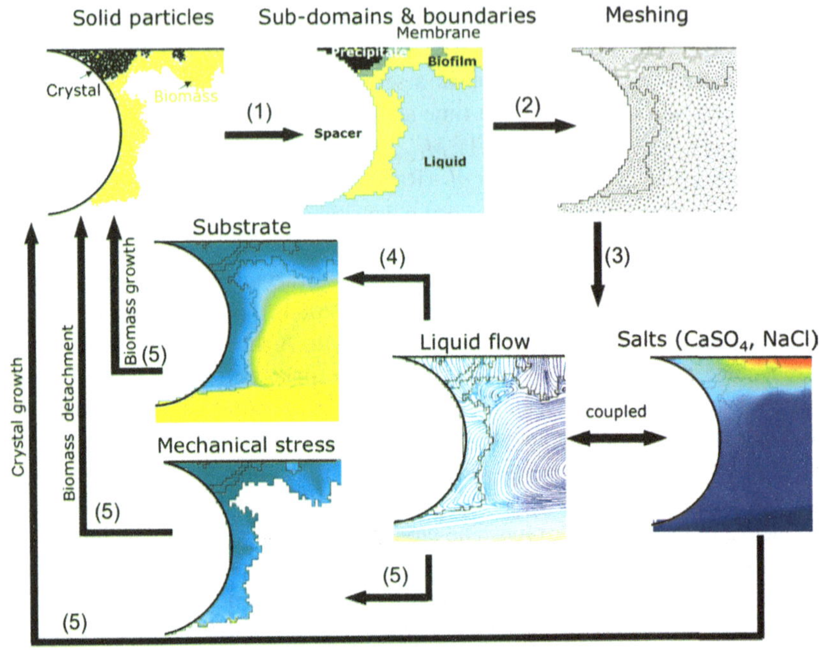

Figure 4.2 Sequence of calculations for modeling foulant development in time: (1) definition of sub-domains and boundaries; (2) construction of finite element mesh; (3) coupled solution of fluid flow and salt transport; (4) calculation of substrate distribution; (5) Particle growth, generation and detachment leading to a new set of particles for the next time iteration.

following sequence of steps is performed (Figure 4.2) at each time step: (1) Based on the current distribution of particles, model geometry and the corresponding computational sub-domains (liquid, biofilm, inner precipitate, outer precipitate) are defined. Specific physical properties are assigned a function of the phase present. A spatially variable function for biomass concentration $c_X(x, y)$ is created and the specific surface for precipitate growth a_v is calculated. (2) A customized finite element mesh is generated to allow accurate calculations of gradients. A finer mesh (maximum size 5 μm) is imposed close to the membrane surface and within the precipitate subdomains. (3) The flow (u, p) and mass balance equations for the two salts (c_1, c_2) are solved together, due to the two-way coupling between local permeate velocity and salt concentration. (4) The mass balance equations for substrate (c_S) are solved separately, using the already calculated velocity field and the mapped biomass concentration c_X. (5) Processes related to foulant development are evaluated as follows: (i) the increase in mass and size of particles is calculated, based on the local degree of saturation obtained in step (3) (for crystal particles) and the local substrate concentration from step (4) (for biomass particles); (ii) new biomass particles attach randomly and new crystal nuclei form based on DS; (iii) new biomass and crystal particles form by division of particles exceeding a threshold mass; (iv) the particles are displaced based on a weighted shoving algorithm, in order to minimize the overlaps between particles; (v) liquid flow and mechanical stress in the foulant layer are computed and particles situated in regions of high stress ($\sigma > \sigma_{det}$) are removed – detachment steps are calculated in a loop, until no particles can be removed (Radu *et al.*, 2010); (vi) this updated set of particles will serve for calculations in the next time iteration starting again from step (1).

RESULTS

Numerical simulations were performed to investigate the combined formation of biofilms and precipitate in the RO/NF feed channel. Idealized cases corresponding to situations in which scaling, biofouling and combined fouling may occur were compared (Table 4.2). The scenarios where only scaling occurs in the channel (cases 3 and 6) imply that there are no bacteria and/or nutrients present in the feed water. For biofouling (cases 2 and 5), it was assumed that the feed water contains the same amount of salts as in the case of scaling (to achieve the same osmotic pressure of the feed), but no nucleation takes place. In the combined fouling (cases 1 and 4), all inlet solute concentrations are identical to the corresponding biofouling case, but nucleation and growth of crystals occurs in supersaturated conditions.

Dynamics of combined fouling in the feed channel

Development of combined biofouling and scaling in a membrane system is governed by the availability of substrate and microbial inoculum, the existence of supersaturated areas and the level of shear stress in the channel. Biofilm formation starts with small, discrete colonies on the membrane and feed spacer. When only a small amount of biofilm is present, the gypsum mainly precipitates around the feed spacer-membrane contact areas (Figure 4.3, day 5). These 'spacer niche' zones have been shown to be prone to gypsum scale formation, both by membrane autopsy analysis (An *et al.*, 2011) and numerical modelling (Radu *et al.*, 2014a). In time, as the biofilm colonies grow and merge into a streamlined layer due to detachment, many new areas of precipitate formation can be observed (Figure 4.3, day 15) because further nucleation occurs within the biofilm. Scanning electron micrographs also revealed crystals embedded in the biofilm (Hou *et al.*, 2010; Thompson *et al.*, 2012). In the presence of feed spacers, both foulants are heterogeneously distributed in the channel: clean membrane areas correspond mostly to the high shear stress zone opposite to a feed spacer, while a mixture of biofilm and precipitate is present in the remaining membrane areas. Nevertheless, the biofilm and the precipitate each occupy different areas within the channel. Uneven distribution of foulants on the membrane was observed also in several membrane autopsy studies (Baker & Dudley, 1998a; Darton *et al.*, 2004;

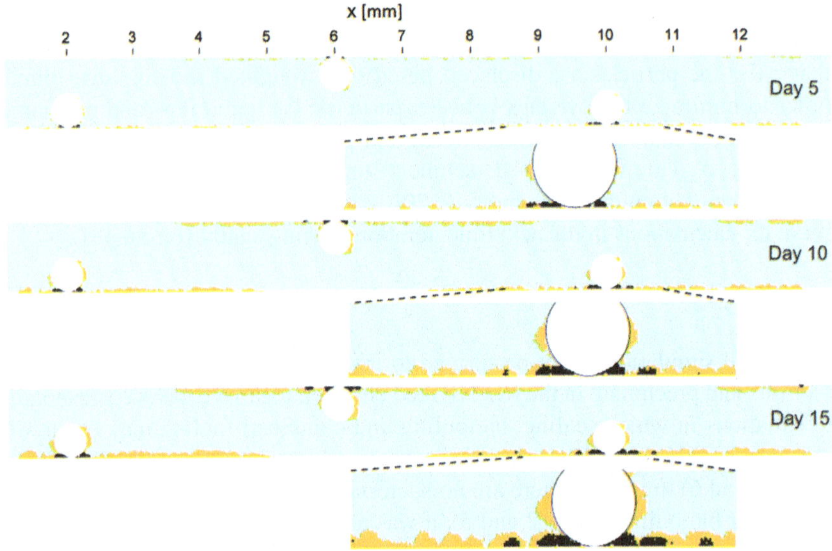

Figure 4.3 Development of combined biofouling and scaling in time (case 4). The yellow areas indicate the biofilm, the black areas indicate the precipitate. The white circles represent the feed spacers. A heterogeneous foulant distribution is observed.

Tran *et al.*, 2007). These findings point to the importance of multi-dimensional simulations for describing fouling in membrane feed channels. Figure 4.4 illustrates the formation of biofilm and precipitate for the three fouling scenarios: only scaling (case 3), only biofouling (case 2) and combined fouling (case 1) after 15 days. For the precipitate, however, there are significant differences between the scaling and

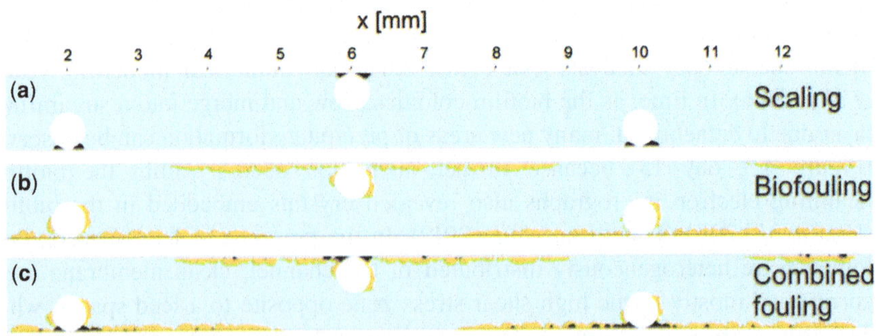

Figure 4.4 Fouling in the feed channel after 15 days for different scenarios: (a) scaling only (case 3), (b) biofouling only (case 2), (c) combined fouling (case 1). The yellow areas indicate the biofilm, the black areas indicate the precipitate.

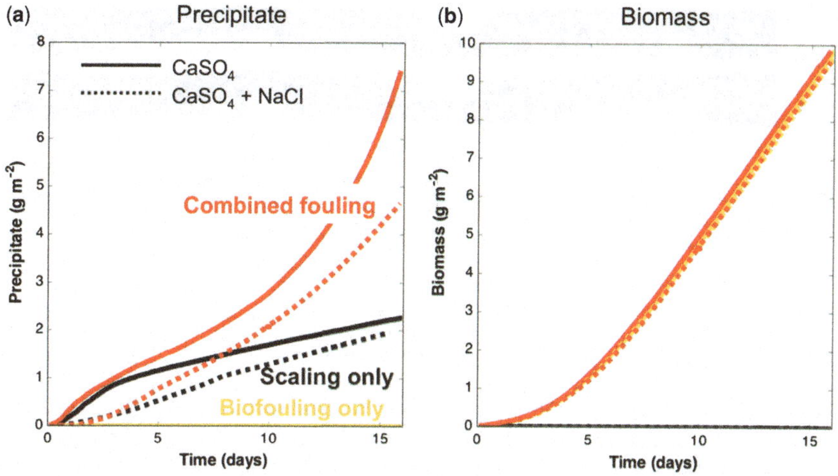

Figure 4.5 Amount of foulant formed in the feed channel for various feed water compositions: (a) Precipitate; (b) Biomass. Red lines correspond to combined fouling, black lines to scaling and yellow lines to biofouling. The solid lines indicate single salt present in the feed (cases 1–3), dashed lines indicate mixture of salts in the feed (cases 4–6).

combined fouling situations, despite the same composition of the feed water. The precipitate is concentrated in the spacer niche in the absence of biofilms, while it is spread in multiple areas on the membrane if biofilms are also formed. Interestingly, when quantifying the amount of foulant in the feed channel (Figure 4.5a), more precipitate was formed in the presence of a biofilm, despite identical feed water composition in the cases of scaling and combined fouling.

In addition, although the feed water has the same degree of saturation, the feed composition (i.e. the presence of NaCl) affects the amount of gypsum that precipitates. The amount of biomass in the biofilm was, however, the same in all simulated cases (Figure 4.5b), since the same attachment rate, identical substrate concentrations in the feed water and constant cross-flow velocity were considered (Table 4.2) (i.e. the precipitation had only a minor effect on biofilm formation).

Biofilm enhanced concentration polarization and scaling

The degree of saturation is the highest at the membrane surface and is the main factor affecting crystal nucleation and growth (Lyster *et al.*, 2009; Radu *et al.*, 2010). In order to understand the cause of more precipitate formation in the presence of biofilm, the degree of saturation distribution in the channel was analysed. The local DS is a result of complex coupling between flow and transport of various solutes in the membrane feed spacer channel. As reported also in previous numerical studies (Radu *et al.*, 2014a; Schwinge *et al.*, 2002a), solute concentrations can vary

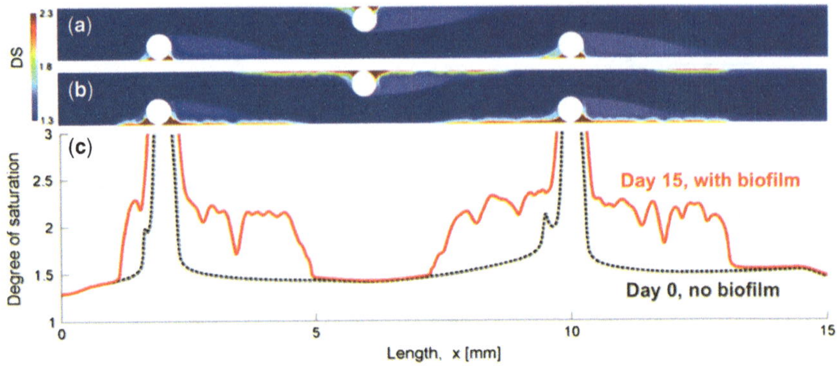

Figure 4.6 Degree of saturation for gypsum in the feed channel (case 2): (a) initial 2-d distribution; (b) altered 2-d distribution due to biofilm presence (day 15 – biofilm structure presented in Figure 4.4b); (c) 1-d comparative profile along the bottom membrane corresponding to cases (a) – black dashed line and (b) – red continuous line.

markedly, as a result of the presence of feed spacers. For a single salt present in the feed water (here $CaSO_4$), an increase in salt concentration means an increase in the degree of saturation. In Figure 4.6a regions of high DS can be observed close to the membrane-spacer contact areas. Biofilm formation alters solute distribution due to biofilm enhanced concentration polarization (Figure 4.6b). This effect has been suggested previously in both experimental (Herzberg & Elimelech, 2007) and numerical studies (Radu *et al.*, 2010). A larger volume reaches high degrees of saturation in the presence of biofilm (Figure 4.6b). In addition, the degree of saturation at the membrane surface is strongly enhanced (Figure 4.6c) as the biofilm reduces solute convection parallel to the membrane. For feed solutions containing mixtures of salts, solute distributions in the channel become even more complicated. Concentration polarization ratios for the two salts considered in this work ($CaSO_4$ and NaCl) are not the same (~20% different), as their diffusion coefficients are not identical. For the slower diffusing $CaSO_4$ the concentration enhancement within the biofilm is greater than for NaCl (Figure 4.7a and 4.7b). Consequently, the DS for gypsum in the presence of the biofilm is not easily predictable, given that it has a non-linear dependency on the solute concentration (i.e. via Pitzer equations used for determining the activity coefficients). For the operational conditions considered in this study, the ratio of DS increase within the biofilm is even more pronounced than the ratio of potentially scaling salt concentrations (Figure 4.7c). However, it should be noted that this observation is very much dependent on the ratio of the two solutes within the feed water, flux, and cross-flow velocity as well as biofilm thickness. In order to quantify the potential for formation of new nuclei in different cases, the nucleation propensity can be defined as the expected number of nuclei that would form within a certain time interval in the channel. Distinct trends can

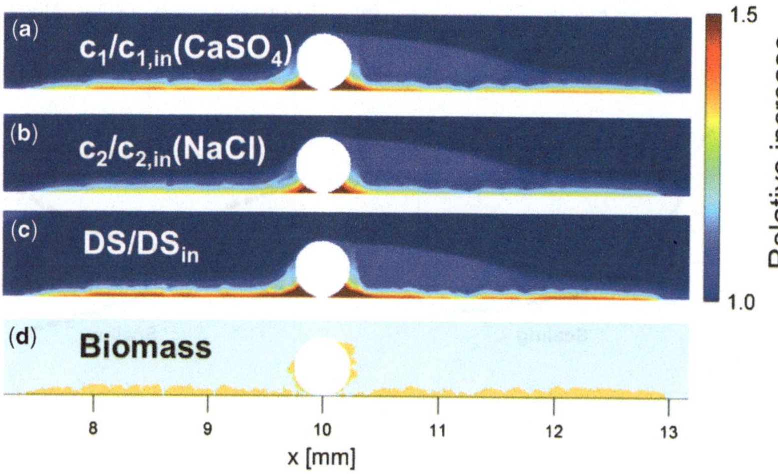

Figure 4.7 The effect of biofilm (day 15, case 5) on solute distribution in the feed channel: (a) Concentration increase of $CaSO_4$ in respect to inlet concentration; (b) Concentration increase of NaCl in respect to inlet concentration; (c) Degree of saturation enhancement ratio for gypsum; (d) Biomass distribution in the feed channel.

be noticed for the scaling only vs the combined fouling scenarios, for both single salt solutions and mixtures (Figure 4.8a and 4.8b). When the precipitate is the only foulant, the expected amount of nucleation declines in time as the feed spacer niche gets filled up with gypsum.

Conversely, for the combined fouling, after an initial decrease in nucleation propensity more nuclei are likely to form as soon as enough biofilm has developed. Optical images taken during an experimental test indeed indicated more nucleation sites and a higher extent of membrane coverage with gypsum in the presence of a biofilm (Thompson *et al.*, 2012). The observation of more precipitate formed in the presence of biofilms is clearly related to the enhanced concentration of solutes in the biofilm. The model results indicate that the presence of a biofilm accelerates gypsum scaling for at least two reasons: (1) enhanced concentration polarization, which may lead to salt concentrations above the solubility product next to the membrane and within the biofilm; and (2) an increased surface with high DS within the biofilm favourable for crystal nucleation.

Effect of combined fouling on process performance

A comparative evaluation of the impact of fouling for the three scenarios in terms of flux decline is shown in Figure 4.9. It can be observed that the combined fouling always leads to the largest flux decline, as also suggested by the experimental findings of Thompson *et al.* (2012). For some stages of the fouling, the normalized

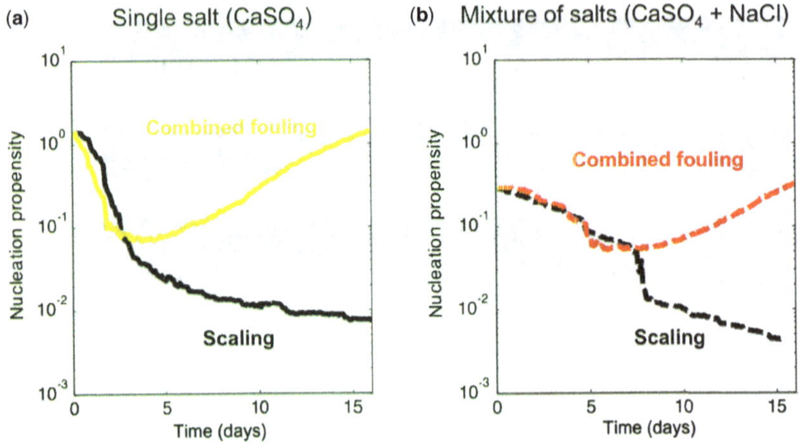

Figure 4.8 Evolution of nucleation propensity in the feed channel for different feed water compositions with identical DS = 1.3, combined fouling (red lines) compared to scaling only (black lines).

flux decline in combined fouling is larger than the summed flux decline for scaling and biofouling alone. This type of synergistic effect due to coupling of two types of fouling has been proposed for gypsum precipitation in the presence of alginate in forward osmosis systems (Liu & Mi, 2012). However, it should be noted that such an observation is mainly dependent on the relative rates of the

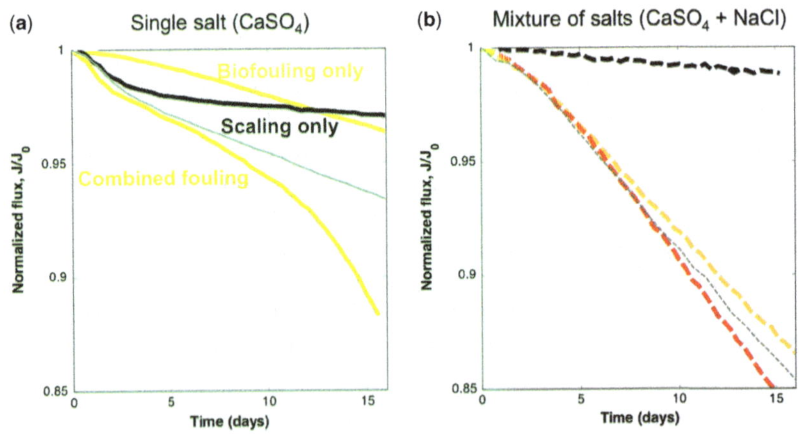

Figure 4.9 Normalized flux decline for different fouling scenarios (red: combined fouling; yellow: biofouling; black: scaling; gray: algebraic sum of flux decline corresponding to individual scaling and biofouling) and feed compositions: (a) single salt ($CaSO_4$); (b) mixture of salts ($CaSO_4 + NaCl$).

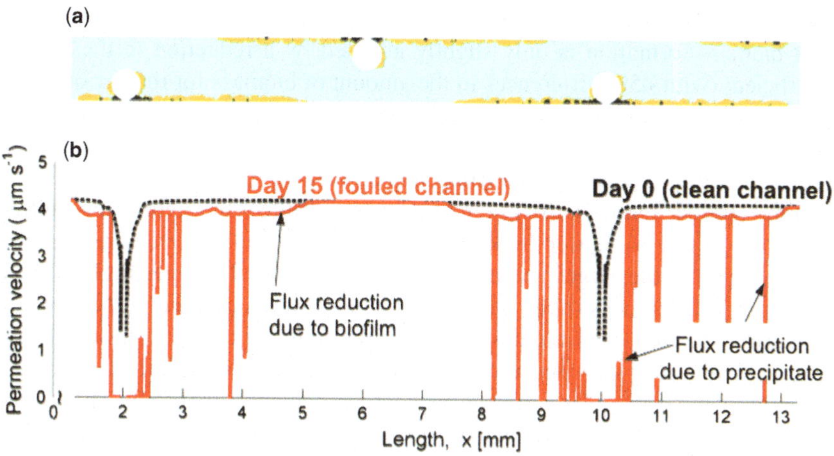

Figure 4.10 (a) Foulant distribution in the feed channel (day 15, case 4). Yellow: biofilm; Black: precipitate. (b) Local permeate flux alteration due to biofilm and precipitate formation. Blue dashed line: day 0; red continuous line: day 15.

various processes involved in foulant development (crystal nucleation and growth, and biomass growth/alginate deposition) and therefore cannot be generalized. The local permeate flux is strongly altered when the two foulants are present, as shown in Figure 4.10. Initially, in the absence of any foulant, the areas with the lowest permeate flux are around the spacer-membrane contact zones, where the solute concentrations (and thus osmotic pressure) are the highest. Once accumulated in the channel, the foulants clearly affect the flux to a different extent: the biofilm reduces the local flux less than mineral scaling; the precipitate is more impermeable, which leads to severe local permeate decline. In RO, the bio- films decrease flux mostly through enhanced concentration polarization (CP), not through their hydraulic resistance (Dreszer *et al.*, 2013; Radu *et al.*, 2010). An even larger adverse effect of CP would be observable at higher operational fluxes or pressures.

Parameter sensitivity studies

The impact of several model parameters on the development of biofilms and precipitates was evaluated in a sensitivity study. Precipitate permeability was increased and decreased by one order of magnitude compared to the initial value derived from Rahardianto *et al.* (2006). An increase in precipitate permeability results in slightly more mineral formation for both scaling only and combined fouling scenarios, as solute transport is enhanced within the outer growing layer. However, the lower permeability does not result in a different amount of precipitate. The importance of such variations for the permeate flux decrease appears to be very small. The diffusion coefficients for both substrate and salt in the biofilm were reduced to 80% and 60% of their values in the bulk liquid. The results indicate that biomass formation is only slightly affected by a reduction in the diffusion coefficient (with <5% differences in the amount of biomass for the various cases), which indicates little diffusion limitation in the evaluated conditions. The amount of precipitate, on the other hand, increased noticeably for the combined fouling scenarios when decreasing the diffusion coefficient.

For salts, a reduced diffusion coefficient resulted in higher concentration polarization due to slower back-diffusion of salts from the membrane surface, which explains the larger amount of precipitate formed. Interestingly, although only minor differences in the amount of biomass were observed in the biofouling scenarios, the impact of biomass on flux decline is noticeable: a lower diffusion rate leads to higher salt concentrations at the membrane surface, thus reducing the effective driving force for permeate production. It appears that for bio-film growth with a reduced substrate diffusion coefficient, the diminished substrate availability was compensated by the intensification of substrate concentration polarization. When doubling the operational pressure, the amount of precipitate in the channel increased substantially. This is due to the greater concentration polarization of salts created by the largely enhanced flux, while the higher substrate concentration polarization has only a minor influence on increasing the amount of biomass. The different flux decline responses to the two types of foulants are due to the different kinetic expressions and mechanisms for crystal and biomass growth, but nevertheless the effect of a doubled pressure on the flux decline is very powerful. It can be concluded that variations in these parameters do not affect the general model outcome: for all cases more precipitate is formed in the presence of biofilm and an accelerated flux decline is observed for combined fouling scenarios. However, while the diffusion and permeability coefficients have only a limited effect, the operational pressure seriously affects the precipitation.

DISCUSSION

In this paper, combined biofouling and mineral scaling in membrane feed spacer channels have been investigated with a two-dimensional mechanistic model, which integrates physical, chemical and biological processes. The numerical results revealed several potential causes for enhanced gypsum precipitation in the presence of biofilms.

Complexity of biofilm–precipitate interactions

A variety of factors are involved in fouling of membrane feed channels. Figure 4.11 illustrates some of the complex interactions between transport processes in RO/NF feed channels (fluid flow and solutes transport/reaction) and fouling (caused by the presence of biofilms and precipitates). Even in the absence of any foulant, there is a clear coupling between the fluid flow and solute transport, inherent to RO/NF membrane separation processes.

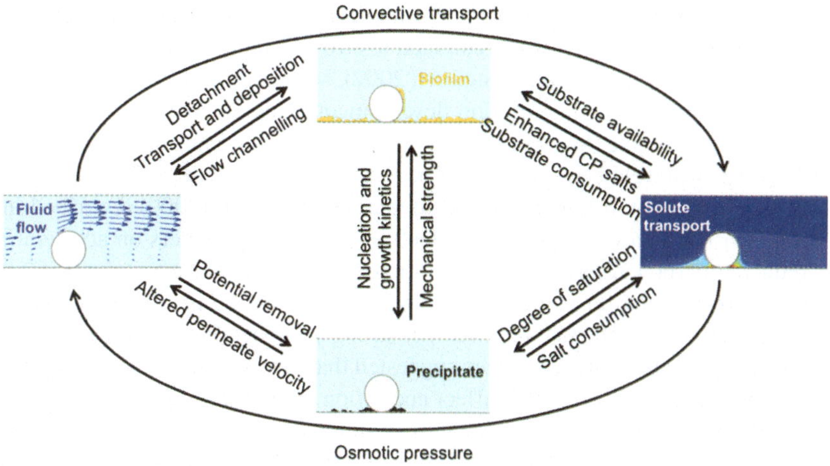

Figure 4.11 Schematic illustration of the interactions between flow and solute transport processes and fouling in RO/NF feed channels due to biofilm formation and mineral precipitation.

That is, local permeation of water through the membrane depends on the local osmotic pressure of the solution (determined by the local salt concentrations), while solute concentrations are in turn influenced by the flow via convective transport (Lyster & Cohen, 2007; Radu *et al.*, 2010). This results in strong local variations in solute concentrations and flow velocity/pressure within the feed spacer channel. After fouling is initiated in the channel, the flow distribution is altered. Biofilm growth leads to a more tortuous flow path, resulting in formation of preferential channels with high liquid velocity and an increased pressure drop (Graf von der Schulenburg *et al.*, 2008; Picioreanu *et al.*, 2009; Radu *et al.*, 2010; Vrouwenvelder *et al.*, 2010b). On the other hand, precipitates are almost impermeable, locally reducing the permeate flux. When operating at constant pressure this results in an overall permeate flux decline, while for operation at constant total flux this may lead to permeate flux enhancement through the clean membrane areas (Radu *et al.*, 2014a). The concentration of solutes varies not only in space, but also in time because the foulants can affect both reactions and transport of solutes. The substrates are consumed for biofilm growth and some salts are consumed when the precipitates are formed. Furthermore, salt concentrations are affected by the biofilm enhanced concentration polarization (Herzberg & Elimelech, 2007; Radu *et al.*, 2010) and flow channelling may lead to stagnant zones with reduced solute transport (Baker & Dudley, 1998a; Picioreanu *et al.*, 2009). Foulant formation itself is influenced by solute concentrations and liquid flow: biofilms grow as a function of substrate availability (Vrouwenvelder *et al.*, 2009b) and detach as a function of mechanical stress caused by the flow (Picioreanu *et al.*, 2001); precipitates are formed in areas of a high degree of saturation (Lyster *et al.*, 2009). The presence of divalent ions in the feed water can lead to stronger biofilms due to ionic cross-linking of biopolymers (Chen & Stewart, 2002). Moreover, dosage of anti-scalants has been suggested to affect biofilm development by enhancing the deposition of cells on the membrane (Sweity *et al.*, 2013) or by providing a nutrient source for the biofilm (Vrouwenvelder *et al.*, 1998). The existence of a biofilm may affect crystal nucleation and growth kinetics. For example, Dupraz *et al.* (2009) described in a comprehensive review the potential effects of microbial polymers on accelerating or reducing carbonate precipitation in microbial mats. In the case of calcite, the influence of the biofilm would be even more complex: the metabolic activity of bacteria may change the pH and this will strongly impact the thermodynamics of calcite precipitation. It has also been suggested that the functional chemical groups on the biofilm and the cells may affect nucleation kinetics (Steiner *et al.*, 2010).

In addition, embedded precipitates may influence the mechanical strength of the biofilm and thus its detachment. Lin *et al.* (2013) has shown with compressive load tests for anammox granules that the presence of solid precipitates (apatite) leads to a reinforced granule structure. These are just a few examples of the many interactive factors of physical, chemical and biological origin that can influence fouling, making the interpretation of the results often challenging. Comprehensive biofilm and precipitate models based on first principles can therefore be helpful for analysing the complex interactions.

Benefits, limitations and extensions of the model

The development of increasingly complex models for fouling in membrane processes has been driven by the aim of better understanding a system that was mostly investigated by a black box approach. The model presented here can adequately describe related experimental observations for combined fouling on a qualitative level and provide a theoretical basis for interpreting the data. The benefit of such a model for studying complex phenomena such as fouling does not reside in prediction capabilities, but rather in the ability to check the relative significance of various effects. The value of this kind of numerical model is to support exploratory laboratory-scale studies on micro-scale variations in relation to macro-scale performance indicators. Through sensitivity analysis, the model can inform experimental efforts. These multi-dimensional mechanistic models may also be applied to interpret the data obtained from in situ studies of combined biofouling and scaling making use of advanced experimental techniques (i.e. optical coherence tomography (Wagner *et al.*, 2010b), nuclear magnetic resonance (Graf von der Schulenburg *et al.*, 2008; Vrouwenvelder *et al.*, 2010b), confocal Raman spectroscopy (Wagner *et al.*, 2009), and X-ray computed tomography (Iltis *et al.*, 2011; Wildenschild & Sheppard, 2013). The model has been developed in the particular context of membrane processes. Nevertheless, this numerical approach can in principle be adjusted for analysing various applications dealing with biofilm–mineral interactions: other membrane processes (e.g. forward osmosis; (Liu & Mi, 2012)), bio-grouting (e.g. soil improvement based on ureolysis followed by carbonate precipitation; (Zhang & Klapper, 2010)), CO_2 sequestration (Mitchell *et al.*, 2010), dental plaque formation (Ilie *et al.*, 2012), bio-corrosion and bioleaching (linking sulphate reducing/sulphur-oxidizing bacteria with mineral precipitation/dissolution processes; (Olivera-Nappa *et al.*, 2010)). Further model development would entail considering individual ions for defining water composition, which, however, implies a large number of soluble species associated with more demanding computations.

CONCLUSIONS

(1) The mechanistic model integrating mineral precipitation and biofilm formation with hydrodynamics and solute transport describes the development of different interacting foulants in membrane feed spacer channels. (2) Micro-scale simulations are important because they can illustrate the heterogeneous foulant distribution in the feed channel. (3) The presence of biofilms enhances gypsum precipitation in the conditions studied. (4) A more severe decline in performance was obtained for combined fouling compared to the sum of the single fouling simulations.

Chapter 5

Effect of different commercial feed spacers on biofouling of reverse osmosis membrane systems: A numerical study

INTRODUCTION

Membrane filtration processes like reverse osmosis (RO) and nanofiltration (NF) have become increasingly important for high quality drinking water production in recent years. The major problem of this technology is biofouling – excessive growth of biomass leading to reduction of produced water quantity and quality, while increasing the operational costs.

Feed spacers are essential parts of the spiral-wound NF and RO modules that keep membranes apart to form the flow channel and to promote mixing of the fluid. Baker *et al.* (Baker *et al.*, 1995) reported that initial deposits of fouling were found to accumulate alongside the membrane feed channel spacer and with time these deposits encroached upon the remaining free membrane area. Van Paasen *et al.* (Van Paassen *et al.*, 1998) observed an exponential increase of the feed channel pressure drop caused by biofouling build up onto the feed spacer of the membrane modules. This biofouling proved to be related with chemicals dosed to the feed water. Tran *et al.* (Tran *et al.*, 2007) found that the vicinity of the feed spacer strands was most affected by fouling. Strategies to reduce feed spacer biofouling have been addressed, e.g. periodic air/water flushing (Cornelissen *et al.*, 2007a) and applying thick feed spacers (Majamaa *et al.*, 2010). Vrouwenvelder *et al.* (Vrouwenvelder *et al.*, 2009a) showed that for fresh water the feed spacer biofouling is much more important than membrane biofouling for feed channel pressure drop increase and thus for overall performance decline. In summary, feed spacers are important for membrane performance and play a key role in biofouling of membrane systems. Therefore, a better understating of the

Author(s): Szilárd Bucs, Joop Kruithof, Mark van Loosdrecht and Johannes Vrouwenvelder
doi: 10.2166/9781780409597_0085

impact of feed spacers on biofouling is of primary importance for an improved design of the spiral-wound membrane modules.

According to Li *et al.* (Li *et al.*, 2002b), feed spacers can be characterized by four parameters: (i) the distance between spacer filaments, (ii) the angle between spacer filaments, (iii) the flow attack angle and (iv) the total spacer thickness. Sablani *et al.* (Sablani *et al.*, 2002) studied experimentally the influence of three feed spacers varying in thickness 20, 28 and 46 mil (1 mil = 25.4 μm; i.e., a 28 mil spacer = 711 μm thick) on concentration polarization. They found a decrease in flux with decreasing spacer thickness, but the highest permeate flow was obtained for the intermediate spacer thickness. Recently, Araújo *et al.* (Araújo *et al.*, 2012a) studied experimentally the effect of different spacer thickness on biofouling. Their findings showed that with the increase of spacer thickness there is a decrease of the feed channel pressure drop due to biofouling.

Computational fluid dynamics (CFD) has become a widely used tool in studying the hydrodynamic behaviour of NF and RO membrane systems (Fimbres-Weihs & Wiley, 2010). Many studies using CFD focused on the effect of feed channel spacer on fluid flow and mass transfer with different type of spacers (Fimbres-Weihs & Wiley, 2007; Geraldes *et al.*, 2002; Koutsou *et al.*, 2007; Saeed *et al.*, 2012; Shakaib *et al.*, 2007). Simplified, cylindrical shapes were used for representation of spacer filaments in most of the numerical studies on the effect of feed spacer geometry on mass transfer and fluid flow. Stereomicroscopic observations of the feed spacer revealed that the spacers used in commercially available spiral-wound membrane modules have more complicated geometry, with filaments varying in thickness and thinning (Vrouwenvelder *et al.*, 2008a). Picioreanu *et al.* (Picioreanu *et al.*, 2009) found by numerical simulations that the feed channel pressure drop for a simplified spacer with cylindrical filaments is significantly different from a more realistic spacer geometry with filament thinnings. Although the effect of feed spacer geometry has been extensively studied, it is still not clear how spacers affect biofouling and the performance parameters.

In this study we examined with a numerical model the impact on feed channel pressure drop of: (i) liquid flow velocity and bacterial cell load; (ii) biomass location on the spacer and/or membrane surfaces and (iii) various feed spacer geometries (28, 31, 34 and 46 mil thick) as applied in commercially available spiral-wound reverse osmosis modules and a modified geometry having a 31 mil thick spacer. To the authors knowledge this is the first paper using a three-dimensional mathematical model on biofouling evaluating several realistic geometries (commercially available) feed spacers as used in full-scale spiral wound membrane modules.

MODEL DESCRIPTION

A three-dimensional numerical model was developed to study the impact of feed spacer geometry on the biofouling of feed channels of spiral-wound membrane modules. The model is based on the work of Picioreanu *et al.* (Picioreanu *et al.*, 2009), implemented here in more efficient computer code coupling COMSOL Multiphysics solvers (COMSOL 4.3a, Comsol Inc., Burlington, MA, www.comsol.com) for fluid dynamics and solute transport with MATLAB (MATLAB, 2011a, MathWorks, Natick, MA, www.mathworks.com) code for biofilm formation.

Spacer geometry

Geometries corresponding to four commercially available feed spacers used in full-scale spiral-wound RO modules were characterized using a calibrated stereomicroscope (Leica M205 FA), followed by measurements using Qwin Pro 3.1.0 software, and then reconstructed in the COMSOL Multiphysics environment. The studied spacers differ in thickness, porosity and their filament shapes. For each filament, several characteristic dimensions were considered, according to Figure 5.1a. Every spacer is constructed from two perpendicular layers of filaments (Figure 5.1b), and the filaments from each layer have a unique shape. The spacer filaments differ mainly in their diameter, with several specific regions identified for all filaments determining a characteristic shape, as shown in Figure 5.1a. The dimensions and the flow channel porosity for each spacer type are presented in Table 5.1. Additionally, a hypothetical 31 mil thick modified spacer geometry was

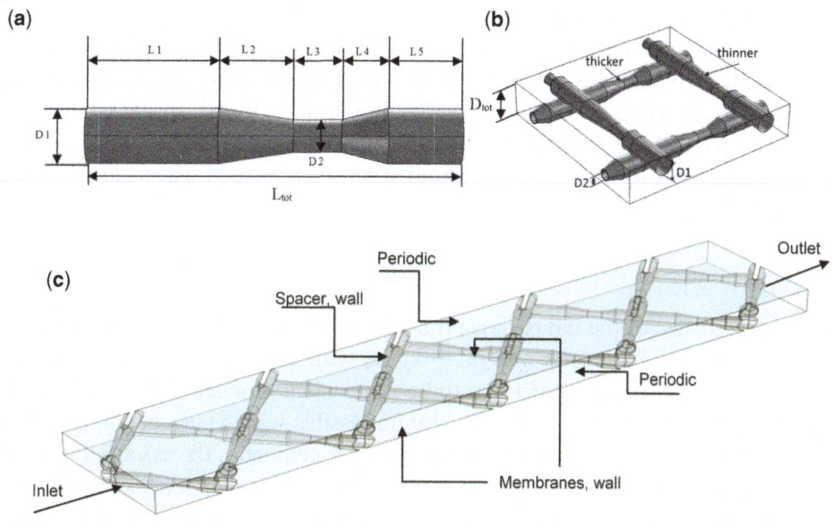

Figure 5.1 Model feed spacer geometry and computational domain. (a) General representation of spacer filaments with measured characteristic dimensions (see Table 5.1); (b) Reconstructed geometry of a representative feed spacer unit varying in filament thickness, as applied in practice in spiral-wound modules for water treatment; (c) Computational domain and types of boundary conditions.

Table 5.1 Dimensions of different types of feed spacers used in this study, according to Figure 5.1. For each spacer type, the filament dimensions (in μm) for the two layers of fibres are presented in columns: (A) thinner filament and (B) thicker filament.

		Spacer Thickness (mil)									
		28 mil		31 mil		31 mil Modified		34 mil		46 mil	
		A	B	A	B	A	B	A	B	A	B
Dimensions (μm)	L1	862	877	983	299	400	400	234	281	364	171
	L2	595	740	557	132	100	100	389	535	346	420
	L3	767	595	358	417	1780	1780	500	92	561	713
	L4	178	252	340	1036	100	100	696	878	526	518
	L5	386	324	542	896	400	400	566	599	614	589
	D1	354	449	447	446	446	446	422	445	563	544
	D2	221	289	263	272	263	263	223	223	315	274
	L_{tot}	2788	2788	2780	2780	2780	2780	2385	2385	2411	2411
	D_{tot}	711		787		787		863		1168	
	Channel porosity	0.90		0.89		0.92		0.88		0.88	

created to study the effect on feed channel porosity of spacers having the same thickness but different filament dimensions.

The spacer geometries used in commercially available RO and NF modules have standard thicknesses, but they can be provided by different manufacturers and produced by different technologies, which may lead to diverse spacer geometries.

These spacers vary especially with respect to the spacer unit length L_{tot} and the flow channel porosity (Table 5.1). All studied commercial feed spacers are displayed Figure 5.2.

Computational domain

The standard size of the industrial spiral-wound RO and NF module is a length of 40 in. (~1 m) and a diameter of 8 in. (~0.2 m) with a total membrane surface area of ~40 m^2. A numerically accurate three dimensional model of flow and biofilm formation in such a large area is virtually impossible within current computational limits. Still, because of the repetitive unit spacer geometry, the essential flow and biofilm formation patterns can be captured within a smaller scale computational domain. The size of the computational domain used in this study is in the range of 10^{-5} m^2. The exact length, width and height of the computational domain differs with the spacer geometry. In all cases, five spacer units in diamond configuration (i.e., 45° rotated against the main flow direction) were placed in the computational domain (Figure 5.1). This is the current compromise between the necessary calculation time and model realism.

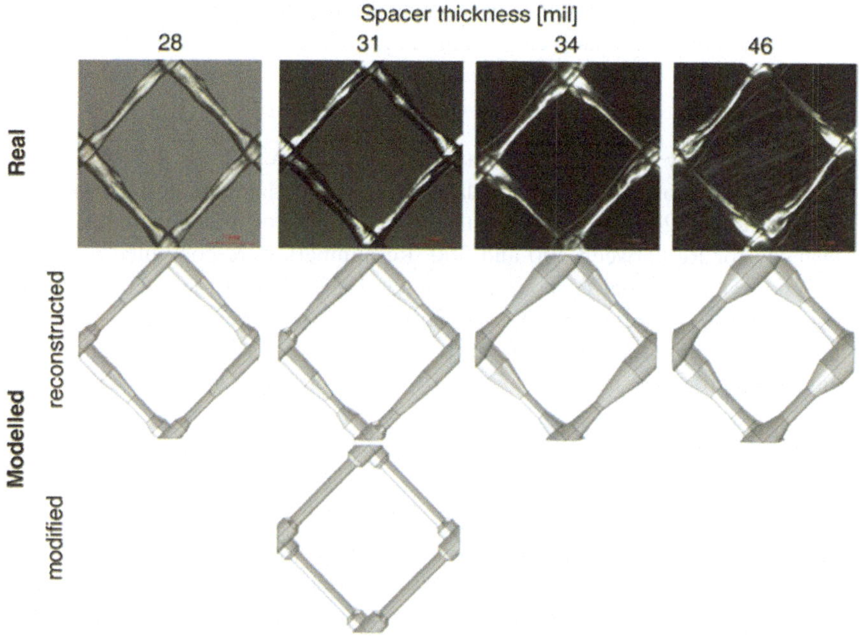

Figure 5.2 Geometries of different commercially available feed spacers: top – microscopic images of spacers as applied in practice, bottom – re-created and modified model geometries. One representative element is shown for each type of spacer.

The computational domain consisted of feed channel volume available for flow and resulted from subtracting the spacer volume from a box bounded by membranes, inlet, outlet and lateral surfaces (Figure 5.1). An imprint of the feed channel spacer on the membranes can be observed during autopsies of spiral-wound RO modules, which suggests that the spacer filaments are actually pressing into the membranes on a sizeable contact area.

Therefore, the flow channel was constructed so that the two membrane planes cut 5 μm from the spacer top and bottom (Fimbres-Weihs & Wiley, 2007). In addition to more realistic model geometry, this construction also avoids very sharp angles in contact areas, which usually lead to computational difficulties and many unnecessarily small mesh elements.

Liquid flow, substrate transport and biofilm formation

Processes influencing biofouling may occur at very different time scales. Therefore, while the slow biofilm formation was evaluated in dynamic conditions (i.e., in time), the fast hydrodynamics and soluble substrate transport were considered stationary at each time step during biofilm development. More details regarding the time scale separation in biofilm models can be found in (Picioreanu *et al.*, 2000).

Liquid flow

It was shown by Fimbres-Weihs and Wiley (Fimbres-Weihs & Wiley, 2010) that the water flow in feed channels with spacers is laminar up to a Reynolds number (Re) of 200 for spacers with 45° orientation of filaments, but becomes unsteady for Re between 200 and 300. Re numbers were computed according to Schock and Miquel (Schock & Miquel, 1987). For all geometries Re was less than 200, therefore the Navier–Stokes equations were used to model the laminar, incompressible and stationary flow which was assumed during the whole simulation process:

$$\begin{aligned} \rho(u \cdot \nabla)u + \nabla p &= \nabla \cdot (\eta \nabla u) \\ \nabla \cdot u &= 0 \end{aligned} \tag{5.1}$$

with $\mathbf{u} = (u_x, u_y, u_z)$ the flow velocity vector, p the pressure, ρ the density and η the dynamic viscosity of water. The boundary conditions were set according to Figure 5.1. Fully developed laminar flow was assumed at the inlet $x = 0$, with the average fluid velocity u_{in}. On the outlet boundary at $x = L_X$, zero pressure (arbitrarily chosen as reference value) and no viscous stress conditions were imposed. The top and bottom membrane surfaces ($z = 0$ and $z = L_Z$), as well as the whole spacer surface were no-slip walls ($\mathbf{u} = 0$). On the lateral boundaries ($y = 0$ and $y = L_Y$) periodic conditions were applied, as frequently used in simulations for geometries with a repeating pattern (Koutsou *et al.*, 2007).

Substrate transport and reaction

Biomass grows in spiral-wound membrane modules as function of local concentrations of biodegradable substrates. The concentration field, C_S, of a single soluble substrate assumed here limiting for biomass growth was calculated from a convection–diffusion–reaction equation:

$$D_S \nabla^2 C_S - u \nabla C_S + r_S = 0 \tag{5.2}$$

with diffusion coefficient of substrate D_S. The substrate is transported by convection and diffusion in the bulk liquid and only by diffusion in the biofilm. The specific growth rate of the microorganisms (μ) was computed according to the simple Monod kinetics:

$$\mu = \mu_M \frac{C_S}{K_S + C_S} \tag{5.3}$$

where μ_M is the maximum specific growth rate of the microorganisms, K_S is the Monod half-saturation coefficient and C_S is the local biodegradable substrate concentration in the biofilm. Monod-type microbial growth rates are commonly applied for biofilm growth (Suidan *et al.*, 1987; Wanner & Gujer, 1986; Wanner *et al.*, 2006). The biomass growth rate r_X was linear with respect to C_X (biomass concentration in the biofilm):

$$r_X = \mu \cdot C_X \tag{5.4}$$

The substrate consumption rate by biomass, r_S, was calculated neglecting the maintenance:

$$r_S = Y_S \cdot r_X \tag{5.5}$$

where Y_S is the yield coefficient (mol substrate consumed per C-mol biomass formed). A constant inlet concentration $C_{S,in}$ was assumed at $x = 0$, while the outlet ($x = L_X$) was a zero-diffusion boundary. Periodic conditions were set for the lateral boundaries at $y = 0$ and $y = L_Y$. The membranes and spacer surfaces were set as impermeable walls.

Biofilm formation

The biofilm development in the feed channel was governed in this model by two processes, microbial attachment and growth. Biomass detachment also plays a role in biofilm development, but mainly at late stages of biofouling development (i.e., thick biofilms).

Effects of biofilm detachment on biofouling will therefore be addressed in follow up studies. Biomass was allowed to attach to randomly chosen places in the computational domain. Three different attachment modes were studied: (i) only on the spacer surface, (ii) only on the membrane surface and (iii) on both spacer and membrane surfaces. The cellular automata (CA) biofilm model developed in (Picioreanu *et al.*, 1998a, b) and modified as in (Picioreanu *et al.*, 2009) was used to simulate the biofilm growth in a spiral-wound membrane system. The biomass growth is calculated in each volume element by a Monod equation linked to the substrate consumption rate:

$$\frac{dC_X}{dt} = r_X \tag{5.6}$$

In every volume element of a 3-d grid a maximum biomass concentration was set to $C_{X\max}$. Once the maximum biomass concentration for a grid element is exceeded, the biomass amount is divided in two parts. One part will remain in place, while the second part will be repositioned in space according to the CA algorithm as explained in (Picioreanu *et al.*, 1998a, b).

Model solution

The model solution algorithm follows the procedure described in detail in (Picioreanu *et al.*, 2009). After the model geometry was created, the biofilm development was followed iteratively in discrete time steps of size $\Delta t = 4$ h. The liquid flow and substrate distributions reach their steady state fast enough (~seconds) compared to the chosen Δt (~hours) to be considered stationary for each time interval of biofilm development. After updating the biomass distribution one time step, equations 5.1 and 5.2 are used again, for calculating the new stationary flow and concentration fields.

For the calculations of liquid flow and mass transport of the soluble substrate a tetrahedral mesh was built with maximum element size of 90 μm. This resulted for 31 mil thick feed spacer geometries in a total of more than 1,300,000 mesh elements. The calculated substrate concentration field was then used in the biomass growth, followed by biomass spreading and attachment of new biomass. The main algorithm was implemented in a MATLAB script split into several steps:

- **A** define model parameters (Table 5.2), create the model geometry, computational domain and build equations in COMSOL (Figure 5.3a)
- **B** create the finite element mesh for flow and mass transport calculations in COMSOL (Figure 5.3b)
- **C** solve the laminar flow equation 5.1 in COMSOL for the flow velocity u and pressure p (Figure 5.3c)
- **D** solve substrate transport-reaction equation 5.2 for concentration C_S (Figure 5.3d)
- **E** evaluate biomass growth from equation 5.4 to get a new C_X; transport biomass by the CA method; attach new biomass (Figure 5.3e). After this biomass update the time is incremented with Δt and the algorithm flow returns to step C.

RESULTS

The model was used to investigate the effect of linear flow velocity, bacterial cell load, biomass attachment location, and different commercially available and modified feed spacer geometries on biofouling and feed channel pressure drop evolution. All simulated cases are listed in Table 5.3.

Effect of linear flow velocity

The linear flow velocity varies along the length of a membrane filtration installation, consisting of several membrane elements installed in series. Also, a variation of linear flow velocities has been reported in the lead membrane elements in practice, from 0.07 to about 0.2 m · s^{-1} (Vrouwenvelder *et al.*, 2009b).

Table 5.2 Model parameter.

Parameter	Symbol	Value	Unit	Source
Maximum biomass concentration in a grid element	C_{Xmax}	1400	$mol \cdot m^{-3}$	Chosen
Biomass maximum specific growth rate	μ_m	1.25×10^{-5}	s^{-1}	Chosen
Substrate half-saturation coefficient	K_S	0.05	$mol \cdot m^{-3}$	Chosen
Substrate concentration in the inlet	$C_{S,in}$	0.4	$mol \cdot m^{-3}$	Chosen
Yield of substrate consumed per C-mol biomass	Y_S	1	$mol \cdot (Cmol)^{-1}$	Corresponding to acetate
Diffusion coefficient of substrate	D_S	1×10^{-9}	$m^2 \cdot s^{-1}$	Small solute in water
Liquid density	P	1000	$kg \cdot m^{-3}$	Water
Liquid viscosity	η	0.001	$Pa \cdot s$	Water

Numerical simulations were run with different linear flow velocities and compared with experimental data (feed channel pressure drop, pressure drop increase) obtained in similar conditions by Vrouwenvelder *et al.* (Vrouwenvelder *et al.*, 2009b) in experimental studies with membrane fouling simulators (MFS).

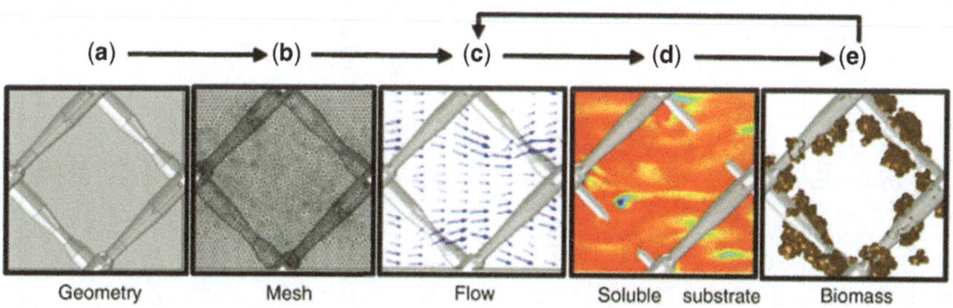

Figure 5.3 Graphical representation of the model algorithm. The arrows indicate the sequence of algorithm steps, with a loop between steps (c) and (e). (a) set model parameters and geometry in MATLAB and COMSOL; (b) create the 3-d tetrahedral mesh and 3-d rectangular biomass grid; (c) calculate fluid flow (blue arrows: local flow velocity in a slice at half the flow channel height); (d) calculate substrate concentration (colour scale: solute distribution in a slice at half the flow channel height; blue: minimum; red: maximum); (e) biomass attachment and growth (brown volume: biomass). In all the top-view figures of the 3-d model system the spacer filaments are shown in grey.

Table 5.3 Variable model parameters in different simulations.

	Spacer Thickness [mil]	[μm]	r_{att}[1] [Cmol · m^{-2} · h^{-1}]	u_{in}[2] [m · s^{-1}]	Q[3] [L · h^{-1}]	Attachment Location
Effect of linear flow velocity	31	787	1.08×10^{-2}	0.163	16.0	Membrane + spacer
				0.082	8.0	
				0.041	4.0	
Effect of bacterial cell load	31	787	1.08×10^{-2}	0.163	16.0	Membrane + spacer
			5.4×10^{-3}	0.082	8.0	
			2.7×10^{-3}	0.041	4.0	
Effect of attachment location	31	787	1.08×10^{-2}	0.163	16.0	Membrane
						spacer
						membrane + spacer
Effect of spacer shape and channel porosity	31 31 modified	787	1.08×10^{-2}	0.163	16.0	Membrane + spacer
Effect of spacer thickness as applied in practice						
At constant flow rate	28	711	1.08×10^{-2}	0.178	16.0	Membrane + spacer
	31	787		0.163		
	34	863		0.153		
	46	1168		0.115		
At constant linear velocity	28	711	1.08×10^{-2}	0.163	14.6	Membrane + spacer
	31	787			16.0	
	34	863			16.9	
	46	1168			22.6	

[1]Biomass attachment rate; [2]average flow velocity in inlet; [3]liquid volumetric flow rate.

For this set of simulations the 31 mil (787 μm) thick feed spacer geometry was used at different linear flow velocities of $u_{\text{in}} = 0.163\ \text{m} \cdot \text{s}^{-1}$ (corresponding to hydrodynamic conditions frequently encountered in industry), $0.082\ \text{m} \cdot \text{s}^{-1}$ and $0.041\ \text{m} \cdot \text{s}^{-1}$, the other model parameters being set as listed in Tables 5.2 and 5.3.

The initial pressure drop in absence of fouling (Figure 5.4a, day 0) showed a small difference (<10%) between the simulation and experimental initial pressure drop. This difference may be due to e.g. the variations in thickness and geometry of the spacer used in the experiments, but also to assumptions in the model calculations.

In time, the development of pressure drop caused by biofilm accumulation resulted in a similar trend for both model simulations and experimental studies. Small dissimilarities between the calculated and measured data are observed. The amount of accumulated biomass was the same at the end of the simulation studies (day 6), since the same biomass attachment rate was used for all simulations and no detachment was considered. With increasing linear flow velocity the effect of the accumulated fouling material (biomass) on pressure drop increased (Figure 5.4b).

These results show that by applying a lower linear velocity the impact of the fouling material on performance can be significantly reduced.

In summary, the biofouling model describes existing experimental observations for pressure drop and pressure drop increase well. However, in reality the linear velocity has more complex implications in terms of substrate load, bacterial transport and deposition rates, as well as biomass detachment rates. All of these

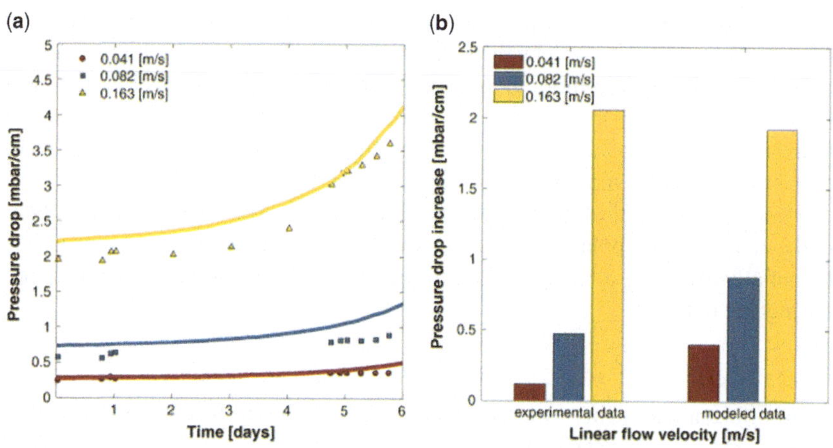

Figure 5.4 Calculated and experimentally determined data at different linear flow velocities, for the 31 mil (787 μm) thick feed spacer. (a) Pressure drop in time during the biofouling process (lines – calculated data, symbols – measured data); (b) pressure drop increase at day 6 relative to the initial pressure drop.

factors impact the amount of accumulated biomass and, consequently, the feed channel pressure drop.

Effect of bacterial cell load

The biofouling onset may also be affected by the bacterial cell load (i.e., cell concentration multiplied by linear flow velocity) in the feed channel. It is expected that when increasing the liquid flow rate, more microorganisms would pass through the channel per unit time (Paris *et al.*, 2007), therefore more cell attachment could take place due to the higher cell supply and higher shear rate. Eventually, a maximum deposition rate exists function of wall shear rate (Margalit *et al.*, 2013). Because the decreased deposition efficiency at high shear rates relates to detachment rates, not included in our model, this effect has to be systematically checked in future studies. For investigating the effect of cell load on performance decline, the 31 mil (787 μm) thick feed spacer was used in numerical simulations at different flow and cell attachment rates (Table 5.3). Three cases were studied when the transported amount of biomass varied with flow rate: the cell attachment rate was reduced proportionally with the reduction in flow rate.

Comparing the results obtained at different linear flow velocities and constant attachment rate (see Table 5.3, effect of linear velocity: i.e., constant cell load) with those at flow-linked attachment rate (see Table 5.3, effect of cell load: i.e., variable cell load) reveals that at lower linear fluid velocities the impact of accumulated biomass is lower. Possibly because of differences in hydrodynamics and less biomass attachment in the flow channel (Figure 5.5). Similar results were obtained experimentally by Vrouwenvelder *et al.* (Vrouwenvelder *et al.*, 2009b): with variable flow rate different amounts of biomass were found during membrane autopsy. However, there can be three causes for this observation: a higher substrate transfer rate leading to more biomass growth, more cell attachment due to more biomass transported and deposited and a combination of more growth and more attachment. These results suggest that the impact of biofouling in spiral-wound membrane modules may be delayed by reducing the linear fluid velocity.

However, in practice, applying lower linear flow velocities leads to an increase of the concentration polarization and to lower shear stress at the membrane surface, potentially affecting other types of fouling (e.g., scaling, colloidal) that may occur during RO operation.

Biomass location

Biofilm formation begins with bacterial transport followed by cell adhesion. When cells become firmly attached to the support surface, growth into micro-colonies takes place, finally leading to a mature biofilm (Pang *et al.*, 2005). The adhesion of microorganisms in the NF and RO membrane systems depends on the hydrodynamics (Stoodley *et al.*, 1998b) and on the surface properties (hydrophobicity, surface charge, etc.) (Kang & Choi, 2005). The preferential places where cells attach in the initial

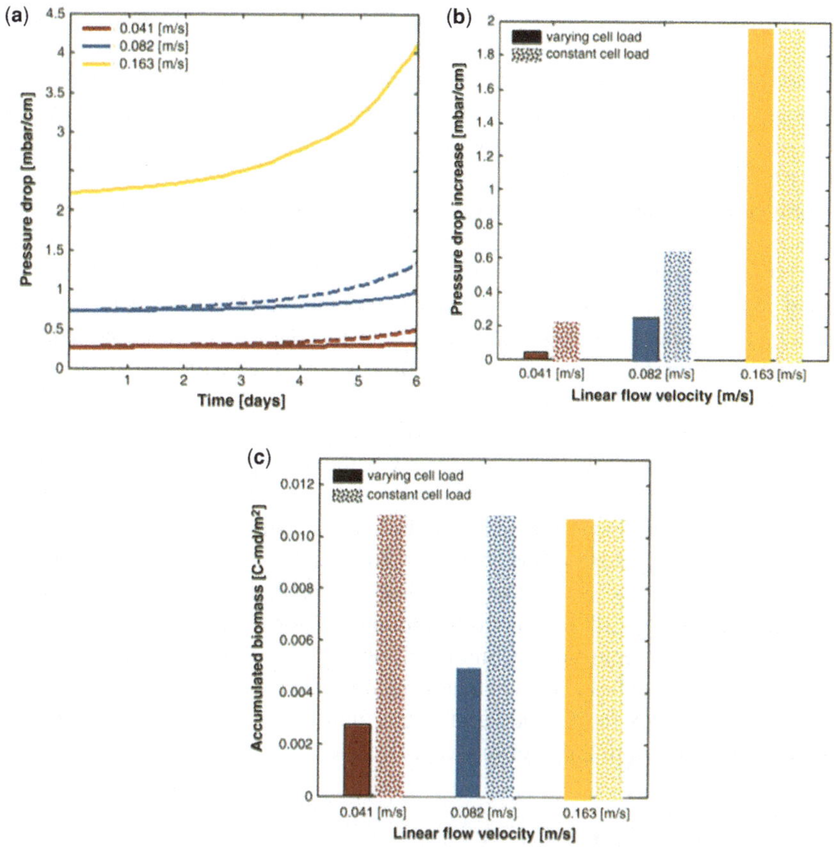

Figure 5.5 Comparing the impact of linear fluid velocity and cell load on feed channel pressure drop. (a) Pressure drop in time (connected line – varying cell load, dashed line for continuous cell load), (b) pressure drop increase due to biomass accumulation (day 6), and (c) accumulated biomass concentration at different linear flow velocities (day 6).

phase of biofilm formation in NF and RO membranes are still unclear, but there are indications (Duddridge *et al.*, 1982) that biofilm formation may be linked to the value of local shear stress induced by the liquid flow. It was reported that biofilm formation is not avoided by high shear stress conditions. On the contrary, biomass developed under high shear was found to be more compact and harder to remove. This is also confirmed by industrial RO membrane module autopsies, for example, Tran *et al.* (Tran *et al.*, 2007) found thicker biofilm layer underneath or in the vicinity of feed spacer. On the other hand, other studies indicate that biofilm formation begins mainly in places behind the spacer filament crossings (Picioreanu *et al.*, 2009; Vrouwenvelder *et al.*, 2009a; Vrouwenvelder *et al.*, 2010c), where the shear stress level is relatively low.

With the numerical model of biofouling the consequences of biomass attachment at different locations on the development of the feed channel pressure drop were investigated. Three different scenarios are presented here, all with the same biomass attachment rate (Table 5.2): (i) attachment only to the membrane surface; (ii) random attachment to both membrane and spacer surface and (iii) attachment only to the spacer surface.

Simulations were run with the same linear flow velocity and inlet substrate concentration, using the 31 mil (787 μm) thick spacer geometry presented in Figure 5.2. In the absence of biomass (at $t = 0$), the highest pressure drop occurs as the fluid passes between the spacer filaments and membrane sheets (Picioreanu *et al.*, 2009). The numerical results indicate that the accumulated biomass has a different influence on the feed channel pressure drop when attached to the spacer surface than when attached to the membrane.

The highest impact of the accumulated biomass on the feed channel pressure drop is observed when the biomass attached only to the feed spacer filaments (Figure 5.6). Presence of biomass on the spacer surface results in narrower flow channels increasing thus the local fluid velocity (Figure 5.7) which leads to higher pressure drop (compare the pressure fields in Figure 5.6c – biomass on spacer only – with those in Figure 5.6a, b). The smallest pressure drop difference was observed when biomass attachment was only allowed on the membranes (Figure 5.8a).

Because the amount of accumulated biomass was similar in all three cases (Figure 5.8c), it can be concluded that there are sensitive locations in the feed spacer channel (especially in the middle of spacer fibres) where even a small amount of biomass could lead to high performance decline.

Spacer geometry

Various feed spacer geometries can be found in commercially available NF and RO membrane modules. The main roles of spacers are to provide an open channel for the flowing feed water and to promote mixing (i.e., intensify mass transfer and reduce concentration polarization (Tran *et al.*, 2007)). The disadvantages of using feed spacers include the increase of feed channel pressure drop (an added resistance to flow) and the enhanced biofouling potential (Rukapan *et al.*, 2012; Tran *et al.*, 2007; Vrouwenvelder *et al.*, 2009a). Feed spacers are most often characterized by the manufacturers just by their thickness. Nevertheless, it was shown that the distance, the angle between crossing filaments and the filament orientation are also important for performance (Guillen & Hoek, 2009; Koutsou *et al.*, 2009; Koutsou & Karabelas, 2010; Subramani & Hoek, 2008).

The main differences of spiral-wound membrane modules are due to the installed membrane type and feed spacers geometry. This paper studies the effect of four different commercially available and one hypothetical feed spacer geometry on the feed channel pressure drop and pressure drop increase due to accumulation of biomass. The spacer geometries used in the industrial modules which are subject of

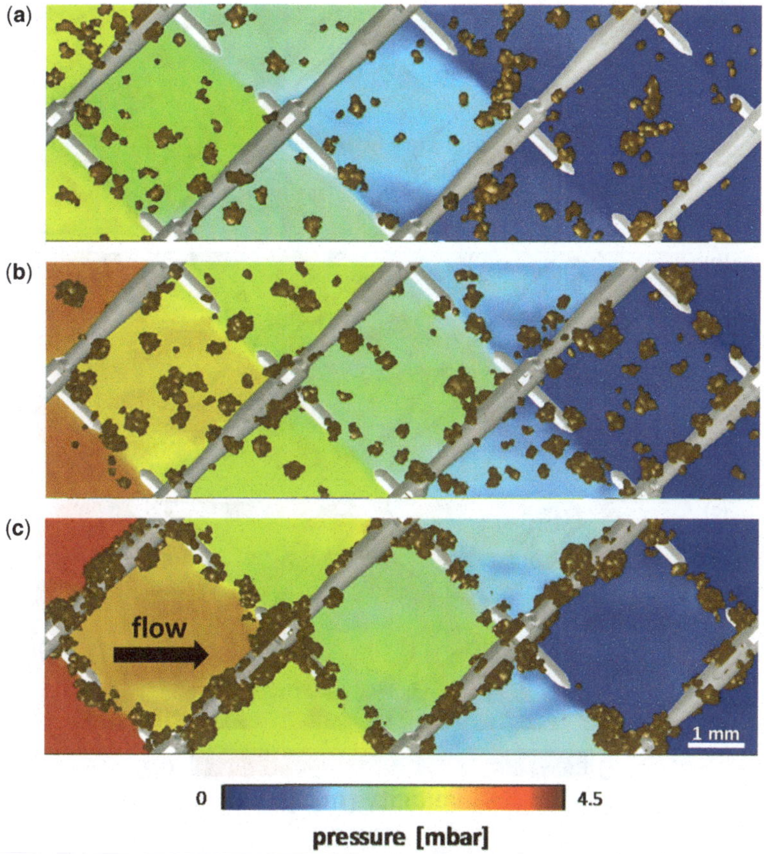

Figure 5.6 Liquid pressure and biofilm developed in the feed channel for different biomass attachment scenarios (day 6): (a) biomass attachment only on the membrane surface, (b) biomass attachment randomly on both membrane and spacer, (c) biomass attachment only on the feed spacer. The grown biomass is shown in brown and feed spacer in grey. The same amount of biomass developed in all three cases. The color scale indicates the pressure in the plane situated at half the channel height, $z = 394$ μm. The arrow indicates the flow direction.

this study differ in their thickness and filaments shape. The simulation conditions, such as biomass attachment rate, substrate inlet concentration and fluid linear flow velocity, were kept constant in all the cases.

Spacer shape and channel porosity

The modified spacer geometry does not represent an ideal feed spacer geometry, but rather a simple hypothetical construct used to investigate the effect of a different

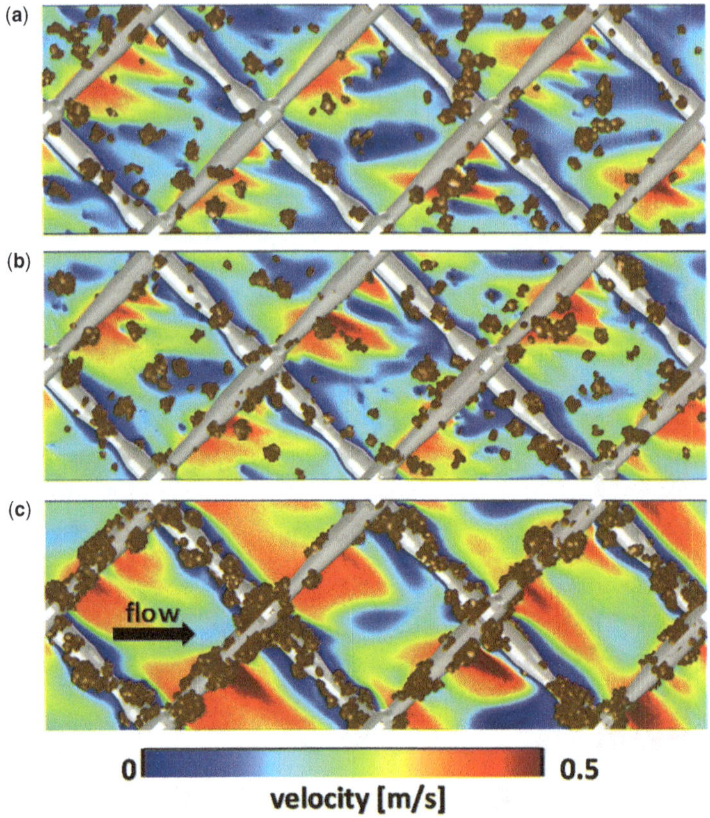

Figure 5.7 Fluid velocity and biofilm developed in the feed channel for different biomass attachment scenarios (day 6). (a) biomass attachment only on the membrane surface, (b) biomass attachment randomly on both membrane and spacer, (c) biomass attachment only on the feed spacer. The grown biomass is shown in brown (day 6) and feed spacer in grey. The same amount of biomass developed in all three cases. The colour scale indicates the magnitude of fluid flow velocity in the plane situated at $z = 197$ μm. The arrow indicates the flow direction.

filament shape with the same thickness on performance decline. The numerical simulations clearly show how the feed spacer resistance to flow can be reduced by increasing channel porosity. Small changes in channel porosity (from 89% for commercial spacer to 92% for the modified spacer) led to significantly lower initial pressure drop (~20% lower) and less pressure drop increase in time in the presence of the same amount of biomass (Figure 5.9). The feed channel spacer has a narrow area of contact with membrane sheets at the cross section of the filaments. The same thickness at the filament crossings and reduced filament thickness elsewhere, result in a higher feed channel spacer porosity and consequently in a lower pressure drop increase.

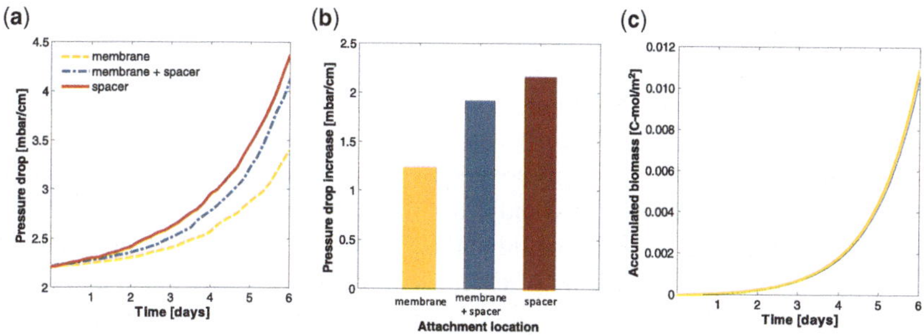

Figure 5.8 Effect of biomass attachment location on the development of feed channel pressure drop due to biofilm formation. (a) Pressure drop development in time; (b) pressure drop increase relative to the clean channel (day 6); (c) accumulated biomass in the feed channel. The biomass attachment was randomly allowed (i) only on the feed spacer filaments (red line), (ii) both on the membrane and spacer surface (blue line), and (iii) only on membrane (yellow line).

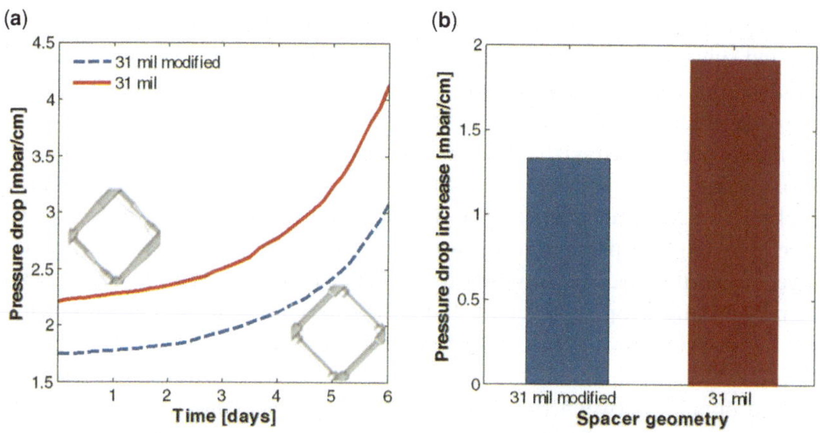

Figure 5.9 Pressure drop for two feed spacer geometries with the same thickness and same accumulated biomass amount, but differing in channel porosity. (a) pressure drop in time; (b) pressure drop increase with biomass relative to the clean channel (day 6). Simulations were performed with a 31 mil thick feed spacer as applied in commercially available spiral-wound modules (connected line) and a 31 mil spacer modified for thinner fibres (dashed line).

These results suggest that beside the spacer thickness (giving channel height), spacer mesh size (distance between filament cross sections) and the fluid flow attack angle, the filament shape (especially its thickness and thickness variation) is also important for feed channel pressure drop and the impact of biomass on pressure drop increase.

Spacer thickness

Industrial RO and NF membrane modules are available with feed spacers differing in thickness. We studied the effect on feed channel pressure drop of four different geometries corresponding to feed spacers used in commercially available spiral-wound RO/NF modules. The studied spacers displayed in Figure 5.2 differed in thickness (28, 31, 34 and 46 mil), porosity and filament shape, and their spatial dimensions are listed in Table 5.1. Two sets of simulations were run: (i) with constant flow rate, $16\ L \cdot h^{-1}$ (i.e., different linear flow velocities for different spacer thickness) and (ii) constant linear flow velocity $0.163\ m \cdot s^{-1}$ (corresponding to various flow rates).

In case of constant flow rate the linear liquid velocity becomes a function of the spacer thickness, with the highest fluid velocity achieved for the narrower 28 mil (711 μm) spacer geometry. As expected, the highest initial pressure drop was found for the thinnest spacer geometry. Both initial pressure drop and pressure drop increase due to biofilm formation show a decreasing trend with increasing feed spacer thickness, as shown in Figure 5.10a, b. The initial pressure drop is caused only by the presence of feed spacer, while the pressure drop increase is due to biomass accumulation. Because the amount of accumulated biomass was similar for each type of spacer ($\sim 7.1 \cdot 10^{-3}$ Cmol $\cdot\ m^{-2}$ after 6 days), due to identical attachment rate and substrate concentration in all cases (with neglecting biomass detachment, which could have also played a role), it was found that the feed channel pressure drop increase is less affected by biofouling when using thicker spacer.

In the second scenario, a constant linear flow velocity was applied at $0.163\ m \cdot s^{-1}$. This resulted in varying flow rates because of different channel thicknesses. Similar trend as for the simulations at constant flow rate were observed. Highest pressure drop was found with the thinnest (28 mil) and the lowest with the thickest (46 mil) feed spacer (Figure 5.10c, d). However, the initial pressure drop is lower for the case where constant linear flow velocity was applied than for the case with constant flow rate. The difference in pressure drop increase for different spacer thicknesses is similar for the two scenarios, but the change between the different thick spacer geometries is smaller for the second scenario. Results for both cases (constant linear flow velocity and constant flow rate) are in good agreement with the experimental data reported in (Araújo *et al.*, 2012a).

DISCUSSION

The objectives of the numerical studies on biofouling development in spiral-wound NF and RO systems were to determine the influence of (i) liquid flow velocity and cell load, (ii) biomass location on the feed spacer and membrane surfaces, and (iii) various feed spacer geometries (28, 31, 34 and 46 mil thick) as applied in practice as well as a modified geometry 31 mil feed spacer. The computer simulations were in agreement with existing experimental data on the effect of liquid flow velocity

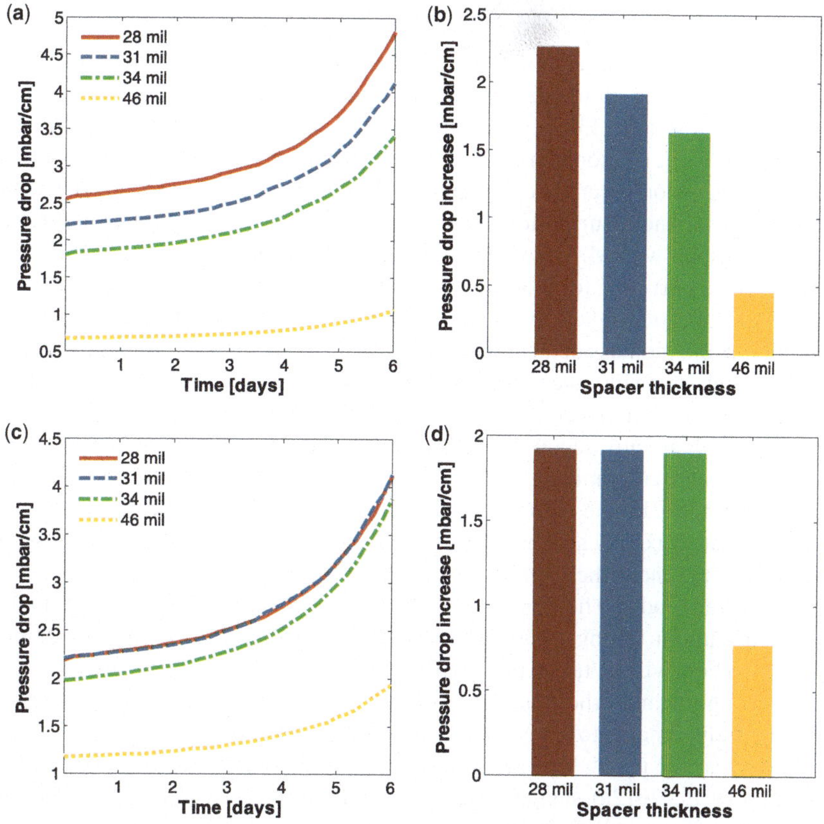

Figure 5.10 Changes in feed channel pressure drop for different feed spacer thicknesses and geometries (28, 31, 34 and 46 mil thick feed spacers as applied in commercially available membrane modules). (a,c) Pressure drop in time, (b,d) pressure drop increase relative to the clean channel (day 0). Results at: (a,b) constant flow rate (16 $L \cdot h^{-1}$); (c,d) constant linear flow velocity (0.163 $m \cdot s^{-1}$).

(Figure 5.4) on feed channel pressure drop and pressure drop increase caused by biofilm formation, indicating that the model describes experimental data well. Biomass attachment on the feed spacer had the highest impact on pressure drop increase compared to attachment to the membrane only and both the membrane and spacer (Figures 5.6 and 5.8). Comparison of an in practice applied 31 mil feed spacer with a modified 31 mil spacer geometry showed that the shape of spacer filaments influences the pressure drop and the pressure drop increase caused by biofilm accumulation (Figure 5.9). A thicker feed spacer resulted in: reduced initial pressure drop and decreased effect of accumulated biomass on pressure drop increase (Figure 5.10).

Numerical model evaluation

The numerical biofouling model describes existing experimental observations well. Compared to earlier numerical biofouling models, a more detailed geometry for the feed spacer was used (Figures 5.1 and 5.2). Also, the numerical model developed by Picioreanu *et al.* (Picioreanu *et al.*, 2009) has been implemented successfully in a more efficient computer code coupling COMSOL Multiphysics solvers (COMSOL 4.3a, Comsol Inc., Burlington, MA, www.comsol.com) for fluid dynamics and solute transport with MATLAB (MATLAB 2011a, MathWorks, Natick, MA, www.mathworks.com) code for biofilm formation.

Possible future implementations into the model can be biodegradable substrate limitation and biomass detachment. The amount of accumulated biomass in an NF and RO membrane module mostly depends on the available biodegradable substrate concentration and hydrodynamics (Vrouwenvelder *et al.*, 2009b). Although formally taken into account, the substrate limitation was not very effective due to the chosen combination of C_S and K_S and, as a result, the biomass grew unlimited. If the biomass growth was substrate-limited, then in certain conditions the permeate flux may also have an effect on the biomass growth rate by providing more substrate near the membrane (Radu *et al.*, 2012a). Biofilm detachment was absent in the model. The biofilm was allowed to develop inside the computational domain without taking in consideration the shearing effect of hydrodynamics. These aspects will be taken into account in the future model versions.

How significantly these model limitations affect the main conclusions reached in this study remains to be evaluated in future investigations.

However, the match between model results and experiments from Araujo *et al.* (Araújo *et al.*, 2012a) suggests that the model already accounted for the main processes affecting the biofouling process.

Importance of feed spacer for biofouling

One of the factors with strong impact on the operational costs of spiral-wound RO and NF membrane water desalination plants is the pressure drop increase along the modules. The feed channel pressure drop is a result of frictional forces when the fluid flows through the module. The initial pressure drop in an RO module when the feed channel is clean (i.e., no fouling material has accumulated on membranes or spacer) is mainly the result of the feed spacer resistance, which depends on the spacer geometry.

It was suggested that by decreasing the linear flow velocity the initial pressure drop can be reduced (Vrouwenvelder *et al.*, 2010). However, by reducing the liquid flow velocity the residence time of the solute increases, which can raise the risk for salt precipitation.

The proposed numerical model indicates that with increasing spacer thickness the effect of biofouling on feed channel pressure drop is decreasing. It was found that with reduced flow rate the amount of accumulated biomass may be lower, which is supported by the experimental findings of Vrouwenvelder *et al.*

(Vrouwenvelder *et al.*, 2009b). However, Ahmad *et al.* (Ahmad *et al.*, 2005) showed in experimental studies that lowering the feed flow velocity results in an increase of concentration polarization. They also studied the effect of different feed spacer geometries on pressure drop and minimization of concentration polarization, recommending cylindrical spacer filaments at higher fluid velocities, but suggesting alternative spacer filaments at lower flow velocities. Mo and Ng (Mo & Ng, 2010) have demonstrated experimentally the importance of feed spacers in reducing the effect of concentration polarization and fouling but also showed that the presence of the feed spacer creates additional resistance on the permeate flux due to the reduction in effective membrane surface area. Therefore, optimizing a spacer for minimal biofouling in practice needs to account also for the complex interactions between fluid dynamics and mass transport. In other words, a spacer and operating conditions that promote less biomass growth and thus less feed channel pressure drop may adversely affect other performance indicators.

Spiral-wound membrane elements can be produced with thicker feed spacers while maintaining the same membrane surface area (Bartels *et al.*, 2008), indicating that using thicker feed spacers should not significantly increase the investment costs for a membrane plant. Results of Bartels *et al.* (Bartels *et al.*, 2008) clearly showed that a thicker feed spacer had a lower initial feed channel pressure drop and, more importantly, a lower feed channel pressure drop increase than a thinner feed spacer. For example, in a membrane module a newly designed 34 mil feed spacer had a 16 percent lower pressure drop than the standard feed spacer of the same thickness. As a result, energy can be saved by using these modules and channel plugging due to biofouling will not rapidly occur. Moreover, the modules with thicker spacer may be more easily cleaned (Araújo *et al.*, 2012a; Bartels *et al.*, 2008; Creber *et al.*, 2010b). In full-scale installations, long term studies were performed with thicker feed spacers at a significantly lower linear flow velocity (Bartels *et al.*, 2008; Majamaa *et al.*, 2009) showing that the lower linear flow velocities can be applied in practice without operational problems related to scaling due to higher concentration polarization. In addition, less biofouling was observed, needing thus a lower cleaning frequency. After cleaning, performances close to the initial values were achieved.

The development of effective biofouling control strategies in spiral-wound NF and RO systems most-likely requires detailed insight in the fouling processes obtained under representative conditions, which can be provided by a tuned combination of numerical modelling with experimental studies. Studies on biofouling control without biocides could include the balancing of feed spacer design, hydrodynamics and concentration polarization.

CONCLUSIONS

The model results emphasise the key importance of feed spacers on the biomass accumulation and performance of NF and RO membrane modules. Numerical simulations indicated that (i) at lower linear flow velocities, the pressure drop increase could be less pronounced because less biomass could accumulate, (ii) biomass on the spacer filaments leads to a higher feed channel pressure drop increase than when developing on the membranes, (iii) the relative feed channel pressure drop increase due to biomass accumulation is much higher at smaller channel porosities and (iv) at same feed flow rate thicker spacers promote a lower pressure drop increase when overgrown with biomass.

Once the model is extended with biomass detachment, flow influenced biomass attachment and with permeation, this mechanistic approach could help to better predict the biofouling process of RO and NF membrane systems and, eventually, to develop and evaluate novel strategies and optimized spacer geometries to reduce and control biofouling.

Chapter 6

In-situ biofilm characterization in membrane systems using Optical Coherence Tomography: Formation, structure, detachment and impact of flux change

INTRODUCTION

Biofouling is the major problem in membrane operation for process and drinking water production (Darton *et al.*, 2004; Flemming *et al.*, 1997; Saeed *et al.*, 2000; Tasaka *et al.*, 1994; Vrouwenvelder *et al.*, 2008b; Winters & Isquith, 1979).

Biofouling occurs when the growth of biofilms in the membrane system negatively impacts membrane performance parameters e.g. feed-channel pressure drop (FCP) and transmembrane pressure drop (TMP), reducing the quantity and quality of produced water. Biofouling is defined as the biofilm formation causing a membrane performance decline exceeding 10–15% of the start-up values under the applied operational conditions. At variations larger than 10–15%, corrective actions are recommended and guarantees are restricted by the manufacturers of membrane elements. Biofouling increases the costs of plant operation strongly and may even be prohibitive for the application of membrane filtration in water treatment.

Historically, biofouling research on spiral wound membrane systems is typically problem solving oriented (Van Loosdrecht *et al.*, 2012). Biofouling is a complicated process in which many factors can influence each other. Fouling is diagnosed on the basis of parameters such as pressure drop and on membrane autopsies in hindsight. To solve or minimize the negative impacts of biofouling, a clear understanding of biofilm development in relation to e.g. membrane module design and operational aspects is needed. '*In-situ*' non-destructive visualization methods in small-scale representative test units should be developed to characterize biofouling, enabling the development of strategies to prevent or minimize the impact of biofilm formation on membrane performance. In other words, suitable tools applied under conditions

Author(s): Szilárd Bucs, Joop Kruithof, Mark van Loosdrecht and Johannes Vrouwenvelder
doi: 10.2166/9781780409597_0107

representative for practice are needed to increase the understanding of biofouling processes, which may lead to effective biofouling control strategies.

A suite of *in-situ* non-destructive imaging tools has been recently proposed for biofouling studies of membrane systems such as e.g. magnetic resonance imaging (Creber *et al.*, 2010a, b; Graf von der Schulenburg *et al.*, 2008; Vrouwenvelder *et al.*, 2009a) and oxygen concentration imaging by use of optodes (Prest *et al.*, 2012; Staal *et al.*, 2011b). Ultrasonic time-domain reflectometry can provide information on biofilm thickness but requires silica dosage to enhance the acoustic signal (Sim *et al.*, 2013). Confocal laser scanning microscopy (Neu & Lawrence, 1997; Schramm *et al.*, 2000) is another suitable technique to obtain information on the biofilm thickness and structure but requires use of fluorescent dyes, possibly affecting the biofilm formation. Besides the application of confocal laser scanning microscopy is limited by the penetration depth. Therefore, with that method the internal structure of thick biofilms cannot be observed.

Non-invasive '*in-situ*' tools/methods offer the possibilities to investigate fouling layer thickness and structure without stopping the operation of the membrane filtration unit. Optical Coherence Tomography (OCT) has the ability to provide volumetric imaging of the biofilm at micron resolution in a non-destructive way (Derlon *et al.*, 2012, 2013; Haisch & Niessner, 2007; Janjaroen *et al.*, 2013; Wagner *et al.*, 2010b; Xi *et al.*, 2006). OCT imaging provides a quantitative, high-resolution, spatially-resolved means to characterize biofilm growth and detachment, biofilm thickness, and structural heterogeneity. It was developed by Huang *et al.* (1991) as a tool for medical imaging in 1991. Since 2006, OCT has been used to investigate biofilm structures in water and membrane filtration systems, in capillary flow cells (Xi *et al.*, 2006), and in crossflow filtration systems under laminar, transient, and turbulent flow conditions (Haisch & Niessner, 2007; Wagner *et al.*, 2010b). Attachment mechanisms of *Escherichia coli* (Janjaroen *et al.*, 2013), and predation of eukaryotes and metazoan organisms in gravity driven membrane filtration systems has been studied as well (Derlon *et al.*, 2012, 2013). So far, no OCT investigations were performed in a filtration system representative for spiral wound membrane filtration in which the thickness and compaction of biofilms was related to performance losses of the membrane system.

In a study on the intrinsic hydraulic resistance of biofilms Dreszer *et al.* (Dreszer *et al.*, 2013) subjected a mature biofilm to changes in permeate flux. This experiment raised the assumption that biofilms react to an increase in permeate flux by compaction. The assumption could not be confirmed due to the lack of *in-situ* biofilm thickness measurements. The application of OCT offers the possibility to provide visual evidence for the compressibility of biofilms. OCT is an optical signal acquisition and processing method able to capture, during contact-free and non-invasive operation, micrometer-resolution images from within optical scattering media (Fercher *et al.*, 2003). OCT is an interferometric technique, typically employing near-infrared light. The use of relatively long wavelength light allows, compared to confocal microscopy, deeper penetration into the biofilm

(Fercher *et al.*, 2003). To the authors knowledge this is the first paper showing *in-situ* biofilm thickness data in a crossflow operated permeate producing membrane system.

The objective of the study was to assess and evaluate the suitability of Optical Coherence Tomography for *in-situ* measurements of biofilm thickness and morphology during (i) biofilm development, (ii) biofilm detachment, and (iii) permeate flux change. OCT data were compared with the hydraulic biofilm resistance, and membrane performance parameters: Transmembrane pressure drop and feed-channel pressure drop. The studies were done with a membrane biofouling monitor (Dreszer *et al.*, 2014a) operated at constant crossflow velocity and permeate flux.

The results were used to discuss potential follow-up OCT research and implications for membrane performance on how to minimize the impact of biofilms.

MODEL DESCRIPTION

Experimental set-up

A flow filtration system with similar hydrodynamic properties as spiral wound membrane modules was employed (Dreszer *et al.*, 2013, 2014a). The transparent membrane biofouling monitor was modified for OCT application by inserting a 4 cm × 15 cm × 0.4 cm sized glass window in the lid of the monitor (Figure 6.1) for imaging at various locations on the membrane (Figure 6.2). The monitor flow channel height was 787 μm.

The application of PES (polyethersulfone) microfiltration membranes (MF) (Nadir MP 005, Microdyn-Nadir GmbH Wiesbaden, Germany) with a pore size of 0.05 μm enabled low pressure operation and the investigation of biofouling/biofilm accumulation without any influence of concentration polarization or other types of fouling. Furthermore, membrane resistance and biofilm resistance could be clearly distinguished according to Dreszer *et al.* (Dreszer *et al.*, 2013, 2014a).

Drinking water prepared from anaerobic groundwater (subsequently treated by aeration, rapid sand filtration, deacidification, softening and rapid sand filtration at treatment plant Spannenburg in The Netherlands) is distributed without primary chemical disinfection and without a disinfectant residual. This drinking water was used as feed water source for the crossflow filtration system experiments. Microscopic investigation of the feed water showed a total bacterial cell number of 3×10^5 cells mL^{-1}.

The number of colony forming units on R_2A media was 2×10^3 CFU mL^{-1} at 25°C after 10 days incubation. A concentrated nutrient solution was dosed into the feed water prior to the filtration cell at a flow rate of 0.12 L h^{-1}, using a peristaltic pump (Masterflex, Cole Palmer, Vernon Hills, USA). To restrict bacterial growth, the pH-value of the nutrient solution was set at 11 by sodium hydroxide addition. The dosing flow rate of the nutrient solution (0.12 L h^{-1}) was low compared to the feed water flow rate (28.3 L h^{-1}). Therefore, the high pH-value of the nutrient solution had no effect on the feed water pH of 7.8. As nutrients, sodium acetate, sodium nitrate and sodium dihydrogen orthophosphate in the mass ratio for

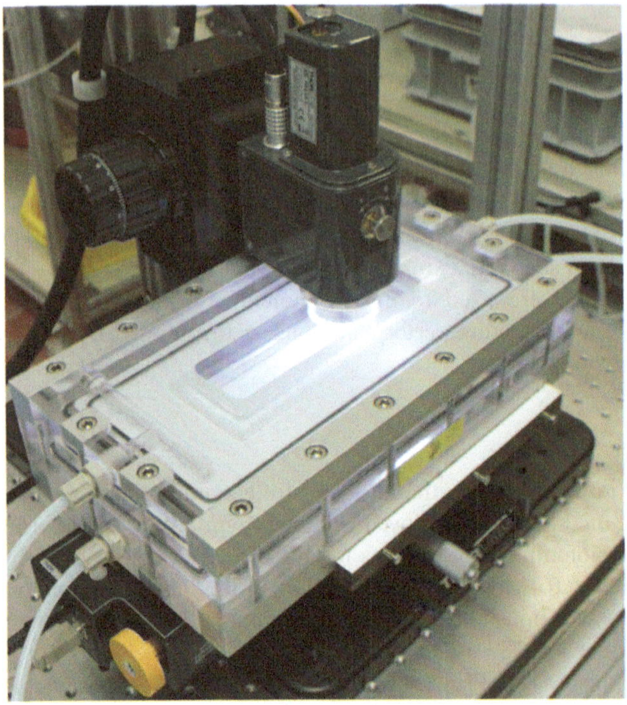

Figure 6.1 Transparent membrane biofouling monitor modified for *in-situ* biofilm thickness measurements with Optical Coherence Tomography (OCT). OCT measurements were performed during monitor operation at constant crossflow velocity (0.1 m s^{-1}) and permeate production (20 or 60 L m^{-2} h^{-1}).

C:N:P of 100:20:10, respectively, were employed (Vrouwenvelder *et al.*, 2009b; Vrouwenvelder *et al.*, 2010a). The concentration of sodium acetate added to the feed water was 1 mg L^{-1} of carbon, the amount of nitrogen and phosphorus was derived from the carbon concentration. Fresh nutrient solutions were installed every 2 days. All chemicals were purchased in analytical grade from Boom B.V. (Meppel, The Netherlands) and were dissolved in deionized water.Feed water was filtered through two 10 μm pore size cartridge filters and the temperature was kept constant at 20°C. A pressure reducer (V782, Vink Kunststoffen B.V., Didam, The Netherlands) enabled a stable feed pressure of 1.7 bar. Before the water entered the filtration cell, nutrients were added. The linear flow velocity of the feed water was monitored by a flow controller (8805/8905, Brooks Instrument, Hatfield, PA, USA) installed behind the outlet of the monitor. The permeate flux was maintained by a peristaltic pump (Masterflex L/S pumps, Cole-Palmer Instrument Company, Vernon Hills, Illinois, USA). The fouling development was monitored by measuring the pressure drop over the feed-channel and over the membrane, using a differential

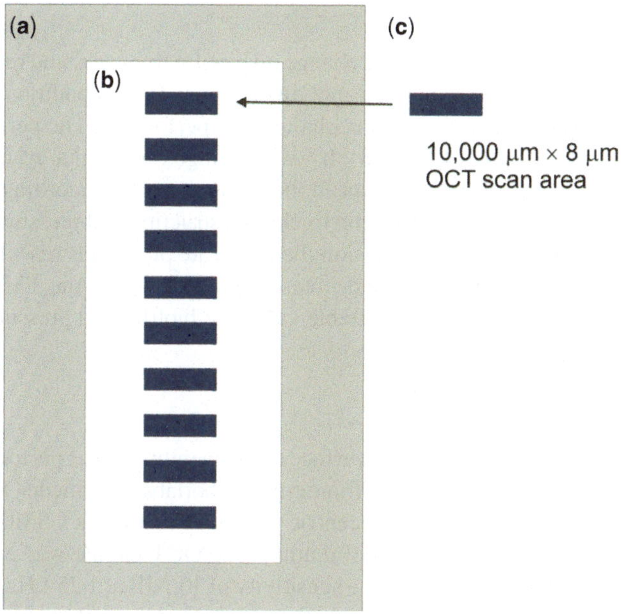

Figure 6.2 Scheme of total membrane surface area (a), imaging window showing the 10 locations for OCT imaging distributed over the monitor window length (b), one OCT imaging area is 10,000 μm × 8 μm (c).

pressure transmitter (Deltabar S PMD70, Endress + Hauser, Maulburg, Germany; (Vrouwenvelder *et al.*, 2009c)). The pressures were measured at the monitor inlet, permeate and concentrate outlet. Temperature, flow velocity, flux, feed-channel pressure drop (FCP), transmembrane pressure drop (TMP) and nutrient supply were measured twice a day. Transmembrane and biofilm resistance data were calculated as described by Dreszer *et al.* (Dreszer *et al.*, 2013).

Experiments and operational conditions

A biofouling monitor was imaged using an OCT microscope in order to investigate biofilm development, detachment and the impact of permeate flux variation in-situ. The system was operated at a crossflow velocity of 0.1 m s^{-1} and a permeate flux of 20 L m^{-2} h^{-1} without the presence of a feed spacer. To enhance biofilm accumulation the feed water was supplemented with a carbon concentration of 1 mg L^{-1}. The monitor with the same hydrodynamic properties as a spiral wound membrane module was operated with a microfiltration membrane in order to avoid concentration polarization. Furthermore, the application of a microfiltration membrane enabled the determination of the biofilm resistance due to a lower clean membrane resistance than the biofilm resistance (Dreszer *et al.*, 2013).

Effect of permeate flux variations

The impact of the permeate flux rate change on biofilm thickness and resistance was investigated. A biofilm was obtained applying the operational conditions from Table 6.1. After 3 days of operation the flux change was performed. The permeate pump which was set to a value of 20 L m^{-2} h^{-1} was changed to a value of 60 L m^{-2} h^{-1}. Along with the increased pumping speed the TMP increased accordingly. After the TMP reached a stable value the biofilm thickness, structure and pressure drops were recorded. To initiate the decompression the permeate pump was reset to a value of 20 L m^{-2} h^{-1} again. Along with the decreased pumping speed the TMP decreased as well. When the TMP reached a stable value the biofilm and pressure drop data were recorded again.

OCT and biofilm image data processing

In-situ imaging of the feed-channel surface of the membrane was performed using a spectral domain Optical Coherence Tomograph (Thorlabs Ganymede OCT System). The OCT was fitted with a 5 × telecentric scan lens (Thorlabs LSM03BB) which provides a maximum scan area of 100 mm^2. The OCT engine was configured to provide high resolution images with a sensitivity of 106 dB at 1.25 kHz A-scan rate. Volumetric images were created using the maximum intensity profile algorithm included in the instrument software (Thorlabs SD-OCT system software version 3.2.1) for a rectangular area 2 mm × 5 mm using 200 B-scans and 500 A-scans of 619 pixels corresponding to a physical depth of 1.1 mm. The axial (i.e. perpendicular to the membrane) resolution for the instrument is below 5.8 μm and the lateral (i.e. parallel to the membrane) resolution is 8 μm.

Table 6.1 Operational parameters and nutrient conditions for the membrane biofouling monitor studies to evaluate the suitability of OCT for *in-situ* measurements of biofilm thickness and structure during biofilm growth, detachment and impact of permeate flux change.

Description	CFV# ($m\ s^{-1}$)	Spacer	Operation (Days)	Dosed Carbon Concentration in Feed Water ($mg\ L^{-1}$ Acetate C)	Permeate Flux ($L\ m^{-2}\ h^{-1}$)
Biofilm development	0.1	no	0–5	1.0	20
Biofilm detachment	0.1	no	4	1.0	20
Impact of permeate flux change	0.1	no	3	1.0	20/60/20

#CFV = linear cross flow velocity.

OCT images were post processed using a custom MATLAB script using the built-in image processing toolbox. Raw images were processed by a multi-step sequence: (1) the red channel from the false color image was converted to grayscale; (2) the resulting image was passed through a threshold filter to locate the membrane surface as this was much brighter than the biofilm signal; (3) the membrane position was stored for each pixel column and used for height calculations; (4) edge detection was performed to determine the outline of the biofilm; (5) morphological filtering was performed to fill in voids due to signal degradation when the biofilm was very thick and to segregate these false voids from real voids in the biofilm; (6) the area between the membrane and biofilm surface was filled and converted to a binary image; (7) spurious nose pixels were removed from the image and (8) the pixel height in each pixel was calculated between the top edge of the biofilm and the upper membrane surface. An absolute size gauge was used within the experimental system (known flow channel height), enabling assessment of the biofilm thickness. From the pixel scaling factor the biofilm thickness in microns was computed. A similar procedure for the generation of biofilm thickness was already applied by Derlon *et al.* (Derlon *et al.*, 2012, 2013).

RESULTS

In this study, the suitability of OCT for in-situ measurements of biofilm thickness and morphology was assessed and evaluated during (i) biofilm development, (ii) biofilm detachment, and (iii) permeate flux change. With OCT, the biofilm thickness and structure was assessed in-situ at (i) a fixed small area for imaging and (ii) 10 locations of the membrane sheet to relate the biofilm thickness, enabling comparison to the hydraulic biofilm resistance and membrane performance parameters: transmembrane pressure drop (TMP) and feed-channel pressure drop (FCP). The studies were carried out in a transparent flow cell (membrane biofouling monitor) containing a microfiltration membrane (0.05 μm pore size), with a clean membrane resistance much lower than the hydraulic resistance of a biofilm. The flow cell was operated without feed spacer at constant linear flow velocity and permeate flux, and at constant water temperature (20°C). The feed water was supplemented with a biodegradable nutrient source (1.0 mg L^{-1} acetate C) to enhance biofilm growth. A summary of the test conditions is shown in Table 6.1.

Biofilm development

Biofilm development and membrane performance were determined over time with a flow cell operated at constant crossflow velocity (0.1 m s^{-1}) and permeate flux (20 L m^{-2} h^{-1}), similar to conditions applied in practice in NF and RO membrane installations.

OCT pictures showing the biofilm development on the membrane over a period of 4 days, while pictures were taken once a day are presented in Figure 6.3. The

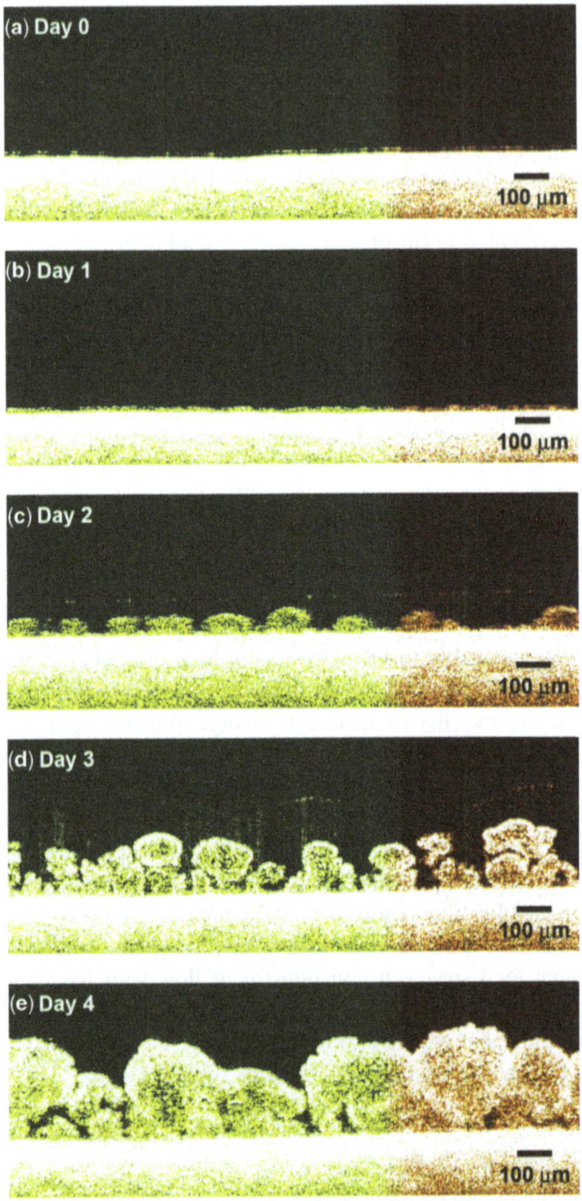

Figure 6.3 *In-situ* OCT observations of biofilm formation on a membrane over a period of 4 days at a fixed position in the monitor. Clean membrane (a), biofilm growth and formation of mushroom-like biofilm structures (b–e). The bars represent the image scale. The monitor was operated at constant crossflow velocity (0.1 m s^{-1}) and permeate production (20 L m^{-2} h^{-1}).

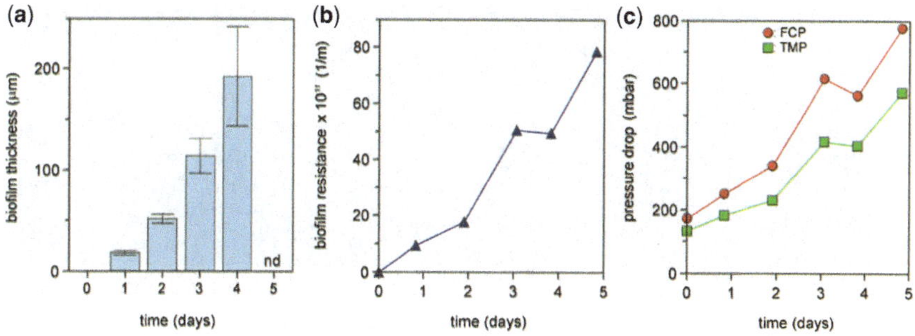

Figure 6.4 Development of biofilm thickness (a), hydraulic biofilm resistance (b) and membrane performance parameters feed channel pressure drop (FCP) and transmembrane pressure (TMP) (c) over time. The biofilm thickness error bars represent the thickness variance.

first OCT picture was taken with a clean membrane (Figure 6.3a), followed by a continuous biofilm growth and the formation of mushroom-like structures (Figure 6.3b–6.3e). The biofilm and the membrane could be clearly distinguished.

In addition to the high resolution pictures of the displayed area, an OCT surface scan along the length of the biofouling monitor was performed (Figure 6.2). Over the whole monitor length, the average biofilm thickness and standard deviation increased over time (Figure 6.4a) and was related to the increase of the hydraulic biofilm resistance (Figure 6.4b). The membrane performance parameters TMP and FCP were monitored over the duration of the experiment and showed the same trend as the biofilm resistance (Figure 6.4c).

OCT imaging allowed non-destructive, in-situ and real-time observations to follow the development of increasing biofilm thickness. Formation of mushroom-like biofilm structures was observed, causing large variations in local biofilm thickness. The increase in biofilm thickness on the membrane caused an increase of the hydraulic biofilm resistance and membrane performance losses.

Biofilm detachment

Local biofilm detachment from the membrane was observed at day 4 of the biofilm development study during the OCT surface scan over the monitor length.

Biofilm detachment occurred at the interface of the biofilm with the membrane (Figure 6.5). Biofilm detachment was found at a small part of the imaged surface: Local biofilm detachment was observed in 4 of the 10 imaged lanes, located at the monitor outlet side. With increasing distance to the monitor feed side, biofilm detachment occurred over a larger surface area. When local biofilm detachment was observed (day 4), the increase of the hydraulic biofilm resistance (Figure 6.4b), TMP and FCP (Figure 6.4c) was temporarily interrupted. No accumulation of detached

Figure 6.5 Biofilm detachment observed locally at a part of the imaged locations after 4 days continuous monitor operation at constant crossflow velocity and permeate flux.

biomass in specific parts of the monitor was found. Biofilm detachment in the monitor could be imaged in-situ with OCT (Figure 6.5).

Effect of permeate flux variation on biofilm thickness and resistance

A recent study showed that the hydraulic biofilm resistance increased when the permeate flux was elevated and returned to the original resistance upon restoring the original flux (Dreszer *et al.*, 2013). To evaluate the influence of flux change on biofilm thickness and resistance, a study with OCT and the transparent membrane biofouling monitor was performed. In a period of three days a mature biofilm was developed in the monitor operated at constant crossflow velocity (0.1 m s^{-1}) and permeate flux (20 L m^{-2} h^{-1}). OCT images were acquired and the biofilm resistance was assessed at the reference permeate flux (20 L m^{-2} h^{-1}). Then the flux was elevated to 60 L m^{-2} h^{-1} followed by a lowering to the original flux (20 L m^{-2} h^{-1}). After each flux change followed by stabilization of the membrane performance parameters, OCT imaging and biofilm resistance measurements were performed. No biofilm detachment was observed. All measurements were carried out in a period of approximately 1 h.

During these three phases the hydraulic biofilm resistance and biofilm thickness were monitored over the total membrane area (Figure 6.7). The biofilm resistance increased at high permeate flux and decreased after restoring the original flux (Figure 6.7a). Comparison of the biofilm resistance at 20 L m^{-2} h^{-1} before and after the temporarily flux elevation showed a 16% higher hydraulic biofilm resistance after the flux changes (Figure 6.7b).

Figure 6.6a shows an OCT image of the biofilm thickness and structure at a permeate flux of 20 L m^{-2} h^{-1}. By elevating the flux to 60 L m^{-2} h^{-1} the biofilm thickness decreased to about half of the original thickness and part of the mushroom-like structure collapsed (Figure 6.6b). After restoring the

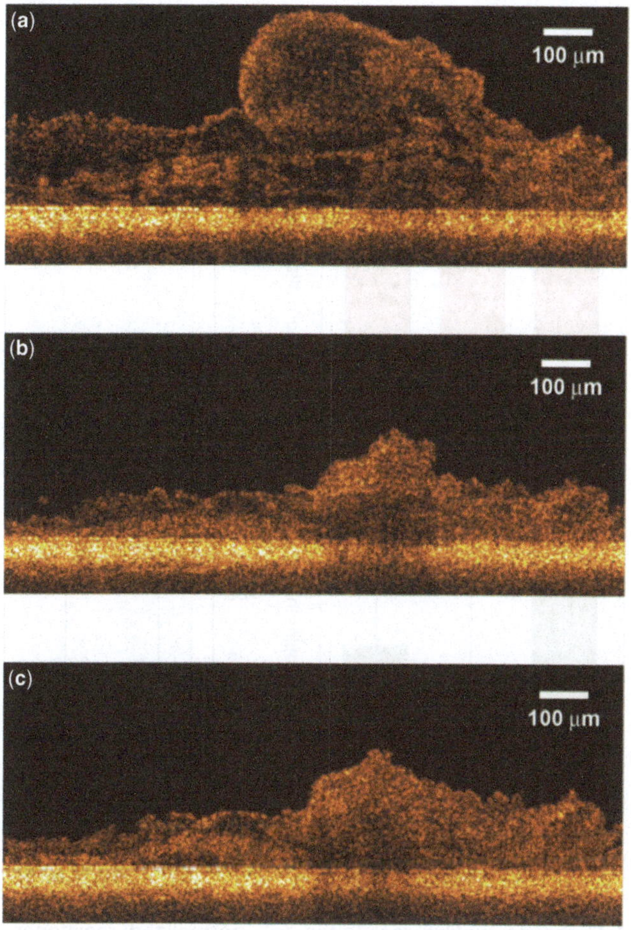

Figure 6.6 OCT observations of the biofilm at permeate flux 20 L m^{-2} h^{-1} (a), after flux increase to 60 L m^{-2} h^{-1} (b) and, after restoring the original flux of 20 L m^{-2} h^{-1} (c). The lateral scan length is 8 μm. The images were made at a stable flux and TMP and constant cross flow velocity (0.1 m s^{-1}).

original flux the biofilm re-gained some of its original thickness and structure (Figure 6.6c).

The biofilm thickness was highest at a permeate flux of 20 L m^{-2} h^{-1} prior to flux increase (Figure 6.7c).

After increasing the flux to 60 L m^{-2} h^{-1} the biofilm thickness decreased, while returning to the original flux caused an increase of the biofilm thickness again. Comparison of the biofilm thickness at 20 L m^{-2} h^{-1} before and after flux elevation showed a 25% lower biofilm thickness after the temporary flux elevation

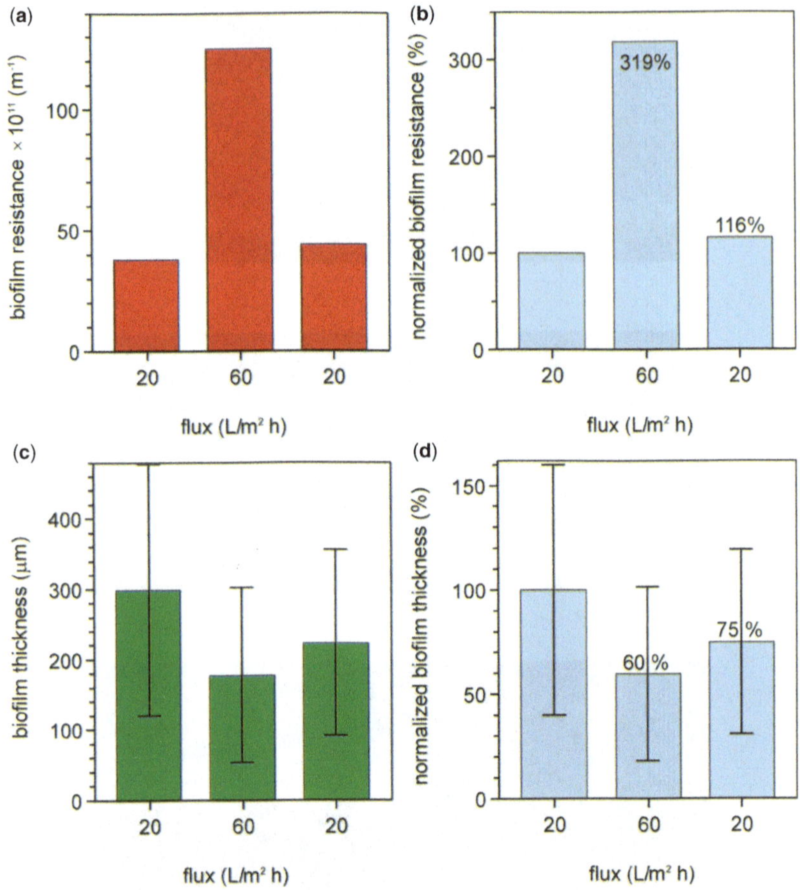

Figure 6.7 Hydraulic biofilm resistance (a), normalized resistance (b), biofilm thickness (c) and normalized thickness (d) at a permeate flux 20 L m^{-2} h^{-1}, after flux increase to 60 L m^{-2} h^{-1} and, after restoring the original flux of 20 L m^{-2} h^{-1}. The bars show the range of biofilm thickness in the total OCT scan area of 100,000 μm.

(Figure 6.7b). A temporarily increase in permeate flux caused a higher hydraulic biofilm resistance and a thinner biofilm. Restoring the original flux the original biofilm resistance and biofilm thickness were not completely restored again. Ten-fold replicated OCT biofilm thickness measurements showed a 4.3% variation in thickness at the initial permeate flux (20 L m^{-2} h^{-1}) and <0.3% variation after elevating and restoring the original flux. At elevated and restored original permeate flux the biofilm structure may be more compact and rigid.

By OCT imaging the impact of permeate flux change on biofilm thickness and structure could be detected. A permeate flux increase caused a thinner biofilm and increased hydraulic resistance. Restoring the original flux did not fully restore the original biofilm thickness and hydraulic resistance. A thinner biofilm can have a higher resistance than a thicker biofilm. Membrane operational aspects like permeate flux can influence the impact of biofilm on the flux.

DISCUSSION

The main objective of this study was to investigate the suitability of Optical Coherence Tomography for in-situ measurements of biofilm thickness and morphology during (i) biofilm development, (ii) biofilm detachment, and (iii) permeate flux change. Biofilms were grown in a crossflow filtration system under the same hydrodynamic conditions (crossflow velocity and permeate flux) as applied in spiral wound nanofiltration (NF) and reverse osmosis (RO) membrane modules. Due to the application of a microfiltration membrane, biofilms without the influence of concentration polarization or other types of fouling were obtained. The OCT images showed an increase in biofilm accumulation with time (Figure 6.3). The biofilm thickness, derived from the OCT images by application of an MATLAB algorithm increased as well. The monitored TMP and FCP, as well as the hydraulic biofilm resistance increased together with the biofilm thickness (Figure 6.4). The intrinsic hydraulic biofilm resistance was related to the membrane performance parameters: TMP and FCP. Biofouling related performance losses in a membrane filtration system are related to biofilm accumulation and biofilm thickness. Local biofilm detachment from the membrane surface was imaged in-situ by OCT in a small part of the test cell (Figure 6.5). A permeate flux increase from 20 to 60 L m^{-2} h^{-1} caused a thinner biofilm with increased hydraulic biofilm resistance. Returning to the original flux (from 60 to 20 L m^{-2} h^{-1}) did not completely restore the biofilm thickness and resistance originally observed at 20 L m^{-2} h^{-1}: A 25% lower biofilm thickness and 16% higher hydraulic biofilm resistance were observed (Figure 6.7).

Suitability of Optical Coherence Tomography

In-situ detection of the development of fouling layers has been investigated applying many different tools. The thickness of cake layers has been determined (i) by a photo interrupt sensor (Tung *et al.*, 2001), (ii) with an optical method (Mendret *et al.*, 2007, 2009), (iii) by an ultrasonic technique (Li *et al.*, 2002b), and (iv) by laser application (Hamachi & Mietton-Peuchot, 2001). Thickness determination of biofilms is hampered by their transparency and similar characteristics as water. Therefore, many methods require additives to the biofilm like stains or signal enhancers. Sim *et al.* (2013) dosed silica to the biofilm to enhance its acoustic signal for ultrasonic time-domain reflectometry (UTDR). Bakke *et al.* (2001) determined the thickness of Pseudomonas aeruginosa biofilms by measuring the optical thickness of the biofilm by light microscopy and translating it into a physical biofilm thickness based on the refractive index of the biofilm. The OCT instrument as configured during these measurements is sensitive to the effective optical path length which is a summation of the refractive index, physical sample thickness, and the total scattering cross-section of the sample. The computed image is greatly affected by these three properties and scattering plays a particularly important role. A good illustration is given the greyscale values of the membrane surface which appear darker in areas where the biofilm is thicker than in areas where the biofilm is thinner. This effect is due to intensity loss due to multiple scattering events in the turbid biofilm. The quantification of biofilm density from OCT images is a future possibility and requires an additional reference arm to measure the refraction index of the sample (Cheng & Liu, 2010) as well as the use of polarization sensitive methods to better resolve images in turbid media (De Boer *et al.*, 1999).

OCT is a new technique for membrane and water treatment applications and enables the investigation of biofilms without addition of stains or signal enhancers. It enables to obtain 2-D and 3-D images of biofilms and has been used already in a number of biofilm studies (Derlon *et al.*, 2012, 2013; Haisch & Niessner, 2007; Janjaroen *et al.*, 2013; Wagner *et al.*, 2010b; Xi *et al.*, 2006). By application of an MATLAB algorithm on an OCT biofilm image the biofilm thickness can be determined (Derlon *et al.*, 2012, 2013). The resolution of the OCT biofilm thickness measurement was below 5.8 μm. The development of a biofilm in crossflow filtration has not been characterized so far by the combination of OCT images, hydraulic biofilm resistance and transmembrane as well as feed-channel pressure drop. The results showed a positive correlation between biofilm thickness and pressure drop; both increasing over the investigated period of time. Compared to a study in which the biofilm thickness was characterized by silica dosage and UTDR similar results were obtained; the biofilm thickness increased over time and showed a non-linear but positive correlation to TMP (Sim *et al.*, 2013).

Biofilms structures have been reported as smooth, fluffy and mushroom-like, in agreement with the OCT observations. Biofilm detachment is not intensively studied on a mesoscopic level, mainly due to the lack of suitable tools. Biofilm detachment is commonly described as detachment of individual microbial cells and/or as sloughing off of biofilm parts. The OCT observations showed detachment of the biofilm base from the membrane surface (Figure 6.5).

The OCT measurements of biofilm thickness and morphology in a membrane test cell operated at constant crossflow and permeate flux showed OCT to be a suitable tool to study biofilms in-situ during e.g. development, detachment and influenced by operational aspects. No additives are needed to obtain quantitative sensitive data on biofilm thickness and structure.

Biofilm compaction

In a previous study (Dreszer *et al.*, 2013) biofilms were grown in a transparent membrane biofouling monitor at two permeate fluxes, 20 and 100 L m^{-2} h^{-1}. After

4 days of operation the biofilm was subjected to changes in permeate flux: 20 to 100 to 20 L m^{-2} h^{-1} and from 100 to 20 to 100 L m^{-2} h^{-1}. The biofilm resistance increased and decreased according to permeate flux variations. In order to study the compressibility of biofilms in more detail, in-situ determination of biofilm thickness under filtration conditions is required. In literature it can be found that biofilms were tested for mechanical compressibility by the application of a film rheometer (Körstgens *et al.*, 2001) or low load compression testing (Paramonova *et al.*, 2007, 2009). In both cases the biofilm was grown on filters placed on agar plates and transferred to the measuring device and compressed between two plates without the influence of flow velocities. The compressibility of biofilms due to a change in permeate flux has not been reported so far.

A change in biofilm thickness can be caused by the biofilm (i) EPS (physical) compression and/or (ii) EPS (chemical) consolidation. The OCT study showed that in a small membrane area a biofilm blob collapsed during the flux increase from 20 to 60 L m^{-2} h^{-1} (Figure 6.6a), suggesting the presence of a local void in the biofilm prior to the flux increase. Biofilm thickness imaging over a large surface area showed that the biofilm completely covered the membrane surface and that the whole measured biofilm thickness varied with the applied permeate flux (Figures 6.7c and 6.7d). A series of repetitive OCT biofilm thickness measurements showed a large variation in biofilm thickness at the initial permeate flux (20 L m^{-2} h^{-1}) and a much smaller variation after elevating and restoring the original flux, indicating that the biofilm became more compact and rigid. The OCT biofilm thickness observations cannot be explained by compaction/decompaction of the biofilm voids or the biofilm EPS only.

The application of OCT for biofilm research enables non-destructive, in-situ detection of biofilm thickness and presents images of the compaction state and decompaction state of a biofilm. An increase in permeate flux causes the biofilm thickness to decrease but the hydraulic biofilm resistance to increase, indicating compaction of the biofilm. The higher resistance is caused by the more compact biofilm containing the same amount of biomass in a smaller volume. It is hypothesized that open spaces (voids) in the biofilm collapse or reduce in volume during biofilm compaction. A decrease in permeate flux and, thus, the lower hydraulic resistance causes the biofilm thickness to increase but the hydraulic biofilm resistance to decrease, indicating the relaxation of the biofilm. The lower hydraulic resistance is caused by a less compact biofilm holding the same content in a larger volume. It is suggested that open voids in the biofilm are (partly) restored during decompaction. The biofilm compaction explains the observation that a thinner biofilm can display higher hydraulic resistance than a thicker biofilm. Thus, biofilm thickness – although intuitively plausible – is not an appropriate parameter to describe the hydraulic permeability. The elasticity of polymers (Flory, 1977) in the EPS (Peppas & Barr-Howell, 1986) may be accountable for this characteristic. The compaction and decompaction behaviour of a biofilm can be compared to the behaviour of a sponge when squeezed and released again.

Biofilm structures

The development of mushroom-like biofilm structures was observed (Figure 6.3 and 6.5). The development of such specific biofilm structures is often reported in literature, in the case of biofilm formation by pure bacterial cultures (Chiang & Burrows, 2003; Sivan *et al.*, 2006), during numerical modelling (Hermanowicz, 2001; Picioreanu *et al.*, 2001, 2007), under specific environmental conditions, e.g. seawater (Mattila *et al.*, 1997) and in dairy industry (Hassan *et al.*, 2010). Possibly, mushroom-like biofilm structures may be a common biofilm structure. However, until now it is not clear how the meso-scale structure of biofilms is influenced by factors such as e.g. feed water microorganism types, biodegradable nutrients, time and conditions of the studies. For example, this study shows that the biofilm structure is affected by variations in the applied permeate flux (Figure 6.6).

The hydraulic biofilm permeability may be influenced by the internal and external biofilm structure (e.g. surface coverage, roughness, voids, and porosity/density), composition (EPS) and a combination thereof. These aspects can be studied in-situ by OCT measurements in combination with the application of the transparent biofouling monitor.

Perspective of OCT studies

OCT imaging offers the advantage of non-invasive, real-time, in-situ characterization of fouling layers. Biofilm development as well as particle deposition can be detected in-situ under real-time operational conditions without the performance of a destructive autopsy. OCT studies enable also the assessment of e.g. biofilm roughness and surface area. The influence of parameters such as flow velocity, pressure, concentration polarization, and fouling control strategies such as chemical cleaning procedures can be measured, documented with images and videos and evaluated.

The combination of OCT and the transparent membrane fouling monitor offers possibilities to investigate biofilms under the same hydrodynamic conditions as present in spiral wound NF and RO membrane systems. Filtration experiments in combination with OCT imaging can be used to gain insights on the effect of crossflow velocity, flow stagnation, stop-start intervals of crossflow, feed spacer presence, and modified anti-biofouling membranes and spacers, as well as the efficiency of cleaning strategies, cleaning solutions and flux enhancers in spiral wound membrane systems. Cleaning of membrane elements is performed at low permeate flux to prevent re-deposition of dirt onto the membrane and spacer. Based on the decompression behaviour of the biofilm as a response to a decreased permeate flux, it may be favourable to lower the flux rate already before applying a cleaning or flux enhancing process to gain a higher efficiency. The diffusion rate of the cleaning agent or flux enhancer through the biofilm might be higher if the fouling layer is less compacted. Experiments are needed to confirm this theory.

This study showed that (i) a thinner biofilm can have a higher hydraulic resistance than a thicker biofilm and that (ii) operational parameters such a permeate flux affect biofilm thickness and compaction. In practice there is a gradient in permeate flux and crossflow velocity over the length of pressure vessels of a membrane installation. Therefore, the biofilm thickness assessed during destructive membrane module studies may not be representative for the fouling layer thickness during operation. Biofilm parameters such as biofilm thickness measured in-situ at various operational conditions such as crossflow velocity and permeate flux should be compared with the same biofilm parameters determined during a destructive study to evaluate the representativeness of the during membrane autopsy determined biofilm parameters.

Application aspects

The impact of an accumulated biofilm on membrane performance can be influenced by plant design and operational aspects. A low and stable permeate flux may reduce the impact of accumulated biomass on membrane performance.

A pressure vessel can contain up to 8 membrane modules in series. The pressure gradient over the pressure vessel length leads to a gradient in permeate flux over the pressure vessel length. The highest permeate flux is observed in the lead membrane module, where commonly biofouling starts. A smaller gradient of permeate flux over the membrane modules in a pressure vessel may be achieved by selecting membrane module orientation on their permeate flux characteristics.

A temporarily increase in permeate flux caused biofilm compaction and increased hydraulic biofilm resistance, even after restoring a lower permeate flux. Therefore, elevated permeate flux should be avoided.

Cleanings aiming at biofouling control may be considered without permeate production: The biofilm thickness is highest and less compact and cleaning chemicals may more easily penetrate the biofilm and high shear cleaning (bubble flow with e.g. air or CO_2, (Cornelissen *et al.*, 2007a, b; Vrouwenvelder *et al.*, 2010c)) may be more effective.

For practical implications, more insight is needed by biofilm research under representative conditions, integrating various fields of expertise (e.g. microbiology, membranes and engineering) to reduce and/or manage the impact of accumulated biomass on membrane performance. The paper gives directions for biofilm research.

CONCLUSIONS

Biofilm development over time was studied with OCT and compared with the hydraulic biofilm resistance and membrane performance parameters:

Transmembrane pressure drop and feed-channel pressure drop. A mature biofilm developed on a membrane in a biofouling monitor was subjected to permeate flux variation while monitored with OCT and pressure drop measurements. Based on the results it can be concluded that:

(1) OCT measurements in combination with the application of the transparent biofouling monitor is a suitable tool for in-situ biofilm investigation.
(2) As to be expected, biofilm thickness increased over time together with the pressure drop and the biofilm resistance.
(3) As not expected to that extent, biofilms responded to changes in permeate flux with compaction and decompaction. The application of a high permeate flux caused the biofilm to compact; the application of a low permeate flux caused the biofilm to relax.
(4) Performance losses in a membrane filtration system were not only related to the biofilm thickness but also to biofilm compaction (density and/or re-arrangement of physical structure) and morphology.

Chapter 7

Early non-destructive biofouling detection and spatial distribution: Application of oxygen sensing optodes

INTRODUCTION

Biofouling in NF/RO membranes

Many areas in the world are experiencing water shortage problems as a result of the increase in human population and water demand (Viljoen, 2010). Reverse osmosis (RO) and nanofiltration (NF) systems are pressure driven, passing water through a semipermeable membrane while retaining particles, ions, and microorganisms (Fritzmann *et al.*, 2007). Production of high quality drinking water depends increasingly on NF and RO membrane processes (Shahan, 1990). A potential drawback of NF and RO is that membranes foul with time, resulting in an increase in pressure drop (PD) and/or a decrease of permeate flux and/or increase in salt passage (Vrouwenvelder *et al.*, 2009a). Major fouling types are scaling, particulate and organic fouling, and biofouling. These different types of fouling may occur simultaneously and can influence each other (Flemming, 2003). Scaling and particulate fouling can be relatively easy to control by physical and chemical pre-treatment, but biofouling and organic fouling seem to be less controllable (Vrouwenvelder *et al.*, 2008b).

Biofouling is referred to as the unwanted accumulation of biofilm by attachment and growth (Flemming, 2002). Biofouling can be operationally defined as the formation of biofilm that results in an unacceptable degree of system performance loss in industrial systems (de Beer *et al.*, 1994). Biofilm formation, which is considered a prerequisite for the occurrence of biofouling, may not always be detected by the present techniques. Detection of fouling is typically done indirectly through measurements of membrane performance decline (Vrouwenvelder *et al.*, 2011a). Existing methods to detect biofouling are mostly based on autopsy and require the destructive opening of a membrane module from the system (Tamachkiarow

Author(s): Szilárd Bucs, Joop Kruithof, Mark van Loosdrecht and Johannes Vrouwenvelder
doi: 10.2166/9781780409597_0125

& Flemming, 2003; Vrouwenvelder *et al.*, 2003; Vrouwenvelder & Van Der Kooij, 2003). The ideal method for biofilm detection should be able to quantify biofouling, and spatial variability in-situ, in real time, non-destructively and at an early stage. Determination of the relevant spatial parameters will result in the formulation of effective biofouling control strategies (Flemming, 2003; Vrouwenvelder *et al.*, 2006, 2007b).

Earlier detection of biofouling enables corrective actions at an earlier stage which may be more effective than cleaning measures at a later stage (Creber *et al.*, 2010b; Vrouwenvelder *et al.*, 2008b, 2011a). In this study, transparent luminescent planar O_2 optodes, in combination with an imaging system, were used to map the two-dimensional distribution of O_2 concentrations and O_2 decrease rates inside a membrane fouling simulator (MFS) (Vrouwenvelder *et al.*, 2007a, b). Planar optode measurements offer a unique opportunity to resolve both spatial and temporal variations on a sub-millimetre scale for a range of analytes (Larsen *et al.*, 2011). So far, the majority of planar optode studies have been focused on O_2 dynamics in marine sediments (Behrens *et al.*, 2007; Glud *et al.*, 1998, 1999; Jensen *et al.*, 2005). This new quantitative and non-destructive method allows direct quantitative assessment of biological activity of fouling in membrane filtration systems, thereby quantifying biofouling.

The primary objective of this study was to evaluate whether biofouling could be detected and quantified with the optode at an early stage, before a significant pressure drop increase is observed (Vrouwenvelder *et al.*, 2009c), providing non-destructive early biofouling detection and diagnosis. This study gives insight into the behaviour of biofilms in membrane systems during their initial developmental phase.

Planar optodes

Oxygen plays a vital role in microbial ecology, and flux of O_2 has been used to quantify primary biomass production, total bacterial community respiration, and mineralization rates (Fenchel & Finlay, 2008). Oxygen concentrations can vary sharply at μm to mm scale, and precise quantification of the O_2 distribution and dynamics is a crucial requirement in understanding spatial and temporal biofilm processes. Planar optodes are a tool that can map spatial distribution of O_2 (Glud *et al.*, 1996). Planar optodes use luminescent O_2 indicators (dyes) immobilized in an oxygen permeable polymeric matrix and can be fixed on foils or glass surfaces. The measuring principle is based on the dynamic collisional quenching of the indicator luminescence by O_2. The dye is excited by an excitation light where after the excited dye can release absorbed energy either by luminescence or collisional quenching by molecular oxygen. The ideal response of optical O_2 sensors is described by Stern–Volmer relation (Bacon & Demas, 1987).

$$\frac{I}{I_0} = \frac{1}{1 + K_{SV}[O_2]} \quad \longleftrightarrow \quad \frac{I_0}{I} = 1 + K_{SV}[O_2] \tag{7.1}$$

I_0: luminescence intensity in the absence of O_2; I: luminescence intensity in the presence of O_2; K_{SV}: constant expressing the quenching efficiency of the fluorophore; $[O_2]$: oxygen concentration.

In planar optodes, only a certain fraction of the O_2 indicator dye remains quenchable upon immobilization thus exhibiting a non-ideal Stern-Volmer response (Carraway *et al.*, 1991). The oxygen dependent quenching can be quantified by a modified Stern-Volmer equation. The relationship can be described by the equation:

$$\frac{I}{I_0} = \frac{1-\alpha}{(1+K_{SV}[O_2])} + \alpha \tag{7.2}$$

where α is the non-quenchable fraction.

For many applications, the most common and simplest form of oxygen sensing involves excitation of a supported dye using an intensity modulated light emitting diode (LED). The excitation light is turned on until the luminescence level of the dye reaches a steady state (Prest *et al.*, 2012). The luminescence intensity is dependent on the O_2 concentration with maximal luminescence intensity reached under anoxic conditions. Recently ratiometric approaches have been published that allow spatial oxygen measurements with low cost, commercially available cameras. The ratiometric method is based on simultaneous measurement of luminescence intensities of the oxygen indicator dye and of the reference (oxygen-insensitive) dye. If the two emissions are spectrally separable, a colour camera can be used for imaging (Larsen *et al.*, 2011). The ratiometric approach allows compensation for the variation in the intensity of the excitation light, sensitivity of the detector and dye density. However, if these parameters are kept constant, reliable measurements are also possible by simply measuring luminescence intensity at anoxic conditions and under different oxygen concentrations (Staal *et al.*, 2011b). Due to its simplicity, this approach was used in our study. Using this method, a calibration curve can be determined by correlating known oxygen concentrations with the ratio of the intensity under anoxic condition to that under different O_2 concentrations. From the calibration curve, any luminescence intensity image can be converted to an O_2 concentration image.

MATERIALS AND METHODS

Experimental setup description

Biofilm was grown in a membrane fouling simulator (MFS) containing a 20 cm × 4 cm coupon of a membrane and a spacer and a transparent window for visual observations (Vrouwenvelder *et al.*, 2006, 2007b). The MFS flow cell was operated without permeate production (Vrouwenvelder *et al.*, 2009d) at a pressure < 4 bar. Hydrodynamic conditions (e.g. flow field distribution and the relationship between cross flow and feed channel pressure drop) in the monitor are similar to spiral wound membrane modules as applied in practice for water treatment. The spacer and membrane coupons were taken from virgin spiral wound membrane elements (4040-TS80-TSF, Trisep Corporation, USA). The feed spacer consisted of a sheet of 31 mil (787 μm) thick diamond-shaped polypropylene spacer (porosity 0.85). The feed spacer was placed in the MFS with the same orientation as in spiral membrane elements (45° contact angle with feed flow). The system was run as a continuous flow reactor and the setup consisted of a large buffer tank (200 L capacity), diaphragm pump, temperature and differential pressure transmitter (Delta bars, Endress + Hauser, PMD75), pressure-reducing valve, substrate dosing diaphragm metering pump, MFS and a flow sensor (Ultra flow sensor ULF01.H.0, POM/FPM) (Figure 7.1). Feed water is filtered by a carbon and cartridge filter (4 μm pore size).

Operating condition

Filtered (carbon and cartridge filter, filter housing model: UPS BB3 [AWF-UPS-3H-20B] Cartridges model: sediment-carbon [AC-SC-10-NL]) tap water ($16\ L \cdot h^{-1}$) continuously refreshed the water in a buffer tank. The tap water is produced from chlorinated RO effluent so in order to have biofilm growth carbon filters were used to remove residual chlorine. Water from the buffer tank was pumped through the MFS at a flow rate of $12\ L \cdot h^{-1}$ equivalent to a linear flow velocity of $0.12\ m \cdot s^{-1}$, which is representative for practice (Vrouwenvelder *et al.*, 2009b). The outflow from the MFS was recirculated into the tank. Water

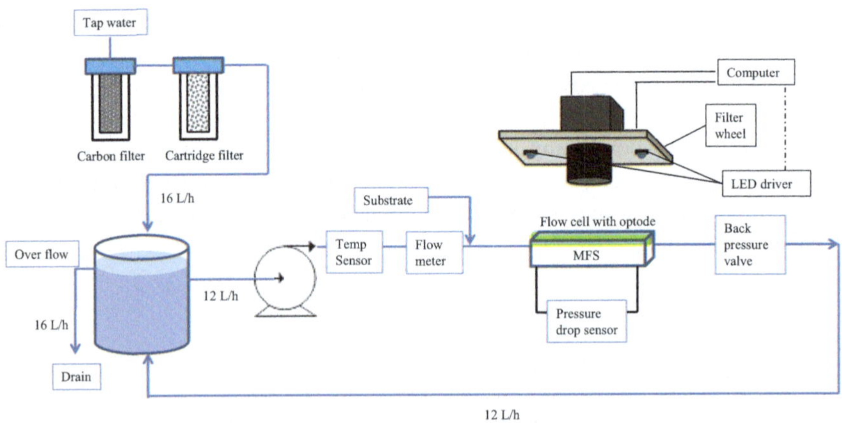

Figure 7.1 Schematic description of the membrane fouling simulator (MFS) including a water mixing system, differential pressure transmitter, flow meter and the imaging setup for O_2 detection. The arrows indicate the water flow direction. The filter wheel gives the possibility to shift between different filters.

recirculation was required to increase bacterial cell numbers in the feed water of the monitor. The buffer tank water was aerated to ensure saturated oxygen conditions in the MFS. Water temperature was maintained at 24°C during the whole experiment. A nutrient stock solution containing sodium acetate, sodium nitrate, and sodium phosphate in a mass ratio C : N : P of 100 : 20 : 10 was added to the feed water up to 1 mg-C · L^{-1} to enhance biofilm formation in the MFS. The stock solution was dosed by a diaphragm metering pump (STEPDOS 03S, Knf NEUBERGER) at a flow rate of 0.05 L · h^{-1}. To restrict bacterial growth in the nutrient dosage vessel, the pH was set at 11 by adding sodium hydroxide at a concentration of 1 mol · L^{-1}. The nutrient dosage flow rate (0.05 L · h^{-1}) was low compared to the feed water flow rate (12 L · h^{-1}). Thus, the effect of the nutrient dosage on the pH of the feed water was insignificant.

Pressure drop was measured over the MFS as an indicator of fouling development (Vrouwenvelder *et al.*, 2009b, c). The development of pressure drop (PD) with time over the monitor length was measured and recorded by a differential pressure transmitter (Delta bars, Endress + Hauser, PMD75). The pressure drop development was normalized by subtracting the pressure drop value at time 0.

Dye and optode description

The planar optode was coated on a Mylar (thin polyethylene terephthalate, thickness equals 0.13 mm) sheet, which was placed into the flow cell under the transparent cover window. The optode sheet was in direct contact with the feed spacer and the water. Biofilm formation occurs on the membrane, the spacer and the optode sheet. The planar optode used in this study was based on the dye PtTPTBPF (excitation 595 nm, emission 775 nm) immobilized in a polystyrene matrix (4% w/w PE/chloroform) (Borisov *et al.*, 2008). Coating the dye directly on the transparent window of the flow cell was not possible since the solvent of the dye also dissolves the MFS window material, making the sight window turbid.

Imaging system

Oxygen imaging was done with an Apogee Imaging Systems Ascent A285 CCD Camera equipped with a monochrome Sony ICX-285 Interline CCD Sensor (1392 × 1040 pixels). A filter wheel (APOGEE CFW31-8R FILTER WHEEL 1.25, 8 POS) was mounted on the camera. A filter (690–850 nm, Astrodon photometrics, sloan photometric filters i'2) was used to take the oxygen images. On top of the filter wheel a Nikon F-mount lens adapter was mounted and a Nikon Nikkor 35 mm f/1.4 lens was used with Nikon macro extension tube Kit resulting in a focal distance of 15 cm, depth of field 1.9–2.2 mm, and a pixel size of 45 μm × 45 μm. Micro-manager 1.4, a freeware microscope-controlling software program based on ImageJ, was used to control the camera and filter wheel settings. The oxygen sensing dye was excited by eight amber LEDs (lumiled type, 595 nm) placed around the lens. The power supply for the LEDs was an AIM-TTI instruments, PL303 power supply,

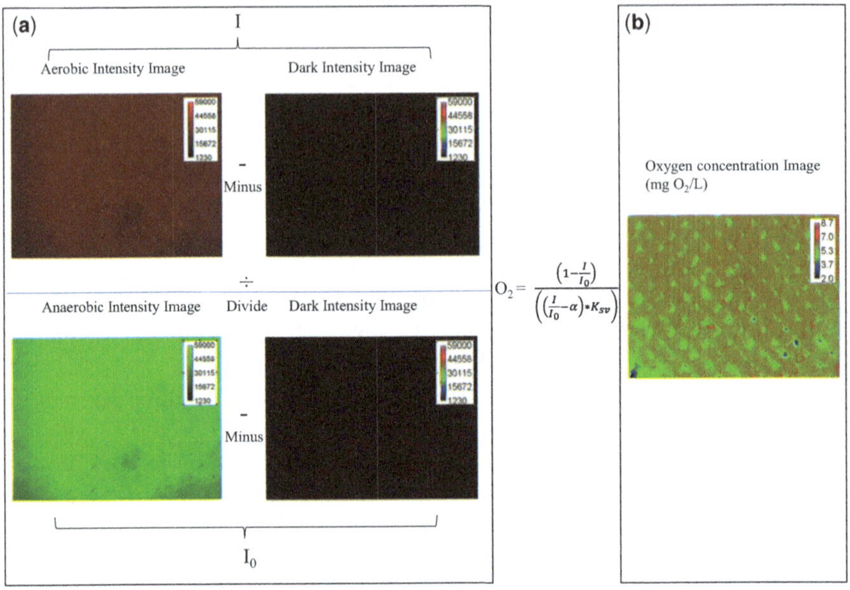

Figure 7.2 Schematic description of the calculation of the spatial oxygen concentration derived from aerobic and anaerobic luminescence intensity images. (a) Scale bar represents intensity values; (b) scale bar represents oxygen concentration (mg · L^{-1}).

1CH, 30V, 3A, providing a stable LED output. The MFS and the camera system were mounted in a light-tight box to minimize the effect of external light on the luminescence images.

Image calculation and analysis

Oxygen distribution was calculated with a luminescence intensity imaging approach (Staal *et al.*, 2011b). In this approach, the spatial O_2 distribution was calculated directly from the ratio of a luminescence image obtained under anoxic conditions (I_0) and a luminescence image (I) at the actual O_2 concentration. A schematic description of the calculation of O_2 concentration from aerobic and anaerobic luminescence intensity images is shown in Figure 7.2. For the conversion from ratio to O_2 concentration a Stern-Volmer equation determined from the calibration of the optode was used.

$$O_2 = \frac{1-(I/I_0)}{\left(\frac{I}{I_0}-\alpha\right)\cdot K_{SV}} \qquad (7.3)$$

α is the non-quenchable fraction of the dye and K_{SV} is the quenching efficiency coefficient of the dye.

Calibration of the planar optode was done by circulating water with different O_2 concentrations through the MFS at 24°C. Different O_2 concentrations were generated using a gas mixing system based on electronic mass flow controllers (Sensor Sense, Nijmegen). Intensity images were taken for each defined O_2 concentration, where after a calibration curve was obtained. K_{SV} and α were determined by fitting the adapted Stern-Volmer equation (eq. 7.2) through the calibration curve.

All post-acquisition image processing was done using the freeware ImageJ version 1.49 g. After the import to ImageJ, all 16 bit images were converted to a 32 bit floating point format. I/I_0 ratio intensity images were then calculated using the 'process > image calculator' option in ImageJ. Conversion of ratio I/I_0 intensity images to O_2 concentration distribution images was done in the process > math > macro menu based on calibration equation linking the intensity ratio values to defined O_2 concentrations. False colouring was applied to visualize the differences in O_2 concentration.

Analysis of spatial indicators

Analysis of spatial indicators of images taken under both cross-flow and stop-flow operation of the MFS was done using the ISA software (Beyenal *et al.*, 2004b). Quantitative parameters describing biofilm physical structure such as homogeneity and contrast to monitor biofilm development were analysed (Yang *et al.*, 2000). Biofilm homogeneity is a textural parameter that measures the similarity of the spatially close image structures: a higher homogeneity indicates a more homogeneous image structure, the homogeneity decreases with increasing numbers of biofilm clusters (Beyenal *et al.*, 2004a). Contrast is determined by the difference in the colour and brightness of an object and other objects within the same field of view.

Cross-flow versus stop-flow imaging

Two different types of oxygen measuring protocols (cross-flow and stop-flow) were tested for their potential to measure biofouling in an MFS. The cross-flow imaging protocol consisted of sequential imaging of luminescence intensity images under (i) dark, (ii) aerobic and (iii) anaerobic conditions for the inlet of the MFS. The dark intensity image was subtracted from all intensity images to eliminate the background effect. The aerobic intensity image was determined under standard cross-flow operation, while the anaerobic luminescence intensity image was made after dosing a concentrated solution of sodium hydrosulphite (20 g · L^{-1}) into the feed water at a rate of 0.125 mL · h^{-1}. The anaerobic luminescence intensity image was taken once a day and sodium hydrosulphite consumed all oxygen present in the monitor. The water flow through the MFS was directly diverted to the drain when sodium hydrosulphite was pumped to prevent contamination of the buffer tank water.

In the stop-flow imaging protocol, a series of 100 intensity images was taken during a 10-minute flow interruption at an interval of 6 seconds. The stop-flow intensity image series was divided by the anaerobic intensity image where after the stack of ratio images is converted into a stack of O_2 concentration image. The stop-flow series gave spatially resolved information about O_2 concentration decrease rates and biofilm development.

RESULTS

In this study, three methods were compared to evaluate their potential for early biofouling detection and diagnosis; (i) pressure drop development with time, (ii) O_2 decrease under MFS cross-flow conditions, and (iii) O_2 decrease under MFS stop-flow conditions. No pressure drop increase was observed during the first 1.5 days of the experiment (Figure 7.3). Both O_2 measuring techniques already showed a decrease in O_2 concentration at day one. Spatially resolved information about oxygen depletion locations was identified from images taken during cross-flow operation (Figure 7.4). These areas indicate local biofilm development. In addition to average O_2 concentration decrease with time, spatial indicators such as image homogeneity and contrast were used as an indicator of biofilm presence (Figure 7.5). Earliest biofilm development detection was based on stop-flow O_2 decrease rates (Figures 7.6–7.8). Furthermore, Figures 7.6–7.8 confirm the oxygen depletion locations identified during cross-flow operation (Figure 7.4) as these locations have

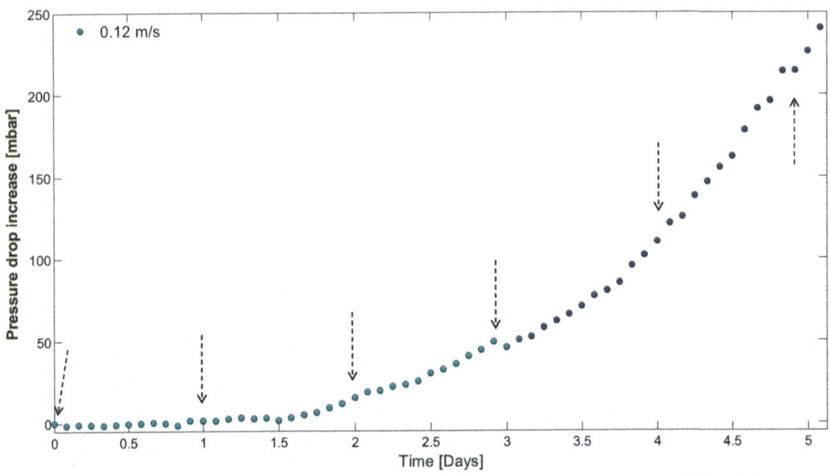

Figure 7.3 Pressure drop increase (PD) with time over the MFS. The MFS feed water was supplemented with 1 mg-C · L^{-1} biodegradable substrate to increase the rate of biofilm accumulation. Arrows indicate the time of O_2 imaging during a short flow interruption (stop-flow) in the MFS.

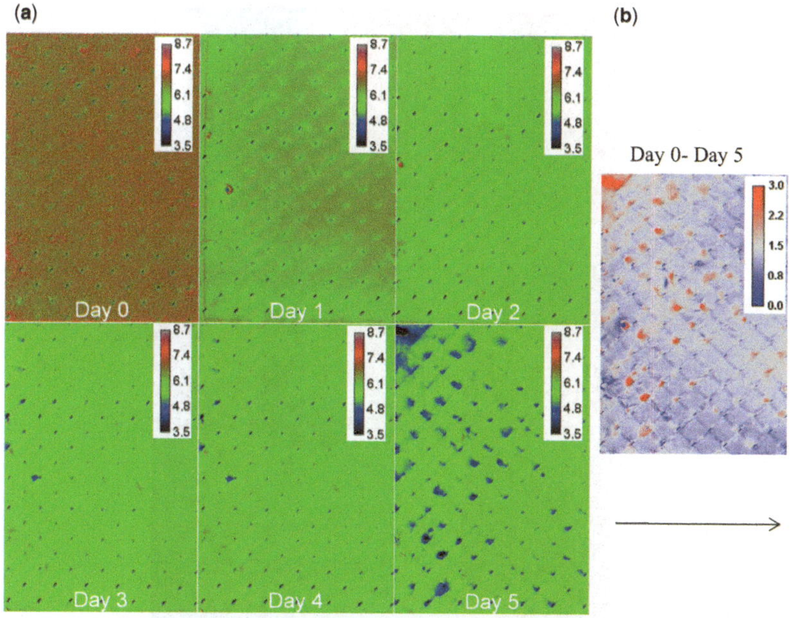

Figure 7.4 (a) Spatial distribution of oxygen concentration (mg · L^{-1}) assessed during cross-flow operation at the inlet side of the MFS (first 2.5 cm) with time (Day 0 to Day 5) at constant cross-flow velocity. (b) Differential image showing oxygen image Day 0 – oxygen image Day 5, indicating the locations where the biofilm is present. The arrow indicates the water flow direction. Scale bar represents oxygen concentration (mg · L^{-1}).

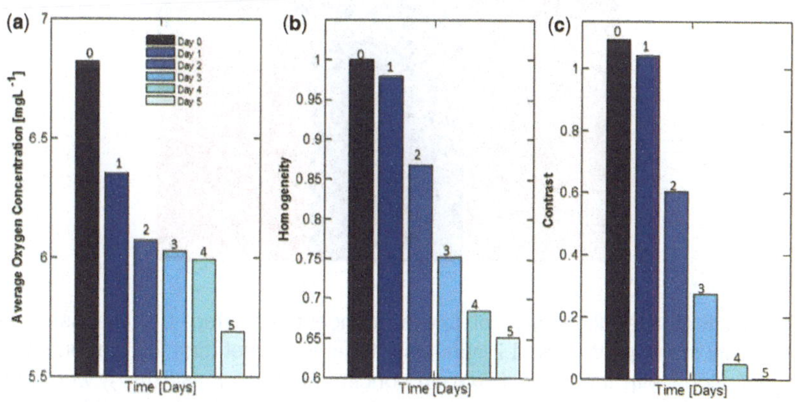

Figure 7.5 (a) Biofilm average oxygen concentration (mg · L^{-1}), and image spatial indicators: (b) homogeneity and (c) contrast change with time determined from images taken at constant cross-flow velocity.

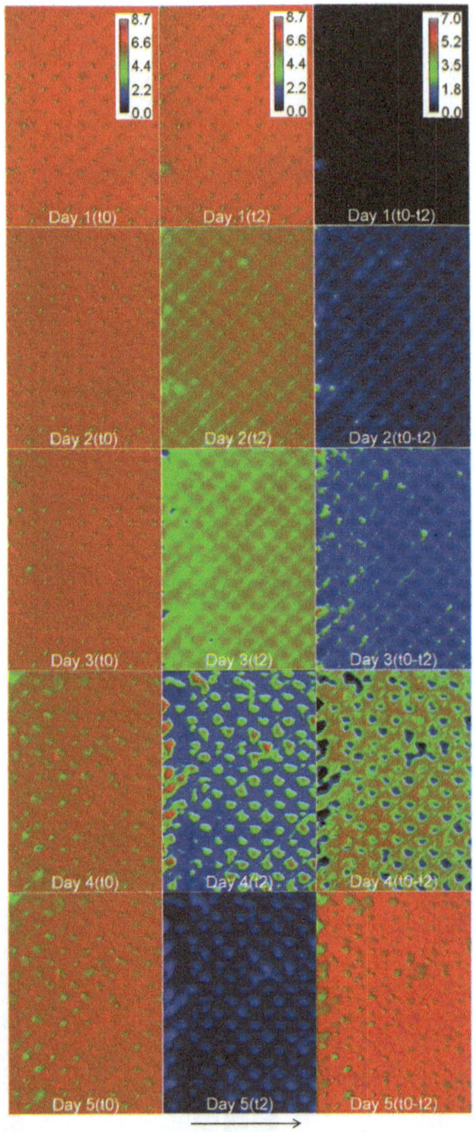

Figure 7.6 Spatial distribution of oxygen concentration (mg · L^{-1}) assessed during water flow interruption (0 and 2 minutes) at the MFS inlet (first 2.5 cm) on day 1–5. The locations with the highest oxygen concentration decrease (t_0–t_2) were derived by subtracting the image after 2 minutes (t_2) from the initial image (t_0). The arrow indicates the water flow direction. Scale bar in first and second column represents oxygen concentration (mg · L^{-1}), Scale bar in third column represents change in oxygen concentration (mg · L^{-1}).

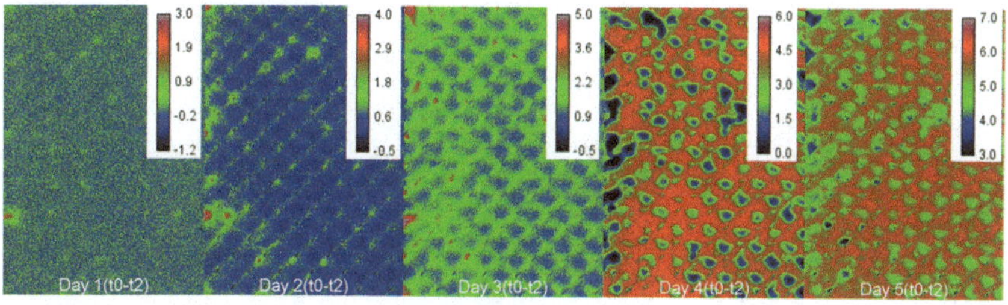

Figure 7.7 Spatial distribution of difference (t_0–t_2) in oxygen concentration (mg · L^{-1}) assessed after water flow interruption (0 and 2 minutes) at the MFS inlet (first 2.5 cm) on day 1–5. Range of O_2 concentration decrease is narrowed to emphasize the oxygen decrease values and locations.

highest depletion rate during stop-flow. Comparison between different indicators for biofouling presence with time is given in Figure 7.9.

Pressure drop

No increase in pressure drop (PD) over the MFS was observed during the first 1.5 days of the experiments (Figure 7.3). After 1.5 days of MFS operation, an exponential increase in PD started as a result of biofilm accumulation in the flow cell. A small effect of flow disruption (stop-flow) on PD can be seen at day three and five (Figure 7.3). Nonetheless, the increasing PD trend recovered within one hour

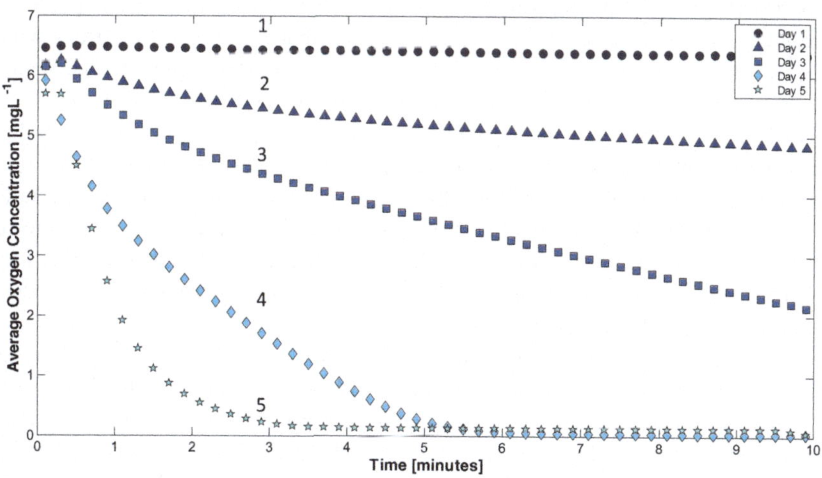

Figure 7.8 Stop-flow plot: Average oxygen concentration (mg · L^{-1}) decrease with time on day 1 to 5 due to a 10 minute water flow interruption of the MFS.

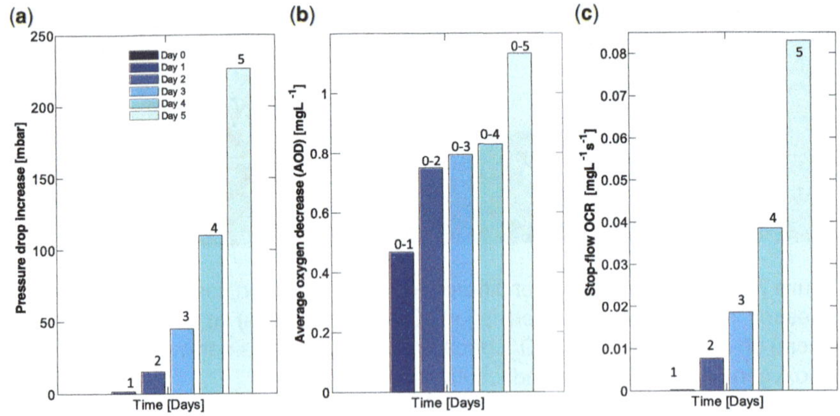

Figure 7.9 (a) pressure drop increase (NPD), (b) average oxygen concentration decrease (AOD), and (c) stop-flow oxygen consumption rate (OCR) with time at constant cross-flow operation of the MFS.

after the original flow ($0.12\ m \cdot s^{-1}$) was restored. The PD increased to 250 mbar within the five day experiment.

Oxygen concentration during cross-flow MFS operation

The spatial distribution of oxygen concentrations ($mg \cdot L^{-1}$) were measured daily under cross-flow conditions for a period of five days at the inlet of the MFS (first 2.5 cm) (Figure 7.4a). Regions with lower oxygen concentrations appeared at the feed spacer crossing on day one, indicating that these were the areas where respiratory activity started and showing initial biofilm development. The regions of lower oxygen concentration become more pronounced with time (Figure 7.4a). The differential image day 0–day 5 (Figure 7.4b) shows the variability in change in oxygen concentration within the imaged area. Almost every pixel shows a decrease in oxygen concentration during the experiment. Compared to day 0, approximately 73% of the area had an O_2 decrease of ~1–1.5 $mg \cdot L^{-1}$ while nearly 16% showed an O_2 decrease of 1.5–2.5 $mg \cdot L^{-1}$. Some small areas (≈1% of the total area) showed even a stronger decrease in oxygen concentration (4–5 $mg \cdot L^{-1}$), occurring predominantly at feed spacer crossings.

The average O_2 concentrations were determined from the images. The highest decrease in average O_2 concentration was found on the first day of the experiment: A reduction of 6% was observed. The O_2 concentration decreased further during the rest of the experiment but at lower daily rates (Figure 7.5a). A suite of parameters were evaluated using ISA software (Beyenal *et al.*, 2004a). Here only image homogeneity and contrast are shown (Figure 7.5b and c). Both homogeneity and contrast decreased with time. The largest decline in image homogeneity was found at day three (≈ 13%); (day one ≈ 2% decline and day two ≈ 11% decline). It was also tested whether the size of the sampled area within the O_2 image affected the results. The O_2 concentration plotted in Figure 7.5a is an average value of the whole inlet area of the MFS (2.5 cm × 4.0 cm). The smallest sampling area giving the same average oxygen concentration values was found to be as small as 0.30 cm × 0.30 cm.

This area is almost as small as one feed spacer frame. Table 7.1 summarizes the minimum surface area needed for representative detection of change in oxygen concentration, where an area as small as 0.30 cm × 0.30 cm can be used to detect O_2 concentration decrease representatively. Table 7.1 shows an increase in the standard deviation of the oxygen values obtained from the different sampled areas as these areas become smaller. The O_2 concentration determined from a smaller area deviated more from the value obtained over a larger area as biofilm spatial distribution becomes more heterogeneous with time.

Oxygen concentration during stop-flow MFS operation

During a stop-flow event, the flow was interrupted for a period of 10 minutes. During this period, a series of 100 intensity images was taken at 6 seconds intervals. In addition to biofilm development locations, average oxygen decrease with time (10 minutes) can be determined from the stop-flow image series. Figure 7.6 shows the spatial distribution of oxygen concentration ($mg \cdot L^{-1}$)

Table 7.1 Minimum surface area needed for representative detection of change in oxygen concentration.

Number of Pixels	Area (cm × cm)	Averagc Oxygen Concentration ($mg \cdot L^{-1}$) ± SD
Biofilm development at an initial stage (Day 1)		
247,390	2.5 × 4.0	6.354
32,900*	1.2 × 1.2	6.246 ± 0.069
10,800**	0.60 × 0.60	6.236 ± 0.086
1,080**	0.30 × 0.30	6.230 ± 0.144
Biofilm development at a late stage (Day 5)		
247,390	2.5 × 4.0	5.695
32,900*	1.2 × 1.2	5.575 ± 0.057
10,800**	0.60 × 0.60	5.556 ± 0.062
1,080**	0.30 × 0.30	5.555 ± 0.091

SD: Standard deviation of the values from the different sampled areas.
*Measurements were done at six different locations of the same surface area.
**Measurements were done at twelve different locations of the same surface area.

assessed after the water flow interruption (0 and 2 minutes) and a difference image (t_0–t_2) derived by subtracting the image at t_2 from the image at t_0 to show locations with the highest oxygen concentration decrease. Figure 7.7 shows the same differential images, but the scale in every individual image is changed so that O_2 decrease values and locations become more pronounced. Figures 7.6 and 7.7 show that the highest oxygen concentration decrease, and thus largest quantity of biofouling present can be seen on and near the feed spacer crossing as early as day two.

On day one only a minor decrease in O_2 concentration was found two minutes after stopping the water-flow through the MFS. The O_2 concentration at t_2 compared to t_0 had an average O_2 decrease of 0.005 mg · L^{-1} and a maximum decrease of 3 mg · L^{-1}. The decrease in O_2 concentration between t_0 and t_2 on day one was relatively homogenous distributed. On day two the O_2 decrease (t_0–t_2) became more evident and more heterogeneous with an average of 0.56 mg · L^{-1} and a maximum of 4 mg · L^{-1}.

The first 1 cm of the flow cell inlet had an O_2 decrease average of 0.65 mg · L^{-1}; the remaining 1.5 cm of the inlet of the flow cell had an O_2 decrease average of 0.50 mg · L^{-1}. More biofilm is present at the first 1 cm (inlet side) of the flow cell. On day three a higher average O_2 decrease (1.25 mg · L^{-1}) over the whole inlet area was seen compared to day two. The largest O_2 decrease was still seen at the first 1 cm of the inlet on day three with a maximum O_2 decrease (t_0–t_2) of 5 mg · L^{-1}. On day four, the first 1 cm of the inlet with the most biofilm growth had a lower O_2 concentration decrease. Overall the biofilm continued to grow with time; the O_2 consumption rate (OCR) during stop-flow increased rapidly until the MFS became completely anoxic within five minutes of stop-flow on day four and within three minutes of stop-flow on day five (Figure 7.6).

Figure 7.8 shows the average oxygen decrease with time after stopping the water flow on days one to five. On day one the OCR was 0.2 µg · L^{-1} · s^{-1} while on day two it increased to 7.65 µg · L^{-1} · s^{-1}; illustrating that respiration increased in the flow cell. Stop-flow images clearly showed regions with lower oxygen concentrations mainly at the feed spacer crossing indicating the initial location of biofilm development and activity (Figure 7.7, Day 1 (t_2), Day 2 (t_2)).

Comparison of the three methods

Three methods were compared in terms of their early biofouling detection potential: pressure drop development with time, O_2 decrease under cross-flow conditions, and O_2 decrease under stop-flow conditions. Figure 7.9 (A) shows the normalized feed channel pressure drop (mbar) development in time, (B) the average oxygen concentration decrease (AOD) (mg · L^{-1}) in time under cross-flow conditions; AOD is calculated by subtracting the average oxygen concentration on any day from the average oxygen concentration on day 0, and (C) the stop-flow OCR (mg · L^{-1} · s^{-1}) with time. Normalized pressure drop development, AOD, and stop-flow OCR all showed an increase with time. The normalized feed channel pressure drop showed an exponential increasing trend. Compared to feed channel pressure drop development, the AOD showed a different trend. The first two days had the highest decrease in O_2 concentration. This was attributed to large clusters of bacterial cells attaching in the MFS and resulting in hotspots of low O_2 concentration. These clusters will already experience diffusion limitation, reducing their O_2 consumption (OCR) locally. Another factor that explains uncoupling of oxygen concentration distribution from OCR is the fact that the oxygen concentration at the base of a biofilm depends on a combination of oxygen consumption and the diffusive boundary layer (DBL) thickness. The DBL thickness is dependent on flow rate and surface roughness. Biofilm development will increase surface roughness and reduce water flow. Both will increase the DBL and therefore biofilms can reduce oxygen concentrations with relative low respiration rates. The stop-flow OCR exhibited an exponentially increasing trend similar to the PD development trend. A decrease in O_2 concentration and an increase in stop-flow OCR were observed within one day after the start of the experiment.

Control experiments conducted with no biodegradable substrate dosing and no biofilm development showed that pressure drop measurements can fluctuate up to ±5 mbar. Thus, the 3 mbar increase observed on day one is still within the sensitivity error of the differential pressure drop transmitter and cannot be used to indicate biofilm development. As for the oxygen concentration, a variation up to $\pm 0.184\ mg \cdot L^{-1}$ was observed when no biodegradable substrate was dosed. The average oxygen concentration was $7.488\ mg \cdot L^{-1}$; resulting in an error of 2.8% of the measured value.

A decrease in average oxygen concentration of $0.184\ mg \cdot L^{-1}$ was considered as the sensitivity error of the imaging methods and any decrease higher than $0.184\ mg \cdot L^{-1}$ was attributed to biofilm growth. Table 7.2 summarizes average time (days) required to detect and quantify the biofouling accumulation inside the MFS.

Earliest biofilm development detection was based on average O_2 decrease and stop-flow OCR as these methods showed a higher signal when compared to pressure drop.

Table 7.2 Average time (days) required to detect biofouling accumulation inside the MFS.

Method	**Detection Period (Days)**
Normalized pressure drop (NPD)	1.5–2
Cross flow	
Oxygen concentration	1
Spatial indicators (Homogeneity, Contrast)	1
Stop flow	≤1

DISCUSSION

Early detection of biofouling development

The recent recognition that membrane biofouling during nanofiltration (NF) and reverse osmosis (RO) processes is inevitable, has led to a paradigm shift in which biofouling management rather than prevention is now becoming a practice being considered (Bucs *et al.*, 2014b; Habimana *et al.*, 2014). Earlier, the key focus of biofouling studies was on developing control strategies that mainly relied on the reduction in the feed water of biodegradable substrates, and the reduction of microorganism's numbers (Kim *et al.*, 2009; Rand *et al.*, 2007; Vrouwenvelder & Van Der Kooij, 2001) to control biofouling. This approach was not successful in controlling biofouling in all cases. Early detection plays an essential role in an adequate biofouling management strategy (Creber *et al.*, 2010a; Vrouwenvelder *et al.*, 2011a). In this study, a sensitive differential pressure drop transmitter was used to monitor the feed channel pressure drop over the MFS length. No increase in feed channel pressure drop was observed during the first 1.5 days of the experiment (Figure 7.3).

Monitoring O_2 concentration changes in time during MFS cross-flow operation was possible using planar optodes. Average O_2 concentration in the MFS decreased with time as a result of biofilm development. This O_2 decrease was seen already one day after starting the study while no feed channel pressure drop increase was recorded. Observing this O_2 concentration decrease under cross-flow operation conditions is considered crucial emphasizing the imaging setup's capability of being an in-situ, non-destructive tool to detect biofouling. Stop-flow oxygen imaging was used to acquire more information on oxygen decrease rates (OCR) and biofouling locations. Usually, stop-flow oxygen imaging can detect biofouling earlier than imaging during cross-flow operation. An oxygen concentration decrease during this 10-minute flow interruption, no matter how small it is, was detected even when no significant decline in O_2 concentration under cross-flow operation was observed (data not shown). In this study, the decrease in O_2 concentration during stop-flow between t_0 and t_2 on day one was relatively homogenous signalling already the start of formation of a thin layer of biofilm (Figures 7.6–7.8). The first 1 cm of the flow cell inlet had the most biofilm development with highest O_2 decrease range .The OCR at the first 1 cm of the flow cell inlet was higher than the remaining 1.5 cm up until day four. On day four, the first 1 cm of the inlet that had the most biofilm growth had a lower OCR and a lower O_2 difference (t_0–t_2). One reason for the reduction in the OCR can be that O_2 respiration rates can become carbon limited in the thicker biofilm due to a lower diffusive transport of carbon into the biofilm compared to O_2 diffusive transport.

Cross-flow and stop-flow imaging of oxygen concentration, as methods to detect biofouling, require imaging skills and are more labour intensive compared to pressure drop measurements. Nevertheless, the superiority of these two O_2 imaging methods is in the detailed information received on biofouling development.

Spatially resolved biofouling development information

In full-scale installations, membrane performance assessment is done according to three operational parameters: feed channel pressure drop, permeate flux, and salt retention. An increase in feed channel pressure drop, decline in flux or increase in salt retention would indicate the existence of fouling. The membrane performance sensors employed are pressure transmitter, flow meter, and conductivity meter. These measured parameters are general and not specific to biofouling (Vrouwenvelder & Van Der Kooij, 2003), they only signal the existence of a problem. Increase in feed channel pressure drop is considered an indirect method to detect biofouling as it can occur due to the accumulation of any fouling type on the membrane and feed spacer. Feed channel pressure drop measurements are based on the resistance that water experiences when passing through the membrane module during operation. Under constant cross flow velocity, feed channel pressure drop increases with an increasing biomass amount, but the location where the biofilm forms will determine the resistance per biofilm volume. The primary roles of feed spacers are to promote water mixing and to provide an open flow channel with little resistance (Araújo *et al.*, 2012a; Bucs *et al.*, 2014a; Tran *et al.*, 2007); however the largest pressure drop along the feed channel occurs in the spacer region, with the narrowest open flow area (Radu *et al.*, 2010). Therefore, biomass present on the spacer will have higher effect on feed channel pressure drop signal than biomass present on the membrane area inside a spacer frame. Measuring the increase in feed channel pressure drop does not provide information about fouling location and distribution. Planar optode imaging specifically detects biofouling through assessment of oxygen concentration as a direct correlation to biofilm activity (de Beer *et al.*, 1994; Staal *et al.*, 2011a). Monitoring the change in O_2 concentration under cross-flow operation conditions is considered very important, emphasizing the imaging setup's capability of being an in-situ, non-destructive tool to detect biofouling.

Spatial distribution of oxygen concentration over the imaged area is the most appealing feature of such a system, enabling identification of the regions where the biofilm starts to grow and how it develops with time. In contrast with differential pressure drop measurements, biomass present on the membrane area inside the feed spacer frame has a higher impact on oxygen consumption signal than biomass present on the spacer since the optode material is directly present under the transparent MFS window. This information about biomass location is mentioned to compare the sensitivity of the measuring methods (pressure drop versus optodes) to show that each method has a different measuring principle, thus illustrating the difference in response with time.

In terms of membrane performance, biomass accumulation on the feed spacer affects the water flow distribution in membrane modules. This flow channelling has a substantial impact on feed channel pressure drop, but also on the permeate flux and salt passage. Flow channelling results in a reduction of the (i) residence time of water in the membrane system and the (ii) effective membrane area. Areas with reduced fresh water flux (stagnant water) experience increased concentration polarization, reducing locally the permeate flux and increasing salt passage. Therefore, the direct consequence of biofilm presence on the membrane and the indirect consequence of biofilm presence on the spacer have a huge effect on membrane performance.

Both cross-flow and stop-flow imaging methods were able to provide spatially resolved information about oxygen depletion locations. The cross-flow method showed local depletion of oxygen concentration mainly starting near the feed spacer crossings. Stop-flow imaging revealed that these same locations (near the feed spacer crossing) had the highest OCR thus confirming the observations seen using cross-flow. Direct biofilm imaging using rhodamine dye (data not shown here) substantiated that the locations with highest O_2 depletion had the highest quantity of biofilm, confirmed also by results seen in (Staal *et al.*, 2011a).

Analysis of the images in terms of spatial parameters such as homogeneity and contrast was also used to determine how the present biofilm is distributed. Thus regardless of the actual pixel values of the image in terms of oxygen, the change in the distribution of these pixels was used to describe biofilm development. Image homogeneity and contrast decreased over time as a result of biofilm development.

Spatially resolved oxygen concentration can be used to determine the minimum surface area required to detect a representative measurement of the change in O_2 concentration. Information on the minimum sample area to get a representative sample of the monitor can be used to implement less labour-intensive non-spatial resolved lifetime oxygen sensing with photodiodes. This minimum sample area information is relevant when spatial distribution is not as important as measuring a decrease in O_2 for biofouling detection. It is essential however for the biofilm bacteria to actively respire O_2; otherwise this method cannot be used.

Effect of stop-flow imaging on biofouling development

Stop-flow imaging of oxygen consumption rate enabled a more evident and earlier decrease in average oxygen concentration, as well as indicating the (the locations of) initialization of biofouling development. Stop-flow images clearly showed regions with lower oxygen concentrations mainly at the feed spacer crossings, showing the initial location of biofilm development and activity at these spots. This observation on the start of biofouling at the spacer is consistent with results obtained by others (Baker *et al.*, 1995; Bucs *et al.*, 2014a; Creber *et al.*, 2010a; Vrouwenvelder *et al.*, 1998).

However, stop-flow imaging, although superior to cross-flow imaging in terms of earlier biofouling detection, resulted in local biofilm detachment. This biofilm detachment drawback was more evident at higher cross-flow velocities (data not shown here) where more biofilm detachment occurred resulting in a reduced biomass amount and lower pressure drop development with time.

Practical applications and future studies

A unique feature of O_2 sensing optode imaging is the quantitative and detailed information this method generates, especially in terms of the spatial distribution of the biofilm. Therefore, the optode method can be used to study the fundamentals of development of biofilms under different conditions such as the effect of different spacer geometries, configurations, materials, and operational aspects on biofouling development. Various cleaning chemicals and protocols can also be evaluated as long as the chemicals do not affect the performance of the optode. O_2 sensing optodes are only relevant for biologically active biofilms that consume O_2. For biofilms under anaerobic conditions, for example, other types of detection dyes can be used such as ones that detect changes in pH or CO_2. In practice, most biofilms in spiral-wound NF and RO membrane systems consume oxygen.

CONCLUSIONS

Biofilm development with time was studied using O_2 optode luminescence imaging and compared with normalized feed channel pressure drop. Surface O_2 imaging was done during cross-flow and during stop-flow operation of membrane fouling simulators. The main objective of these experiments was to evaluate whether O_2 optode imaging can detect biofouling presence at an early stage providing a non-destructive early biofouling diagnosis. Based on the results it can be concluded that:

(1) An increase in normalized feed channel pressure drop started to occur after 1.5–2 days from the beginning of the experiment signalling the fouling development.
(2) A decrease in the average oxygen concentration during cross-flow operation, and an increase in the stop-flow OCR were observed as early as day one from the experimental start indicating earlier detection of fouling and diagnosis of biofilm activity.
(3) O_2 optode imaging, in addition to earlier biofouling detection, provided quantitatively the spatial distribution of oxygen concentration over the imaged surface area, identifying with high resolution the regions where the biofilm starts to grow and how it develops with time.
(4) O_2 optode imaging provided detailed quantitative information on biofouling development therefore it can be successfully implemented in evaluating control strategies or/and practical applications.

Chapter 8

Chemical cleaning of biofouling in reverse osmosis membranes evaluated using magnetic resonance imaging

INTRODUCTION

Increasing global demand for freshwater has pressed the need for effective, efficient water treatment technologies. Desalination is a common process used for treating seawater for drinking, industrial, and agricultural applications. Current desalination capacity is in excess of 55 million m^3/day and is expected to continue to grow steadily (Viljoen, 2010). Of the two main desalination technologies, thermal desalination and membrane based desalination, membrane based technologies have emerged as the market leader (Fritzmann *et al.*, 2007). Reverse osmosis (RO) membranes, able to reject 99% of salt ions, is the predominant membrane desalination technology (Greenlee *et al.*, 2009). RO membrane modules are typically constructed as spiral-wound systems, consisting of multiple layers of membrane sandwiches separated by spacer sheets and coiled around a central perforated tube. The modules operate in cross-flow, whereby pressurised salty water enters at one end, flows across the spacer sheets and over the membrane layers, with (i) water leaving the membrane as salty concentrate and (ii) permeate passing into the membrane sandwiches and travelling spirally toward the central tube. The RO membrane schematic in Figure 8.1 depicts this process.

RO membrane performance can be significantly reduced due to fouling from different sources, including dissolved inorganic material, particulate matter, and micro-organisms (Potts *et al.*, 1981). The micro-organisms feed off of nutrients available in the feed water to form biofilms, complex structures of microbial cells and polymer matrices, within the RO module. This biofouling can lead to a multitude of operational problems, including increased membrane resistance, decreased permeate production, membrane degradation, and decreased product

Author(s): Szilárd Bucs, Joop Kruithof, Mark van Loosdrecht and Johannes Vrouwenvelder
doi: 10.2166/9781780409597_0145

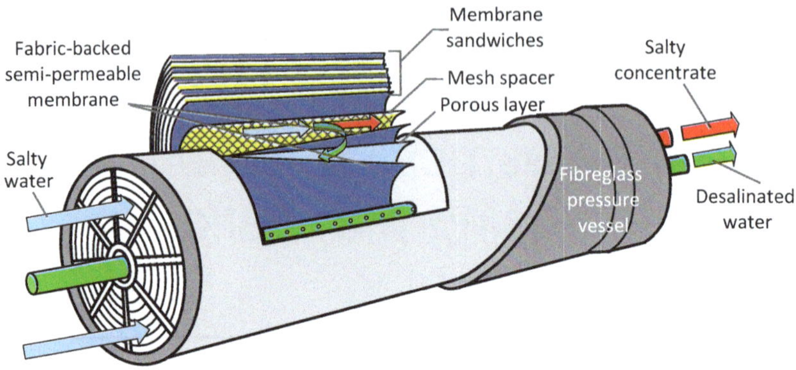

Figure 8.1 Schematic of an industrial spiral-wound RO membrane module.

quality (Flemming, 1997). Despite extensive pre-treatment of source water before it is fed to an RO installation, biofouling persists. Thus, additional monitoring and cleaning of the RO modules are used to control biofouling (Van der Bruggen & Vandecasteele, 2002). Different cleaning methods for biofilm removal may be used, including chemical cleaning, air or water flushing, and electrotechnology (Cornelissen *et al.*, 2007a; Romo & Pitts, 1999; Zondervan & Roffel, 2007). Chemical cleaning solutions are most commonly applied to RO modules for performance restoration with a wide range available commercially; appropriate chemical cleaning solution selection depends on the nature of the membrane and biofilm among other factors (Al-Amoudi & Lovitt, 2007). Several studies have examined the efficiencies of different cleaning solutions on biofilm removal, with alkaline and alkaline coupled with a surfactant solution showing the greatest cleaning potentials (Al-Amoudi & Lovitt, 2007; Ang *et al.*, 2006; Liikanen *et al.*, 2002). Typically in these studies cleaning effectiveness is related to pressure drop, permeate flux, permeate turbidity, and pH and salt passage measurements. These are indirect measures for biomass presence and removal, whereas direct quantitative analysis of the cleaning efficiency requires module disassembly and presents no dynamic measurement ability. Magnetic resonance imaging (MRI) has recently been applied to biofouling studies of RO membranes, identifying both biofilm accumulation within the module and the resultant effects on water flow (Graf von der Schulenburg *et al.*, 2008). MRI thus offers an in situ and non-destructive method to quantify biofilm directly.

The specific objectives of the work presented here were to (i) demonstrate MRI as a tool for direct studies of chemical cleaning of biofouled RO membranes, (ii) to use the resultant imaging data to extract out quantitative measures of cleaning effectiveness and its relationship to more conventional (and more easily employable) measurements, and (iii) to provide preliminary measurements of the effect of fouling extent and time on cleaning effectiveness using these established MRI protocols.

MATERIAL AND METHODS

Flow cells

A model 2D membrane flow cell, known as the Membrane Fouling Simulator (MFS), was developed by Vrouwenvelder *et al.* (2007b) to study biofouling of spiral-wound membrane systems (UF, NF and RO) on a laboratory scale. The MFS provides a convenient means to study biofouling at a spacer and membrane level on a planar surface. The MFSs have an internal cavity fitted with two membrane sheets separated by a membrane spacer. Water enters from one end of the MFS, flows across spacer between membranes, and exits at the far end of the cell. The MFSs are operated without permeate production for simplicity which has a negligible effect on observed biofouling (Vrouwenvelder *et al.*, 2009a). Two sizes of MFSs were used for the experiments; the dimensions of both are shown in Figure 8.2. The larger flow cell is referred to as the MFS and the smaller one as the S-MFS (Spacer Membrane Fouling Simulator). The different cell sizes permitted analysis of different regions. The S-MFS fits entirely within the MRI radiofrequency coil (imaging region) which facilitates study of the entire system, this potentially will be useful in future for comparison with simulation efforts as in this case boundary conditions are well defined. The MFS has a greater surface area, allowing more area for biofilm growth, and providing greater variability within each sample, thus better representing the very large surface area contained within an industrial RO module. The MFS however is longer than the radiofrequency coil, and thus only a section could be imaged at a time.

Fouling

The RO (S)-MFSs were fouled by feeding a continuous stream of dechlorinated (unless otherwise stated) tap water (Cambridge, UK) supplemented with a nutrient solution through the cells for one or two weeks. The nutrient solution was composed of sodium acetate (CH_3COONa), sodium nitrate ($NaNO_3$), and sodium phosphate ($NaH_2PO_4 \cdot 2H_2O$) at concentrations of 1000 μg C/L, 200 μg N/L, and 100 μg P/L respectively. The total water flow through the cells during fouling was 16 L/h for the MFS and 12 L/h for the S-MFS. A differential pressure transmitter

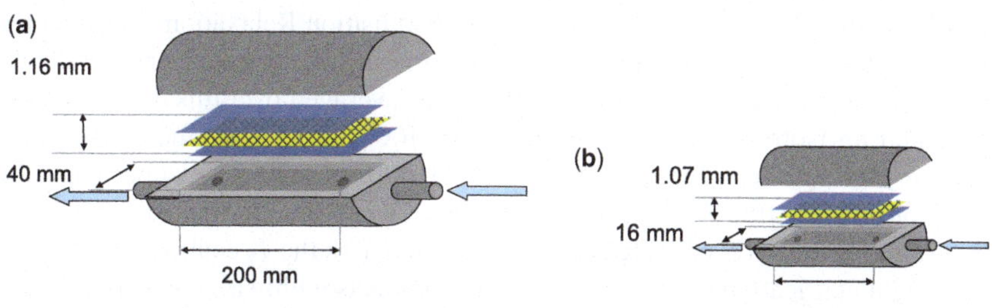

Figure 8.2 Schematics of the (a) MFS and (b) S-MFS cells used for the studies.

(Endress + Hauser, Deltabar S PMD70) was connected to either side of the MFS, allowing measurement of the pressure change across the MFS as a result of fouling (hereafter referred to as feed channel pressure drop). The MFSs were fouled outside of the MRI magnet.

Cleaning

As discussed above, different chemicals may be used to clean RO modules. In the current studies, a selection of cleaning agents was tested at different flow rates and temperatures. In all cases, the cleaning was performed while the cells were placed inside the magnet for imaging. Several different cleaning experiments were performed under different conditions:

Set 1. The first set of experiments focused on the MFS. These experiments examined a greater surface area and were aimed at developing a better idea of how cleaning progresses within the cell and also evaluated different cleaning solutions. In these cases, the cleaning solutions were heated to 45°C and pumped through the RO cell for 90 minutes at a flow rate of 100 mL/min. The cleaning solutions used were (i) NaOH at pH 12, (ii) 10 mM SDS, and (iii) 10 mM SDS and NaOH at pH12.

Set 2. The second set of experiments examined two S-MFS cells and the effect of extent of fouling. One cell was fouled for 7 days, and the other for 14 days. Both were cleaned using NaOH at pH 12 heated to 45°C at 100 mL/min for 90 minutes.

MRI

The MRI measurements were made using a Bruker DMX 200 spectrometer, with a vertical bore and a 4.7 T superconducting magnet. ^{1}H radiofrequency coils of diameter 64 mm and 38 mm for the MFS and S-MFS respectively were used. ^{1}H signal was detected predominately from the water content of the (S) MFS. Two types of imaging measurements were used, 2D structural images and 2D velocity images.

The 2D structural images were used to identify biofilm distribution within the (S)MFS cells. This was done using a Rapid Acquisition Relaxation Enhancement (RARE) pulse sequence (Hennig *et al.*, 1986). Images were orientated perpendicular to the narrowest dimension of the (S)MFS cells, averaged over this dimension and presented an isotropic in-plane resolution of 100 μm/pixel for the S-MFS cells and non-isotropic in-plane resolution of 200 μm/pixel (along the superficial flow direction) by 100 μm/pixel (along the direction transverse to the superficial flow direction) for the MFS cells. RARE imaging increases the T_2 contrast of a sample, allowing identification of different local environments within a sample, in this case biofilm containing regions and surrounding free water.

The 2D velocity images were used to capture the flow field within the (S)MFS cells, producing a 2D structural image of pixel-averaged velocities along the superficial flow direction. The observation time (required to encode the signal for velocity) was 10 ms. Velocity images were acquired using a phase shift velocity encoding pulse sequence (Callaghan & Xia, 1991). Again, images were averaged over the narrowest dimension of the (S)MFS cells, with an isotropic in-plane resolution of 200 μm/pixel. Flow through the cells was reduced to between 2 and 5 mL/min for the velocity imaging experiments. The superficial flow velocity (i.e. the average velocity through the cell in the flow direction in the absence of fouling) used is noted in the relevant text.

RESULTS AND DISCUSSIONS

Fouling and image analysis

Characterisation of an unfouled and fouled S-MFS cell are presented here to familiarise the reader with the MRI and accompanying techniques used to identify biofilm. The structural images for an unfouled and fouled cell are shown in Figure 8.3a. Looking first at the unfouled cell (i) the bright regions are those of strongest (water) signal intensity. The dark grid pattern is the solid spacer sheet. Contrasted to this, image (ii) of the fouled cell contains several dark regions not visible in the unfouled cell (i) which can be attributed to biofilm growth (highlighted by the red ovals). The contrast between water trapped in the biofilm and the surrounding water is the result of the shorter T_2 NMR relaxation parameter value of the trapped water. Velocity images for the unfouled and fouled cell are shown in Figure 8.3b, acquired with a superficial flow velocity 6 mm/s. The unfouled image (i) exhibits a relatively homogeneous flow distribution, influenced by the presence of the membrane spacer sheet. The fouled cell (ii) however shows significant flow disruption as a result of the biofilm growth, with narrow flow channels formed and distinct large regions of high and low (or zero) flow.

RO modules operate in and require a cross-flow regime and hence the greater membrane surface accessible to flow, the greater the potential for permeate production. The effective surface area (ESA) of the module or cell is thus defined here as the % of the total membrane surface area with significant flow across it (and thus reduced concentration polarisation and hence significant local permeate production). For ESA values calculated based on structural image analysis (as in Figure 8.3a), the ESA is referred to as ESA_{img}; likewise for ESA values calculated based on velocity image analysis (as in Figure 8.3b), the ESA is referred to as ESA_{vel}. Calculation of the ESA in both cases requires selection of a gating or critical value to differentiate what we term permeate active (which contributes to ESA) and permeate inactive (which does not contribute to ESA) surface area. With respect to ESA_{img} selection of the gating value is based on image intensity that differentiates biofilm or spacer containing regions from free space. With respect to ESA_{vel} the gating level is selected based on the magnitude of flow velocity, with zero or

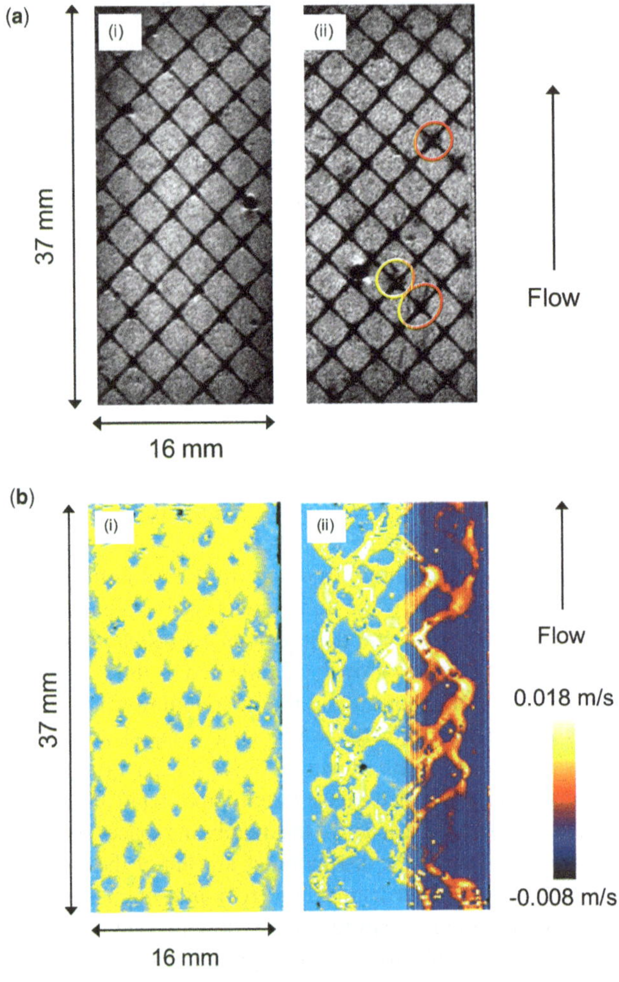

Figure 8.3 (a) MRI structural images of (i) an unfouled and (ii) a fouled S-MFS. (b) MRI velocity images of (i) an unfouled and (ii) a fouled S-MFS.

virtually zero velocity regions being assigned to be inactive and thus not contribute to ESA_{vel}. Selection of the gating level is partially subjective as the image intensities or velocity values form a continuum, however the level chosen is consistently applied to all images for comparative purposes. Figure 8.4a shows the resultant ESA_{img} images for the unfouled and fouled S-MFS respectively (extracted from Figure 8.3a). White indicates regions that contribute to ESA_{img}. An ESA_{img} of 100% would represent a membrane surface area covered entirely in free water, whereas an ESA_{img} of 0% represents a membrane surface area covered entirely with spacer

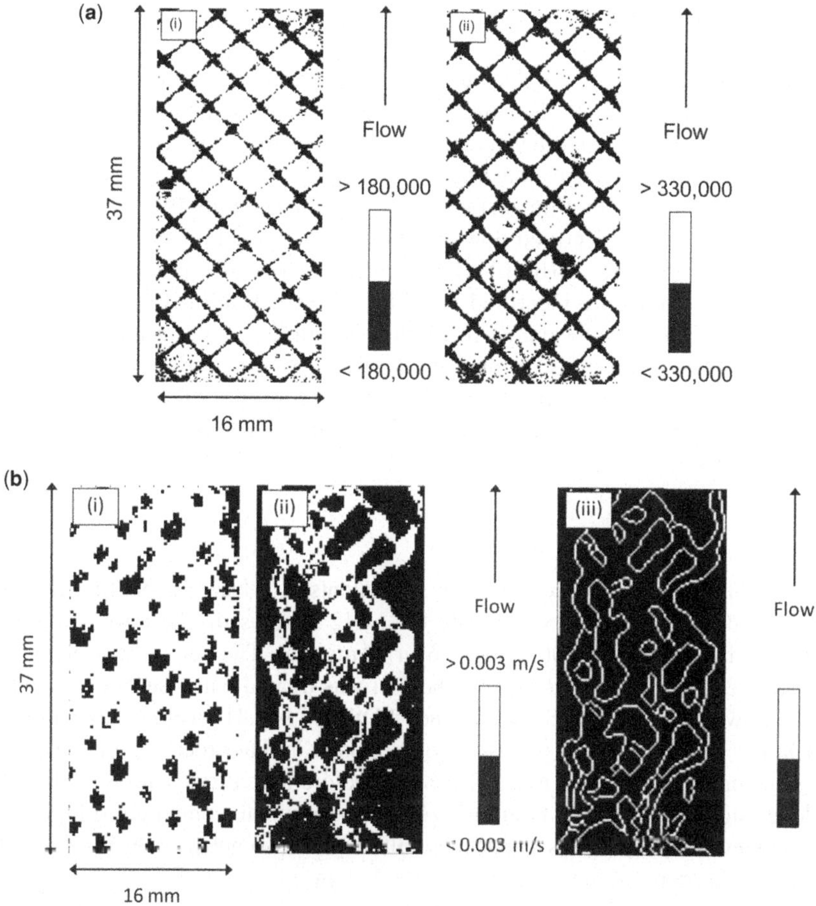

Figure 8.4 (a) Gated structural images of (i) an unfouled and (ii) a fouled S-MFS. (b) Gated velocity images of (i) an unfouled and (ii) a fouled S-MFS. White shows the regions that contribute to the respective ESA. Edge identification for ABS analysis from (b, ii) is shown in (b, iii).

sheets and biofilm. In Figure 8.4a, ESA_{img} is seen to reduce from 74% to 70% upon fouling. Figure 8.4b shows the corresponding images for ESA_{vel} for the unfouled and fouled S-MFS respectively. In this case an ESA_{vel} of 100% indicates flow across the entire membrane surface, whereas an ESA_{vel} of 0% indicates no flow across the membrane. The ESA_{vel} for the unfouled cell is 81%, and reduces to 35% for the fouled cell. The spacer sheet is largely responsible for the zero-flow regions in the unfouled cell. However there are significant regions of stagnation upon fouling that cause a significant reduction in ESA_{vel}, which is clearly more sensitive to biofilm growth and more accurately reflects the potential loss of permeate production from stagnant regions.

This quantitatively highlights our previous qualitative observation (Graf von der Schulenburg *et al.*, 2008) that limited biofilm growth (predominately attached to the spacer sheet) has a significant effect on system hydrodynamics resulting in large stagnant and unproductive areas/regions.

Furthermore, the biofilm surfaces accessible to the bulk fluid can be defined by the edges of the identified flow channels in the fouled images. This is shown graphically in Figure 8.4(b, iii) where the edges from the gated velocity image in Figure 8.4(b, ii) are identified in white. The fraction of pixels attributed to surfaces divided by the image resolution is defined as the accessible biofilm surface (ABS), with units of inverse length and is essentially a 2D calculation of a surface to volume ratio. Hence, a higher ABS represents a greater biofilm surface area in contact with bulk cleaning solution. The ABS of Figure 8.4(b, iii) is 0.567 mm^{-1}.

Different cleaning solutions

For the multiple cleaning solution tests, treatments of (i) NaOH at pH 12, (ii) 10 mM SDS, and (iii) 10 mM SDS and NaOH at pH12 were used. These were applied at 45°C with a flow rate of 100 mL/min over 90 minutes. The treatments were applied to two very heavily fouled MFSs. The first was treated with NaOH only, and the second was treated with SDS, and then SDS and NaOH together (after minimal cleaning was revealed by the use of SDS alone), in that order. Figure 8.5a shows a structural image of the fouled MFS cell before and after cleaning with NaOH alone. Significantly more biofilm is seen to accumulate compared to Figure 8.3. Following cleaning there is a noticeable, although not complete, reduction in biofilm content. Appropriate image analysis revealed an increase in ESA_{img} from 25% to 44% after cleaning. Figure 8.5b shows the structural images acquired for the MFS cell fouled and subsequently treated with the series of treatments (SDS, SDS and NaOH). Contrary to the first cell and despite the more aggressive cleaning regime, this second cell did not experience a significant degree of cleaning; visually, very little biofilm was removed. Appropriate image analysis showed negligible changes in ESA_{img} from a value of 12%.

The corresponding velocity images for the cell are displayed in Figure 8.6a and 8.6b respectively, with a superficial flow velocity of 0.8 mm/s. The first MFS cell (Figure 8.6a), cleaned using NaOH, presents a significant recovery of flow in previously stagnant regions following cleaning.

Quantitatively, this equates to a change in ESA_{vel} from 3% to 23% after cleaning, again this parameter proved to be more sensitive to biofilm accumulation. The second MFS cell (Figure 8.6b) however did not show a significant recovery of flow channels as a result of cleaning. The ESA_{vel} did not change significantly, ranging from 7% when fouled to 9% after cleaning.

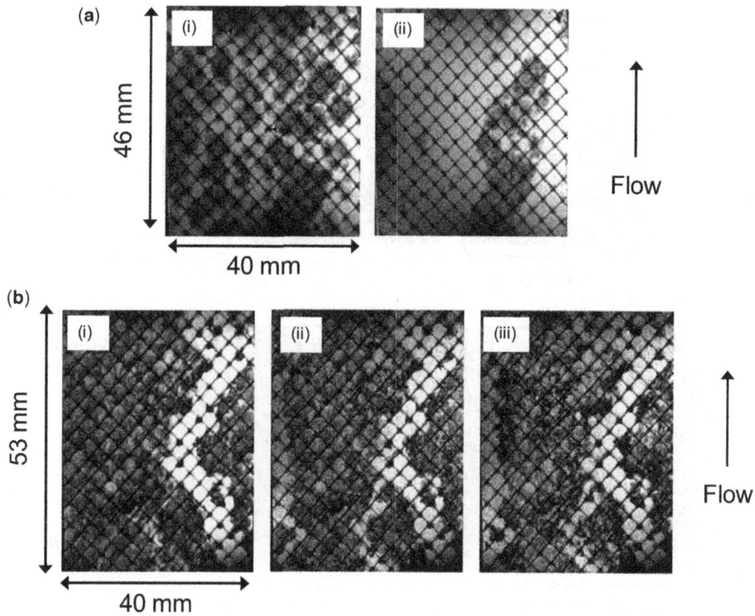

Figure 8.5 (a) Structural images of MFS (i) fouled and (ii) cleaned using NaOH. (b) Structural images of MFS (i) fouled and cleaned using (ii) SDS, and (iii) SDS + NaOH.

The feed channel pressure drop was recorded during the cleaning experiments and is presented in Figure 8.6c. The pressure data for both MFSs display a sharp decrease as a result of cleaning, and suggest significant cleaning progress. However, as the MRI data showed, this is not the case. Clearly a strong drop in pressure drop should not be too easily interpreted as an effective cleaning of the cell. The significant cleaning efficiency differences between the two MFS cells may be due in part to the different cleaning solutions applied; however, considering that NaOH was used on both cells, this is unlikely to be the cause. What is significant in the initial level of fouling with the second cell presenting significantly more fouling initially. The ABS values for the fouled cells are 0.433 mm^{-1} for the first cell and 0.305 mm^{-1} for the second cell. Hence, in addition to having a greater amount of initial biofilm, the second cell also presented less contact area between the biofilm and the bulk cleaning solution.

This possibly indicates that there is a critical extent of fouling beyond which the chemical treatments employed here are ineffective. This will be explored further in future. None of the chemical cleaning solutions tested were however able to clean the RO cells completely from biofilm, despite marked decreases in feed channel cell pressure across the cells.

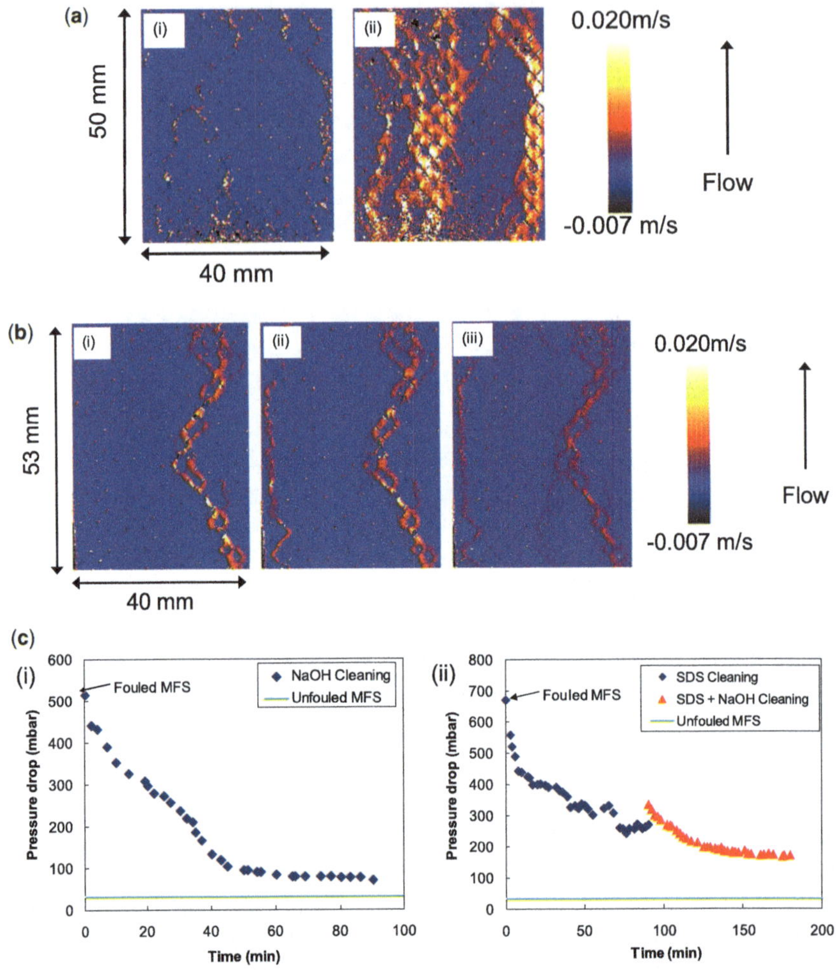

Figure 8.6 (a) Velocity images of MFS (i) fouled and (ii) cleaned using NaOH. (b) Velocity images of MFS (i) fouled and cleaned using (ii) SDS, and (iii) SDS + NaOH. (c) Pressure drop measurements recorded during cleaning of the MFS with (i) NaOH and (ii) a series of cleaning solutions.

Cleaning mechanism

The structural images presented in Figures 8.3 and 8.5 were acquired under static conditions. We now explore the nature of these images under cleaning and hence flowing conditions to explore what additional insights can be provided and to quantify the development of cleaning over the 90 minute application window. In this context, two different chemical cleaning solutions were investigated to determine

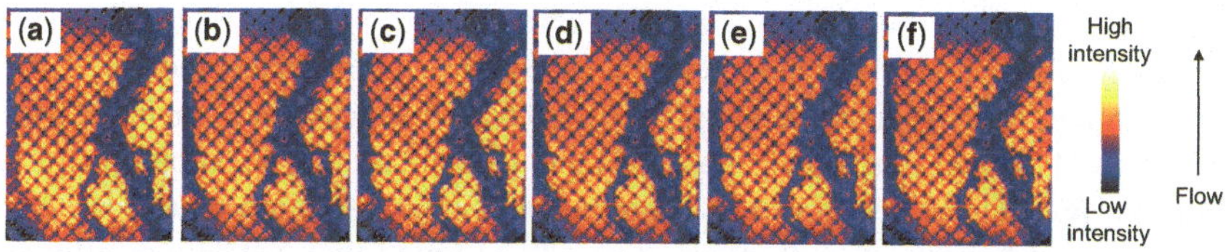

Figure 8.7 Time series of RARE structural images of the MFS cleaned using SDS taken under flow at times (a) 15 min, (b) 30 min, (c) 45 min, (d) 60 min, (e) 75 min, and (f) 90 min. Signal is detected predominately from stagnant or slow flowing water.

if the difference in the cleaning mechanism is visible as applied to a MFS cells. The cell is the same as was used above in Section 3.2, whereby first, a solution of SDS alone is applied for 90 minutes, followed by a solution of SDS and NaOH for 90 minutes at 100 mL/min and 45°C. In this case rapid RARE imaging was performed with a total acquisition time of 4 minutes per image. However as signal is acquired under rapid flow (100 ml/min) contrast in the images is different from that above. Regions of the cell experiencing high velocity will present reduced signal intensity due to both cell flush out and enhanced signal relaxation due to flow through imposed and inherent magnetic field gradients. Conversely stagnant regions will present comparatively high signal intensities.

The series of RARE images taken for cleaning with SDS alone are shown in Figure 8.7, acquired with a superficial flow velocity of at a flow of 42 mm/s. The regions of high intensity are regions of stagnant water, trapped predominately within the biofilm structure. The regions of low intensity (dark regions) are the flowing water along the narrow flow channels. There is little visible change in the images over the 90 minutes of observation, with only moderate broadening of the single flow channel along the left-hand side. Next, the series of RARE images taken for cleaning with SDS in combination with NaOH are shown in Figure 8.8, also acquired at superficial flow velocity of 42 mm/s. In this case, there is a visible change in image intensity over time. This clearly indicates penetration of the

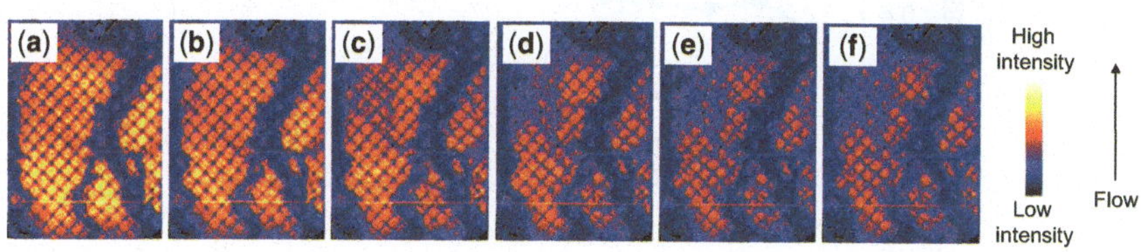

Figure 8.8 Time series of RARE structural images of the MFS cleaned using SDS + NaOH taken under flow at times (a) 15 min, (b) 30 min, (c) 45 min, (d) 60 min, (e) 75 min, and (f) 90 min. Signal is detected predominately from stagnant or slow flowing water.

cleaning fluid into the biofilm regions. There is clearly a reduction in the local T_2 NMR signal relaxation parameter. This could originate from increased flow rate through these regions (although Figure 8.6(bii)) suggests not) or more likely a structural change in the biofilm such as the accumulation of local paramagnetic ions. At present the origin of this reproducible signal reduction is unclear. Note however that Simões *et al.* (Simões *et al.*, 2005) found that NaOH was effective at promoting biofilm removal whereas SDS was not, broadly consistent with these results. Figure 8.9a shows how the total signal intensity varies across the 12 images presented in Figures 8.7 and 8.8. The change in signal intensity during cleaning with SDS in combination with NaOH is relatively linear, qualitatively suggesting a constant cleaning rate. Relevant to the rate of cleaning is the change in ABS. This was calculated for the data in Figures 8.7 and 8.8 and is presented in Figure 8.9b as

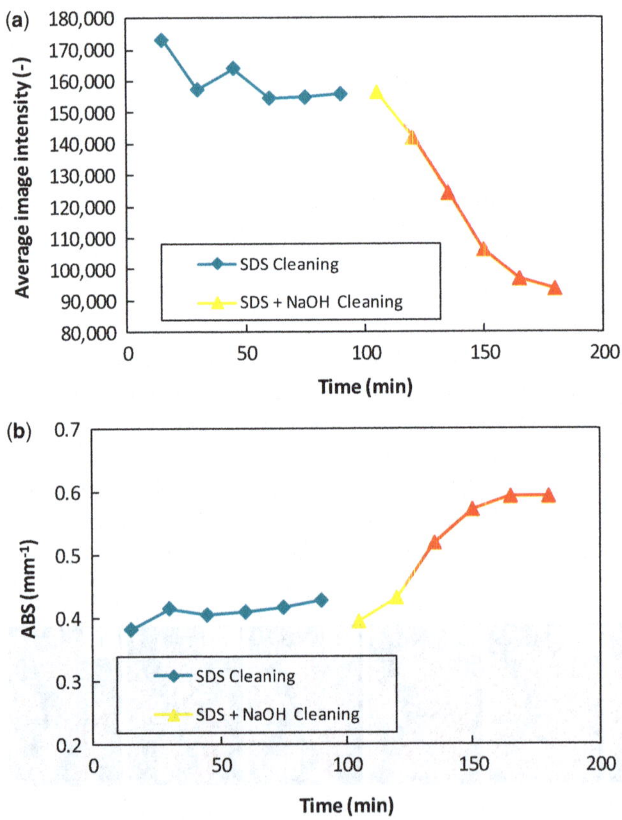

Figure 8.9 (a) Total signal intensity as a function of time for the series of images in Figures 8.7 and 8.8. (b) Change in biofilm perimeter as a function of time, as extracted from the images in Figures 8.7 and 8.8.

a function of time. For the case of the SDS + NaOH solution, there is a noticeable increase in this parameter with time.

Effect of fouling time

In order to study the effect fouling time has on cleaning efficiency, two S-MFS cells were studied: one was fouled for one week and one was fouled for two weeks. They were fouled in parallel to minimise the influence of seasonal and other variations. Both were cleaned with NaOH at pH 12 at 45°C for 90 minutes at 100 mL/min. Figure 8.10a displays the structural images for the S-MFS cells fouled for (i) one week and (ii) two weeks, and subsequently cleaned. Examining the images qualitatively, there

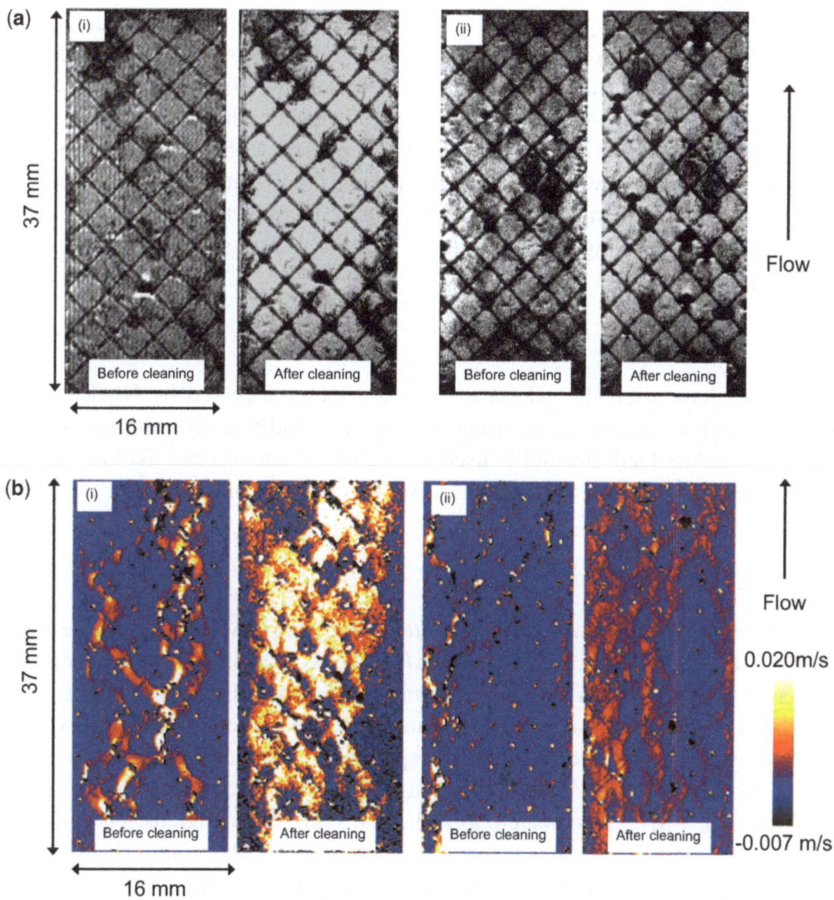

Figure 8.10 (a) Structural images and (b) velocity images for the S-MFS cells fouled for (i) one week and (ii) two weeks, then cleaned using NaOH.

is evidence of greater initial biofilm deposit on the cell fouled for longer, which is expected. In both cells, the structural images after cleaning show some but not complete reduction in biofilm accumulation. For the S-MFS cell fouled for one week, the ESA_{img} increased from 44% to 54% after cleaning and for the S-MFS cell fouled for two weeks, it increased from 35% to 55% after cleaning. The velocity images for the RO cells are shown in Figure 8.10b, acquired at a superficial flow velocity of 2 mm/s. There are fewer flowing channels in the cell fouled for two weeks than that fouled for one week, as visible in the velocity images before cleaning. More so than the structural images, the velocity images after cleaning show broadened flow channels and increased effective membrane surface area.

For the cell fouled for one week, ESA_{vel} increased from 12% to 53%, whereas for the cell fouled for two weeks, ESA_{vel} increased only from 5% to 12%. Thus it appears that the cleaning was significantly more successful for the system fouled for one week by these criteria. The results of this study thus demonstrated quantitatively that a longer fouling time resulted in reduced cleaning efficiency of the RO cell. This observation is linked with the maturity and extent of the biofilm growth within the module or cell limiting cleaning agent access. The ABS values for the S-MFSs fouled for one and two weeks are 0.457 mm^{-1} and 0.340 mm^{-1} respectively, thus the more mature biofilm has a reduced surface area in contact with the bulk cleaning solution and supporting the theory that limited cleaning agent contact reduces cleaning efficiency.

Link to feed channel pressure drop

Above we have used the effective surface area as accessed by NMR imaging to evaluate the success of cleaning operations. Traditionally the pressure drop change over the feed channel is used to evaluate such success. To link the ESA measurements with pressure drop data, ΔP, we can use Darcy's law to calculate the fouling cell permeability, κ:

$$\kappa = Q\mu L / A\Delta P \tag{8.1}$$

Where Q is the volumetric flowrate, μ is the dynamic viscosity, L is the length, A is the total cross-sectional area, and ΔP is the feed channel pressure drop. Thus different cell dimensions are compared on an equal basis. ESA effectively accounts for the change in cross-sectional area integrated along the flow channel axis, also effectively integrating over the channel depth. We collated our data from the various experiments and plot ESA_{vel} and ESA_{img} against κ in Figure 8.11. The actual fouling cell, MFS or S-MFS is indicated, as is the cell state – unfouled, fouled or cleaned. What is immediately obvious is that the cleaning protocols employed only effect a modest improvement in the measurements ESA_{img}, ESA_{vel}, and κ. There is a coarse linear relationship between ESA_{vel} and κ (with significant scatter) but not between ESA_{img} and κ. Linear regression coefficients of determination, R^2, for the two sets of data support this analysis and are 0.83 and 0.75 respectively.

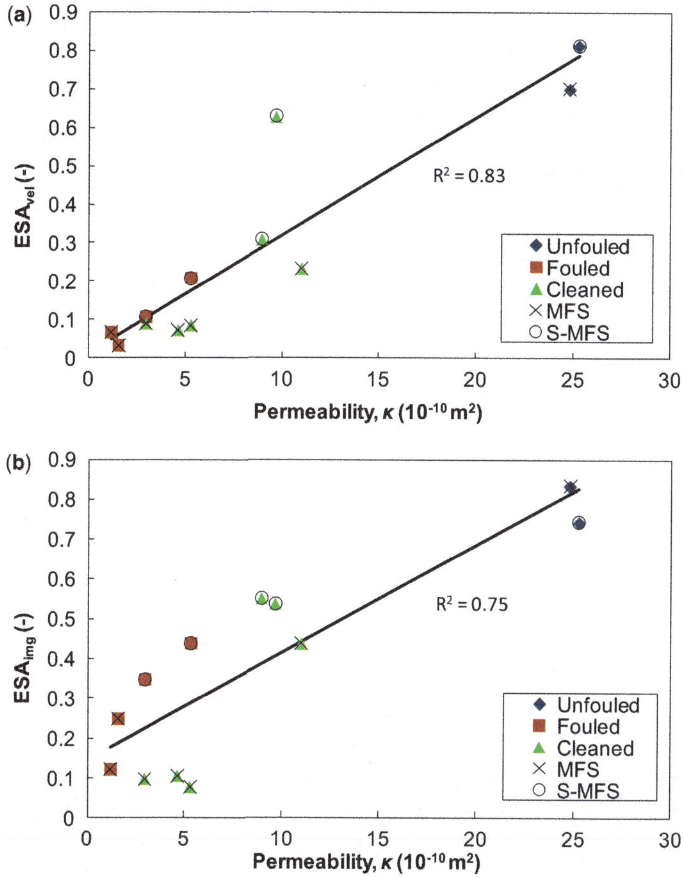

Figure 8.11 Comparison of (a) ESA_{vel} and (b) ESA_{img} and permeability for (S)MFS cells.

CONCLUSIONS

We have successfully employed MRI to visualise and quantify the effectiveness of various membrane cleaning operations on the effective membrane surface areas experiencing cross-flow. The MRI structural and velocity images showed that chemical cleaning acts to broaden existing flow channels by 'attacking' biofilm in contact with the cleaning solution. However of the multiple cleaning strategies employed, none were able to remove completely the biofilm present with effectiveness decreasing as the fouling was either more extensive or had aged more. We were able to show a coarse correlation between feed channel pressure drop and effective surface area determined from velocity images, ESA_{vel}. Significant deviations from this linear dependency were however also observed.

Preliminary results were presented and substantial scope remains for additional cleaning studies of RO membranes employing these MRI techniques both in the 2D MFS analogues used in this work and in actual spiral wound membrane systems. In the future we will also explore the relationship between our definitions of effective surface area and local permeate flux – this will be measured by employing velocity maps through the membrane.

Chapter 9

Early non-destructive biofouling detection in spiral wound RO membranes using a mobile earth's field NMR

INTRODUCTION

Highly populated areas and arid zones around the world suffer from vulnerability to freshwater availability. It has been projected (in 2013) by the Food and Agricultural Organization of the United Nations that by 2025 there will be 1.8 billion people living in countries or regions with absolute water scarcity, and two-thirds of the world population could be under water stress conditions. Traditional water sources in the future will need to be supplemented in order to meet the projected increase in water consumption. Reverse osmosis (RO) membrane technology has developed over the past 40 years to account for 80% of the total desalination plants installed worldwide (Greenlee *et al.*, 2009) with strong growth expected into the future to meet further demand (Lawler *et al.*, 2012). Although many advances have been made, membrane fouling continues to be a major concern for the desalination industry (e.g. Shannon *et al.*, 2008; Tasaka *et al.*, 1994; Vrijenhoek *et al.*, 2001; Vrouwenvelder *et al.*, 2008b) as it decreases membrane performance, requires frequent chemical cleaning and shortens the membrane lifespan. It poses a major economic burden on plant operation, estimated at 25–50% of total plant operating costs (Vrouwenvelder *et al.*, 2011a).

Of the four main types of fouling (Vrouwenvelder *et al.*, 2011a): Inorganic, organic, particulate and microbiological, only microbiological (or biofouling) can arguably be significantly reduced by pre-treatment alone (Al-Juboori & Yusaf, 2012; Goosen *et al.*, 2004). Biofouling occurs extensively, even with significant pre-treatment (Flemming *et al.*, 1997). This is attributed to the formation of biofilms on membrane and feed spacer sheet surfaces within the spiral wound membrane modules, which once established are capable of re-emergence even after intensive

Author(s): Szilárd Bucs, Joop Kruithof, Mark van Loosdrecht and Johannes Vrouwenvelder
doi: 10.2166/9781780409597_0161

chemical treatment (Costerton *et al.*, 1995). Biofilms are an accumulation of microorganisms and extracellular material which forms a matrix which is irreversibly bonded onto the membrane and/or feed spacer surface, with the main mechanism for cohesion and adhesion of the biofilm matrix due to the extracellular polymeric substances (EPS), which makes up 50–80% of the organic matter (Costerton *et al.*, 1995; Guo *et al.*, 2012). Biofilms go through three stages of growth: the induction phase, growth phase and plateau phase; it is crucial to detect and stop biofilm growth during the adhesion-dominated induction phase to effectively prevent biofouling (Matin *et al.*, 2011). During the growth phase the biofilm increases predominantly by bacterial cell multiplication with this phase being typically associated with biofouling (or the threshold of interference) where up to 25% reduction in permeate flux has been observed due to hydraulic resistance from the EPS matrix and concentration polarisation across the membrane surface (Guo *et al.*, 2012). The type of flow regime within the RO membrane affects the stability, adhesive strength, density and structure of the biofilm (Vrouwenvelder *et al.*, 2011b), with "fluffy" biofilm growth occurring at low Reynolds numbers (Re) while high Re induces the formation of thin, compact, harder to remove biofilms (Vrouwenvelder *et al.*, 2011a). The biofilm growth inhibits convective transport (Pintelon *et al.*, 2012), increasing flow resistance which causes a back-bone (or telescoping) flow regime to emerge (Creber *et al.*, 2010b; Flemming, 1997; Manz *et al.*, 2005).

As membrane biofouling causes reduced separation efficiency, permeate contamination, membrane deterioration and increased energy costs it is a common reason for early disposal of RO membrane modules (Matin *et al.*, 2011). Vrouwenvelder *et al.* (Vrouwenvelder *et al.*, 2010c, 2011a) estimate that a warning detection system sufficiently sensitive to detect early stages of biofouling (i.e. before the biofilm becomes established within the membrane) could reduce energy costs by up to 5%, and decrease the annual membrane replacement and cleaning costs by up to 10%. Vrouwenvelder *et al.* (Vrouwenvelder *et al.*, 2009b, 2010b) showed that biofilm accumulation in RO membrane elements increased the feed side pressure drop between the inlet and the salty discharge outlet of the module (hereafter referred to as the feed-channel pressure drop). According to Flemming *et al.* (Flemming, 1997) commonly used measurement techniques such as monitoring permeate flux decrease, colony count and membrane autopsy are not sensitive to detect early stages of biofouling or do not provide in-situ non-invasive quantitative information about the level of fouling through membrane module or pressure vessel containing membrane modules. Direct observation technique (Marselina *et al.*, 2009) and impedance spectroscopy (Chilcott *et al.*, 2013) have recently been used to characterise fouling. A membrane fouling simulator (MFS) can also be operated in parallel to the actual RO membrane system (Vrouwenvelder *et al.*, 2006); previous research (Vrouwenvelder *et al.*, 2011a) has found that MFS closely follow the biofouling effects inside the RO membrane modules.

Here we consider the use of a low cost, portable non-invasive spectroscopic technique, namely nuclear magnetic resonance (NMR), to sensitively and non-destructively detect the early stages of biofouling in spiral wound membrane modules. Specifically we consider the use of Earth's field NMR (EF NMR) which offers a comparatively cheap, mobile and non-invasive measurement. We explore the sensitivity of various readily accessible EF NMR parameters (signal relaxation values T_1 and T_2 as well as suitably modified total NMR signal) to the extent of biofouling. We also consider the ability for such measurements to be performed under operational conditions (i.e. with flow through the module) as well as the sensitivity of the measurements to the relative location of the EF NMR apparatus along the spiral wound RO membrane module length.

BACKGROUND

Relevant nuclear magnetic resonance (NMR) theory

Nuclear magnetic resonance (NMR) is a well-known phenomenon in which the magnetic moment of NMR active nuclei (e.g. ^{1}H) can be analysed using electromagnetic (radio frequency (r.f.)) pulses and applied magnetic fields. Information on the basic principles of NMR can be found elsewhere while useful sources for understanding NMR under flow conditions can be found in Callaghan's texts on magnetic resonance microscopy (Callaghan & Xia, 1991) and transport measurements (Callaghan & Xia, 1991). In this section we will briefly describe some basic features of NMR, relevant to our measurements.

When a substance containing NMR active nuclei (exclusively ^{1}H in the work conducted here) is placed in a static magnetic field ($\boldsymbol{B}_0$) the bulk magnetization ($\boldsymbol{M}$) of the substance varies with time (Pendlebury *et al.*, 1979) as follows:

$$M_Z = M_{Z,0}(1 - e^{-t/T_1}) \tag{9.1}$$

where M_Z is the magnitude of the magnetization along the direction (Z) of the net magnetic field vector, $M_{Z,0}$ is the magnetization at time zero and t is time. The magnetization ($\boldsymbol{M}$) is proportional to the ^{1}H density in the sample and the static magnetic field strength as follows: $\boldsymbol{M}_0 \propto \boldsymbol{B}_0^{7/4}$ (Fukushima, 1999). The spin-lattice relaxation time (T_1) characterises the time it takes the system to reach thermal equilibrium, i.e. for the net magnetization vector (M_Z) to reach $M_{Z,0}$ after the sample is placed within magnetic field $\boldsymbol{B}_0$. The NMR signal is also characterised by the spin-spin relaxation time (T_2) which characterises the loss of signal phase coherence, i.e. the time it takes the NMR signal in the transverse plane (X-Y) to decay towards zero, following radio frequency (r.f.) excitation and can be described by,

$$M_{x,y} = M_{x,y,0}e^{-t/T_2} \tag{9.2}$$

where $M_{x,y}$ is the magnitude of the magnetization along the transverse plane (X-Y) relative to the longitudinal direction (Z) of the net magnetic field vector, $M_{x,y,0}$ is the magnetization at time zero. T_1 and T_2 are material properties and can be used to distinguish different phases, e.g. biofilm matrix and bulk fluid in our case. It has been observed by multiple researchers (e.g. Beuling *et al.*, 1998; Codd *et al.*, 2011; Renslow *et al.*, 2010; Vogt *et al.*, 2013) that the T_1 and T_2 relaxation times for NMR signal from water in a biofilm matrix are up to an order of magnitude lower than that of the bulk water phase.

Measured values of T_1 and T_2 are also however (erroneously in the intended application here) influenced by flow, predominately flow into and out of the measurement volume (defined as the regions where r.f. excitation occurs) during the acquisition sequence. This can be approximated as follows (Mueller *et al.*, 1986):

$$\frac{1}{T_{1\text{f or }2\text{f}}} \approx \frac{1}{T_{1\text{S or }2\text{S}}} + \frac{U}{l_d} \tag{9.3}$$

where $T_{(1\text{f or }2\text{f})}$ is the relevant relaxation time of flowing water, $T_{(1\text{S or }2\text{S})}$ is the relaxation time of stationary water, U is the fluid velocity and l_d is the length of the excited region (approximately the length of the r.f. coil employed for NMR signal excitation). The effect summarised by equation 9.3 is obviously complicated when the sample being measured contains a distribution of flow rates or velocities.

NMR and biofilms

The use of NMR to detect biofilms has a long history. Lewandowski *et al.* (1992) showed that the T_1 relaxation time of flowing water in a bioreactor will decrease with the presence of biofilm. Hoskins *et al.* (1999) measured the T_1 and T_2 relaxation times of water in a packed bed reactor and observed T_1 and T_2 relaxation times decreasing in the presence of biofilms. Seymour *et al.* (2004) measured the effect of biofilm growth in a porous medium and observed anomalous transport dynamics using NMR displacement propagator experiments. Graf von der Schulenburg *et al.* (Graf von der Schulenburg *et al.*, 2008) used NMR imaging and velocity imaging techniques to demonstrate that it is a suitable technique for detecting stagnant zones in a reverse osmosis membrane module caused by biofouling. Creber *et al.* (Creber *et al.*, 2010b) used NMR to study biofouling in a MFS and observed a decrease in T_2 relaxation time with increased biofilm growth. Vogt *et al.* (2013) used NMR correlation techniques to observe the effects of changing T_2 time and changing flow field. These previous NMR studies of biofouling have demonstrated some of the detection capabilities of NMR; however these NMR tests have typically been made using relatively small samples and large magnetic fields (generated using superconducting magnets) on the order of a few Tesla (Blümich *et al.*, 2009).

In terms of the application of comparatively low magnetic field NMR measurements (employing permanent magnets) to the study of membranes, biofilms and biofouling, Sanderlin *et al.* (2013) recently used a low-field (^{1}H resonance frequency of 280 kHz) NMR well-logging instrument in a laboratory setting to show a reduction in T_2 caused by biofouling. Yang *et al.* (2014) demonstrated the usefulness of a comparatively low-field bench top NMR (^{1}H resonance frequency of 12.9 MHz) apparatus in the investigation of hollow fibre membrane modules as employed in membrane distillation processes in desalination. Here we consider the use of the Earth's magnetic field, which dispenses with the need for a strong homogeneous magnet for signal detection; this is significantly cheaper and enables a portable comparatively easy to handle system.

MATERIALS AND METHODOLOGY

NMR methods employed

Measurements were performed using a Terranova-MRI (Magnetic Resonance Imaging) Earth's field NMR instrument purchased from Magritek, New Zealand. A photo of this instrument is shown in Figure 9.1a. This instrument enhances the NMR signal using a coarse electromagnet ($B_{polarising}$: 18.8 mT) as a pre-polariser, and then detects at the local earth's magnetic field (52 μT with $B_{polarising}$ turned off) which corresponds to a ^{1}H (Larmor) resonance frequency of 2240 Hz which is detected by a solenoid r.f. coil of an effective internal diameter (i.d.) of 76 mm. Using the pre-polarisation coil, it is possible to substantially increase the signal to noise ratio (SNR) (Béné, 1980) of the detected NMR signal. This instrument has previously been employed to determine the mean droplet size of various water in oil emulsions (Fridjonsson *et al.*, 2012) as well as included into a novel flow meter (Fridjonsson *et al.*, 2014). The RO membrane module employed is readily incorporated into the EF NMR apparatus as shown in Figure 9.1b.

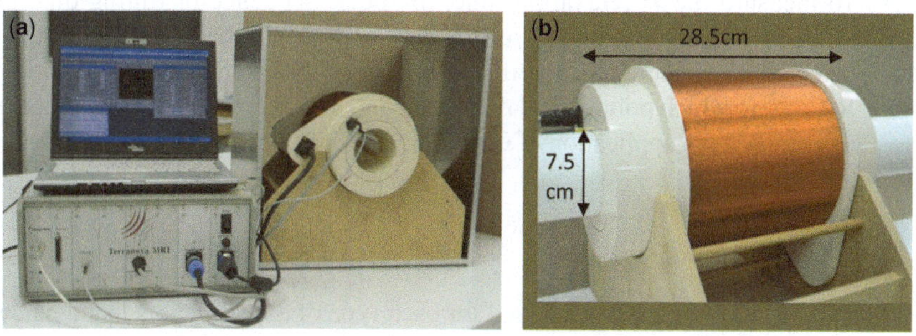

Figure 9.1 (a) Photograph of the Earth Field (EF) Nuclear Magnetic Resonance (NMR) instrument components (Laptop computer, Terranova MRI spectrometer and EF NMR probe) and (b) photograph of the EF NMR probe showing its dimensions; shown inserted into the probe is the spiral wound RO membrane module housing (pressure vessel).

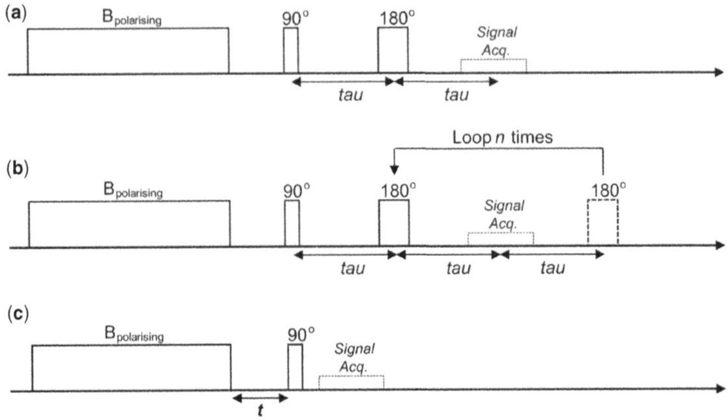

Figure 9.2 Timing diagrams for (a) spin echo, (b) T_2 measurements – Carr–Purcell–Meiboom–Gill (CPMG) (looped 180° pulses n times) and (c) T_1 measurements, where t is a variable wait time.

The following NMR measurements were performed:

- To measure the total NMR signal (S), a spin echo pulse sequence was employed, the timing diagram of which is shown in Figure 9.2a. The *tau* waiting time was increased such that the signal detected was only from the stationary fluid, which is discussed in more detail below. $B_{\text{polarising}}$ was applied for 4 s.
- To measure the T_2 relaxation time, a Carr–Purcell–Meiboom–Gill (CPMG) pulse sequence was employed – see Figure 9.2b for the relevant timing diagram. The data is fitted with equation 9.2. $B_{\text{polarising}}$ was applied for 4 s and a *tau* value of 100 ms was used.
- To measure the T_1 relaxation time, the pulse sequence or timing diagram schematically shown in Figure 9.2c was used. Here the NMR signal (S) is detected as a function of the time duration (t) following a 4 s application of $B_{\text{polarising}}$ at which point signal excitation and detection occurs. T_1 can then be determined via application of equation 9.4:

$$S = S_0 \exp\left(-\frac{t}{T_1}\right) \tag{9.4}$$

Specific EF NMR parameters used for each type of experiment is listed below:

(1) Spin Echo – Polarising current: 6A; polarising duration: 4 s; B1 gain: 2.5; 90 pulse duration: 1.1; 180 pulse duration: 2.3; echo time: 1000 ms; receiver gain: 2; number data points: 16384; 180-acq. delay (ms): 25; acquisition time: 0.5 s; repetition time: 10 s; number of scans: 4; display range: 50 Hz; phase cycling: CYCLOPS.

(2) Carr–Purcell–Meiboom–Gill (CPMG) – Polarising current: 6A; polarising duration: 4 s; B1 gain: 2.5; 90 pulse duration: 1.1; 180 pulse duration: 2.3; receiver gain: 2; number of echoes: 16; echo time: 200 ms; number data points: 16384; dwell time: 200 us; number of scans: 16; repetition time: 10 s; number of scans: 4; integration width: 20; display range: 100 Hz.
(3) T_1-Be – Polarising current: 6A; polarising duration: 4 s; B1 gain: 2.5; receiver gain: 2; 90 pulse duration: 1.1; pre-90 minimum delay: 0 ms; pre-90 delay step size: 200 ms; number of steps: 16; 90-acquisition delay: 25 ms; number of data points: 16384; acquisition time: 1 s; repetition time: 15 s; number of scans: 4; integration width: 50 Hz; display range: 50 Hz.

All NMR relaxation measurements were performed at superficial velocities (U) of 0, 0.64 and 1.28 cm/s, each experiment was run three times; spin echo experiments were conducted at U of 0, 0.64, 1.28, 1.92 and 2.56 cm/s, with each experiment conducted four times. Measurements were performed with the EF NMR r.f. coil and electromagnet (length = 17 cm) assembly positioned at the following locations along the RO membrane module axis (total length = 53 cm): (i) near the entrance (6 to 23 cm), in the middle (18 to 35 cm, as shown in Figure 9.1b) and near the end (30 to 47 cm) respectively. The central permeate channel of the RO membrane module was filled with water during all measurements.

RO membrane module and test conditions

Tests were conducted on a commercial 61 mm diameter spiral wound RO membrane module (purchased from Dow FILMTEC™ Model XLE-2521, FilmTec Corporation, USA). This was snuggly accommodated in a 75 mm outer diameter PVC pipe housing with the module inlet and outlet connected to the fouling circuit using the end caps indicated in the full fouling circuit shown in Figure 9.3. During fouling, the feed flow rate was 120 L/day (U = 0.07 cm/sec), which was dosed with 1 L/day of nutrient solution (i.e. 16 g sodium acetate, 3.2 g sodium nitrate and 1.6 g sodium phosphate in 1 L of DI water) to enhance biofouling. Sodium phosphate was dosed to avoid phosphate limitation (Vrouwenvelder *et al.*, 2010a). Nutrients were added to the fouling circuit using a peristaltic pump (Masterflex L/S, Cole-Parmer Instrument Company, Australia). Mains water monitored with a manual pressure gauge (~5 barg) was used during the experiments and an activated carbon filter (M Series Carbon Filter AQP-RFM1, Aquaport, Australia) was used to remove contaminants (particularly chlorine) which cause membrane degradation (Simon *et al.*, 2009). The feed-channel pressure drop between the module inlet and (usually concentrated solution) outlet was monitored using a high precision differential pressure gauge (Cerabar PMC131, Endress Hauser, Germany) (Vrouwenvelder *et al.*, 2009c). This fouling procedure was consistent with what we have employed in previous RO membrane module biofouling studies and has been generally applied in research on membrane biofouling (Creber *et al.*, 2010b, 2010a; Dreszer *et al.*, 2013;

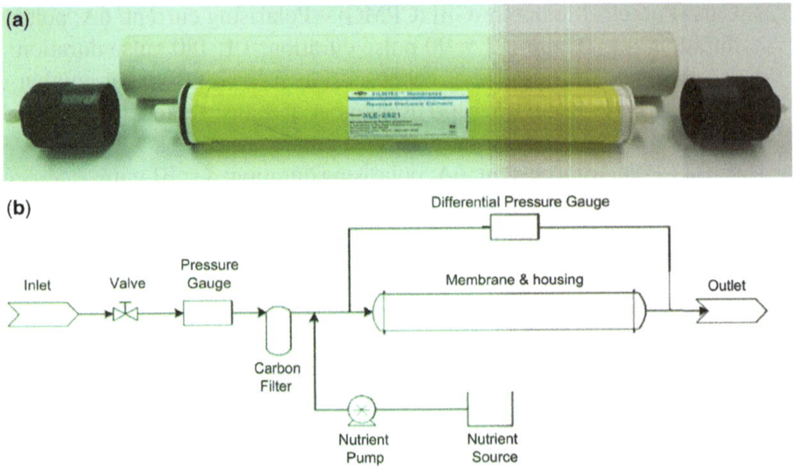

Figure 9.3 (a) RO membrane module, pressure vessel and end caps and (b) schematic of the experimental flow loop setup.

Graf von der Schulenburg *et al.*, 2008; Kappelhof *et al.*, 2003; Miller *et al.*, 2012; Vrouwenvelder *et al.*, 2009a).

Fouling was performed over a collective period of 72 days. The suite of NMR measurements discussed above were performed on the initial membrane prior to fouling, and subsequently at three time points during fouling, hereafter referred to as the three stages of biofouling. The extent of fouling at each successive stage was chosen to be more severe than the previous as indicated by the feed-channel pressure drop.

Figure 9.4 shows this pressure drop evolution with time over the 72 day period and the location of the three fouling stages where the NMR measurement suite

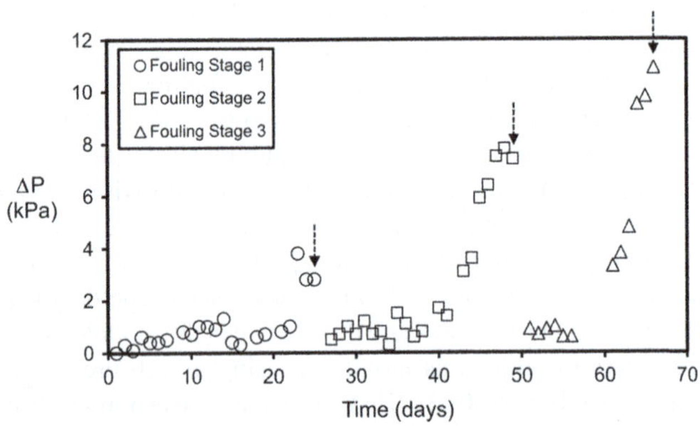

Figure 9.4 Feed channel pressure drop during fouling stages; EF NMR experiments are conducted at the end of each fouling stage (indicated by arrows).

was performed. An induction phase is evident in all stages. Following stages 1 and 2, and on completion of the NMR measurements, the membrane was flushed at a high value of U in excess of 10 cms^{-1} until the feed-channel pressure drop was restored to its original value, as is evident in Figure 9.4. However the reduction in induction time between successive stages suggests that some residual biofouling remained. Supplementary X-ray CT images of the final fouled module were also acquired using a Siemens medical CT scanner (SOMATOM Definition AS), and high field magnetic resonance images were acquired of the same module using a Bruker BioSpec 9.4 T scanner.

RESULTS AND DISCUSSION

Visualisation of final fouled module

A three-dimensional (3D) X-ray image of the final fouled RO membrane module after Stage 3 is shown in Figure 9.5. Cross-sectional images are also presented, as acquired using both 3D X-ray imaging (which will reflect the distribution of

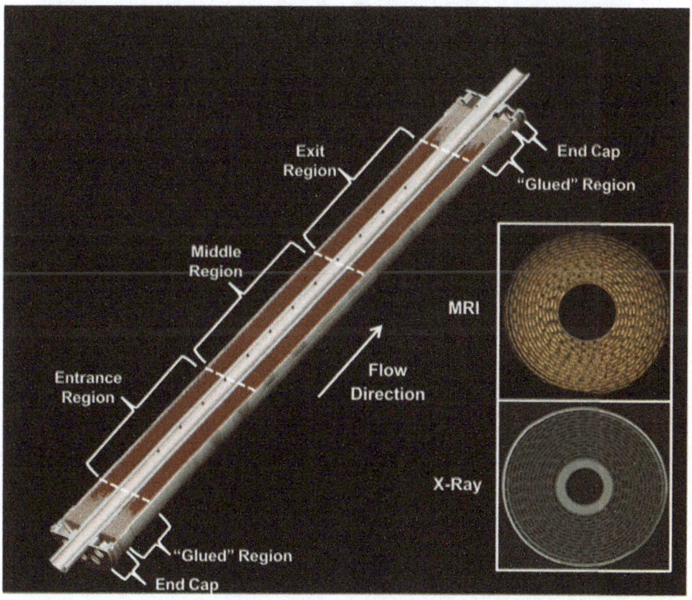

Figure 9.5 CT scan of the reverse osmosis (RO) membrane module, with the location of the entrance, middle and exit regions shown. Insert pictures shows examples of high field MRI and X-ray images obtained. Also indicate are the RO membrane end caps and "glued" regions. CT scan parameters: 120 kV/500 mAs, 0.5 mm resolution. MRI parameters: slice thickness 1 mm, 1024 × 1024 pixels, field of view 60 mm × 60 mm, in-plane resolution 59 μm × 59 μm.

density) and high magnetic field 3D MRI. The membrane module was filled with water during these experiments. The external PVC casing of the module is evident in the X-ray images as are the inlet and outlet fittings. Also evident in Figure 9.5 is the presence of glue sealing the ends of the membrane sheets; the EF NMR measurement zones were fortuitously selected so as to avoid these regions. Figure 9.6 shows photos of the membrane module following an autopsy – clear visual evidence is presented of the accumulation of biofouling material near the entrance of the module, which was not present in the middle or exit regions of the module. The presence of this fouling material is not immediately evident in either the X-ray

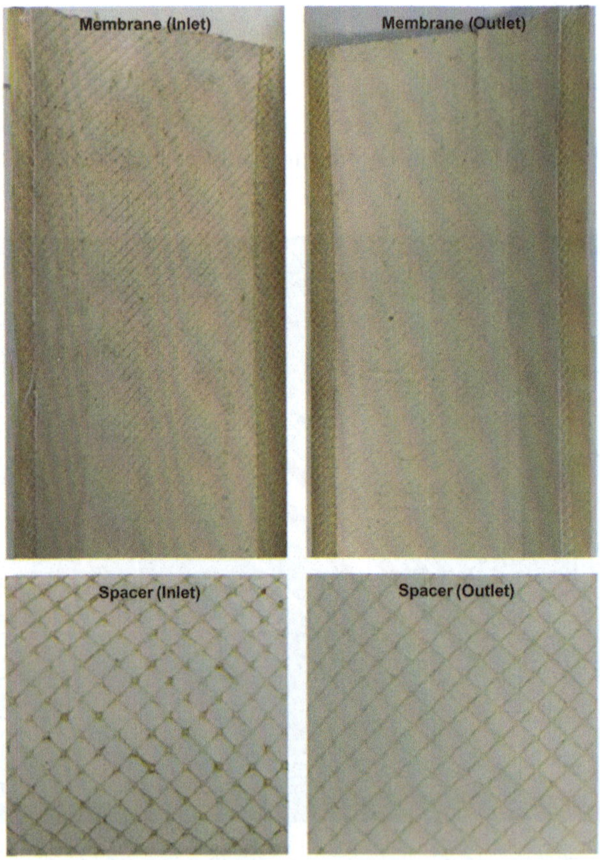

Figure 9.6 Photographs of membrane and feed spacer at the membrane module inlet and outlet sides after RO membrane module dissection, performed at the end of stage 3 (Figure 9.4). A significant amount of biomass was found at the RO membrane module inlet, compared with limited amount of biomass at the outlet.

or high field MRI images in Figure 9.5. It can however be detected based on imaging the distribution of T_2 values when performing the 3D high field MRI acquisition. Figure 9.7 shows the measured T_2 distribution maps/images for the entrance and middle regions, along with the corresponding T_2 probability distributions. Clearly a reduced value of T_2 is measured in the entry region where biofouling was visually evident, consistent with literature. The high field MRI is able to detect biofouling

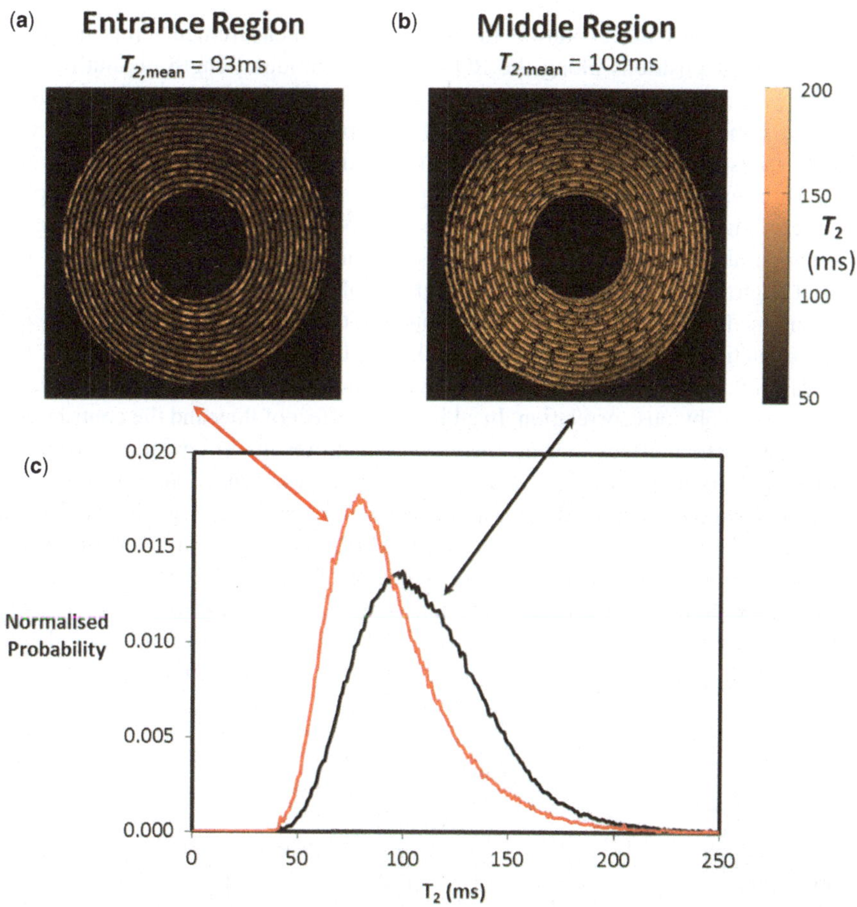

Figure 9.7 T_2 maps obtained at (a) entrance and (b) middle region of the fouled RO membrane module, with the mean T_2 value shown and (c) shows the T_2 distributions obtained from these T_2 maps; measured using the high field MRI instrument. MRI parameters: Echo time (TE) = 23.8 ms, 10 echo images acquired, slice thickness = 1 mm, 1024 × 1024 pixels, field of view 60 mm × 60 mm, in-plane resolution 59 μm × 59 μm. The probability distributions have been normalised such that each equals unity (i.e. $\sum P(T_2) = 1$), this is for ease of comparison.

using T_2 relaxation times. It is worth highlighting that the quality and resolution of images and data in Figures 9.5 and 9.6 using high field MRI are not readily accessible using EF NMR (Fridjonsson *et al.*, 2015a); nevertheless very useful non-invasive data is able to be detected using EF NMR regards biofouling, as will be subsequently demonstrated.

EF NMR relaxation measurements

Figure 9.8 shows the evolution in the T_1 and T_2 measurements for the three measurement positions along the RO membrane module length (as outlined in Figure 9.5) for the clean module and as a function of fouling stage. The error bars are the standard deviations produced by multiple repeats of the measurements. These measurements were performed without flow through the module. T_1, and to a lesser extent T_2, decreased as fouling progressed distinctly in the inlet region of the module, which is consistent with the observed biofouling in this region discussed above. The effect is much less pronounced for the middle region and outlet regions. Furthermore once the effect of flow ($U \neq 0$) is introduced (as summarised in Figure 9.9) it becomes apparent that a simple T_1 and T_2 relaxation time reduction as biofouling progresses is obscured by the effect of flow on the relevant relaxation parameter; at the highest flow rate ($U = 1.28$ cm/s) studied, there was no obvious correlation. In addition the effect of flow and the contribution of inflow-outflow effects during the measurement results in deviations from the exponential assumption of the T_1 and T_2 relaxation time fits. This effect may be de-convolved and quantified but requires detailed knowledge of the flow field (U). A simple robust correlation between the EF NMR relaxation parameters that could be routinely used as an early warning indicator of the onset of biofouling in the RO membrane modules is clearly precluded by the complex effects of flow. Even in the absence of flow, the ability of EF NMR relaxation measurements to detect biofouling requires deployment of the measurement in the vicinity of the fouling

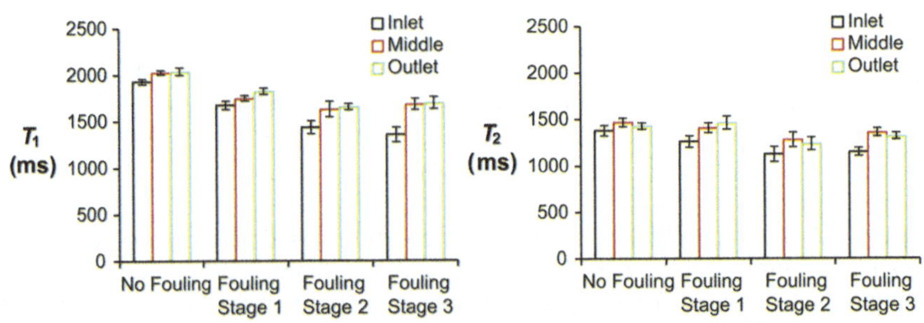

Figure 9.8 T_1 and T_2 relaxation time of the fluid at the membrane module inlet, middle and outlet regions, at different stages of biofouling (no fouling, stage 1, stage 2 and stage 3) during no flow conditions; measured using EF NMR instrument.

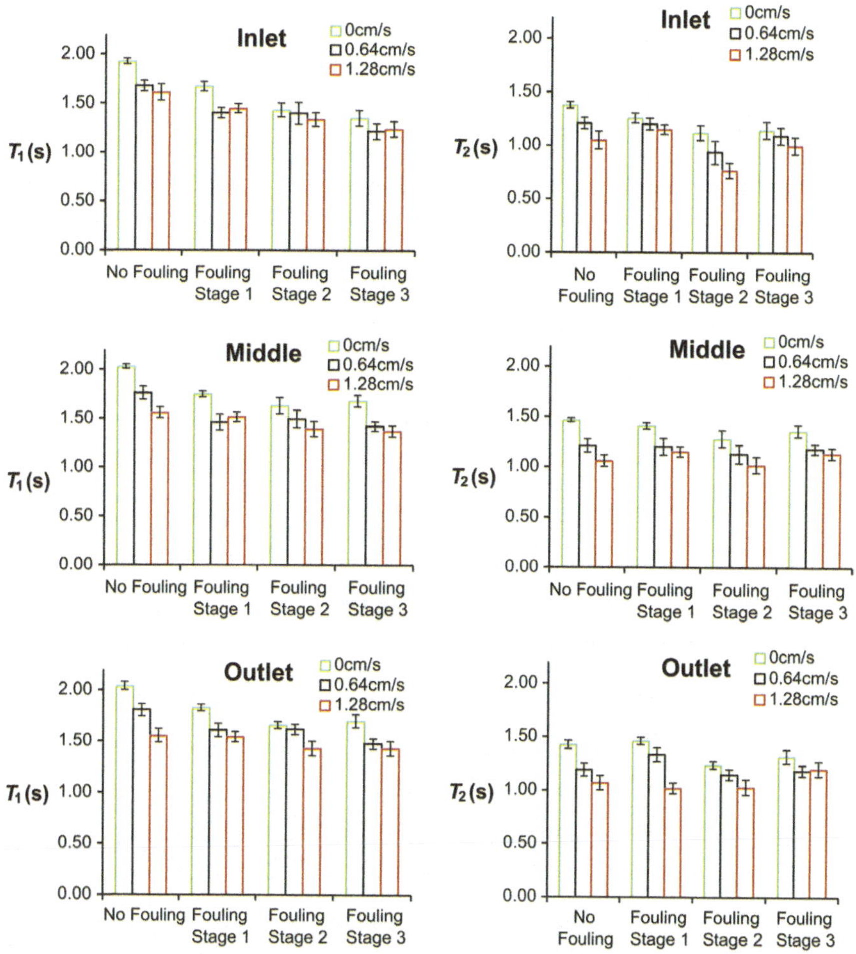

Figure 9.9 T_1 and T_2 relaxation times at three different flow rates ($U = 0$, 0.64 and 1.28 cm/s), at different stages of biofouling (no fouling, stage 1, stage 2 and stage 3); measured using EF NMR instrument.

along the RO membrane module axis (in this case near the module entry). While useful information is available from the T_1 and T_2 relaxation time measurements, due to the complicated relationship between fluid flow distribution and relaxation times and their effects on the measured EF NMR signal, these measurements were found to be less suitable at detecting biofouling under flowing conditions. However, the portable EF NMR derived T_1 and T_2 relaxation times are suitable to non-destructively detect the location of biofouling in spiral wound membrane elements in the absence of fluid flow.

Spin-echo measurements of total NMR signal

Figure 9.10 shows the measured EF NMR signal as a function of velocity, fouling stage and position along the module length, as acquired using the simple spin-echo pulse sequence shown in Figure 9.2a. The signal intensities are normalised relative to that detected in the absence of any flow or fouling. It has been shown in the literature that biofouling within complex flow structures (e.g. porous media and membranes) results in dramatic transition of the flow-field, characterised by the

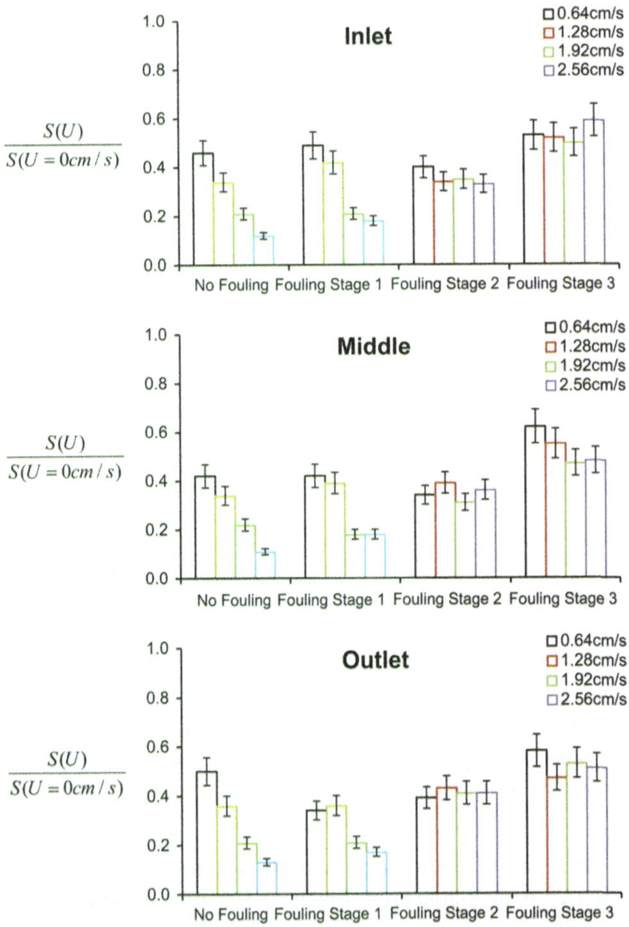

Figure 9.10 Normalised total NMR signal measured with the EF NMR at different stages of biofouling (no fouling, stage 1, stage 2 and stage 3) for increasing mean flow rates (0.64 cm/s, 1.28 cm/s, 1.92 cm/s and 2.56 cm/s) non-destructively determined at the spiral wound RO membrane module inlet, middle and outlet regions.

development of stagnant regions and skeletal rapid flow structure (Creber *et al.*, 2010b; Graf von der Schulenburg et al., 2008; Kim & Fogler, 2000; Seymour *et al.*, 2007; Sharp *et al.*, 2005; Thullner *et al.*, 2002; Vogt *et al.*, 2013). As *tau* and *U* are increased, signal intensity from this skeletal flowing fluid is reduced and eventually eliminated; this essentially occurs when all fluid that has been pre-polarised has been flushed out of the module when signal acquisition occurs. This can be approximated as occurring when:

$$\frac{2tauU}{\sigma} > L \tag{9.5}$$

where σ is the porosity of the membrane/spacer assembly and L is the length of the polarised region.

The *tau* time in Figure 9.2a was increased (to a value of 500 ms) such that for the highest flow rate systematically studied here ($U = 2.56$ cm/s), signal was essentially received only from stagnant fluid inside the module; this can be checked by repeating the measurements at a higher flow rate and checking that the signal intensity detected is unaffected (this of course assumes the fraction of fluid that is stationary remains constant as velocity increases). Constraints on selecting *tau* and *U* in practice will naturally be the prevailing operational velocity and a reduction in SNR as *tau* is increased due to T_2 relaxation effects. It is expected that this methodology will be more sensitive at higher velocities such as at typical operational range up to 0.2 m/s, as the *tau* values consequently required are shorter.

With reference to Figure 9.10, for the un-fouled membrane, the effect of flow is clearly evident as $S/S_{(\text{no flow})}$ reduces systematically with flow for all three positions. The largest flow rate employed ($U = 2.56$ cm/s) corresponds to a plateau in signal and represents the stagnant fluid in the module. This plateau value of $S/S_{(\text{no flow})}$ is 11%; based on geometric considerations, the contribution from the permeate channel (which was flooded during all EF NMR measurements and which was stationary) is 10%. This indicates that for no fouling the vast majority of the detected signal originated from the permeate channel and that the flow field through the module is homogeneous. Data for $U = 2.56$ cm/s thus acts as a semi-quantitative indicator of stagnant water fraction in the module.

As fouling progresses, what is immediately obvious is that the reduction in $S/S_{(\text{no flow})}$ with increasing flow rate is gradually eliminated. This reflects the change in internal hydrodynamics within the membrane module. For fouling stages 2 and 3 (Figure 9.4), the value of $S/S_{(\text{no flow})}$ is essentially independent of flow rate; this reflects the portion of the module that is essentially stagnant. The mean EF NMR signal ($S/S_{(\text{no flow})}$) for the inlet, middle and outlet sections of the RO membrane module is respectively 0.54, 0.53 and 0.52. This means that approximately half the fluid volume of the total module is stagnant, while the other half is available for fluid transport. If we focus on the measurement of $S/S_{(\text{no flow})}$ for $U = 2.56$ cms^{-1}, which we know in the unfouled state reflects predominately on stagnant fluid, we see a significant increase as fouling progresses. Of equal interest is that this increase is only partially reduced between the inlet (where fouling is visually evident upon autopsy) and the middle and outlet regions. Fouling near the entrance of the spiral wound membrane module significantly disturbed the flow field, as measured with the EF NMR, through the whole spiral membrane module. These results indicate that the total NMR signal from the EF NMR, as reported in Figure 9.10, allows for the early non-destructive detection of biofouling in spiral wound membrane elements during cross flow operation in a pressure vessel, and that this measurement can be performed at any position along the module axis.

Practical implications and further studies

Potential field of EF NMR use

The mobile EF NMR can be used to non-destructively (i) detect early fouling, as demonstrated here, and (ii) hence also evaluate the efficiency of cleaning directly during the cleaning process. Most likely the EF NMR can also be useful to evaluate the production process of membrane modules via its sensitivity to the internal flow field. The total NMR signal (as measured using a simple spin echo pulse sequence) is suitable for detecting biofouling during cross flow operation, and the sensitivity is improved at linear flow velocities applied in practice. In the absence of fluid flow, the EF NMR T_2 relaxation time is suitable to non-destructively detect the location of biofouling along the axis of the spiral wound membrane elements. Use of the EF NMR during cleaning will reduce cleaning chemical consumption and down time since the cleaning can be halted when fouling is removed. The cleaning process can also be prolonged or modified in the case of the fouling not being removed.

Measurement of membrane element fouling in pressure vessels during operation

Pressure vessels contain in practice up to 8 spiral wound membrane elements in series. These pressure vessels are made of non-metal material (glass fibre). Therefore, it is anticipated that the mobile EF NMR can be used to non-destructively detect fouling in membrane modules while placed in pressure vessels during operation. In this study, linear cross flow velocities of fluid in membrane modules up to 2.56 cm/s were used to detect fouling. The detection of accumulated fouling will be better at higher cross flow velocities, indicating that fouling detection at practice operational conditions (e.g. 0.2 m/s) will provide even more sensitive and enable early fouling detection. These pressure vessels contain metal connections at their beginning and end for the feed, concentrate and permeate water flows and connection to the frame.

These would not be directly in the measurement field of view, however they would prevent ready placement of the EF NMR system on the pressure vessel during operations. An additional complication is that the modules in industry practice are typically 4 or 8 inches in diameter.

Further research

Consequently, the development of an EF NMR system for larger diameter membrane elements is currently being considered, as is reconfiguration of the polarising coil such that it is able to be clamped on to the pressure vessel during operation (Windt *et al.*, 2011), in the form of a Cuff. We note that there is no obvious physical limitation preventing scale-up of the apparatus to these larger dimensions. In addition possibilities for non-destructive discrimination between fouling types: particulate, biofouling and inorganic are being explored. This features a composition of the current demonstrated simple signal measurement that correlates with the extent of fluid stagnation, with measurements of NMR signal relaxation and diffusion/transport. Other significant challenges in terms of the practical application of EF NMR to these modules include: (i) the module axis needs to be perpendicular to the Earth's magnetic field for optimal SNR and would deliver no signal if parallel and (ii) in practice flow will occur in the central permeate channel. (i) could be eliminated by detection in the presence of a weak magnet. Regarding (ii), this needs to be researched further as to the effect of flow in the central permeate channel on the NMR signal detected; if at a sufficiently rapid velocity, it would have no effect and actually increase the sensitivity of the instrument relative to the data shown in Figure 9.10. We will also in future look to benchmark our approach more widely against other potential non-destructive (and preferably truly non-invasive) options for early fouling detection.

CONCLUSIONS

This work has shown the use of a relatively simple and robust Earth's field (EF) NMR based measurement technique for the early non-destructive detection of fouling within a spiral wound reverse osmosis and nanofiltration membrane modules. This is based on measuring the total EF NMR signal under flow conditions, with increasing NMR signal detected due to fluid stagnation, which correlates with biofouling within the reverse osmosis membrane module. The proposed non-destructive and direct measurement technique can be performed under operational conditions and at any location along the membrane module or, in practice the pressure vessel. The proposed technique is suitable for early fouling detection as the time to evaluate the presence/absence of biofouling is fast (i.e. on the order of seconds). There are no predicted issues with applying this technique at higher operational pressures, and whilst the above study has focussed on biofouling, it could equally be applied to other fouling sources which change the flow field inside the membrane module.

Chapter 10

Experimental and numerical characterization of the water flow in spacer-filled channels of spiral-wound membranes

INTRODUCTION

Membrane filtration processes for seawater desalination and wastewater reuse are becoming increasingly important. In a spiral-wound membrane module, most commonly used in these water treatment processes, the membrane sheets are rolled around an inner tube. To keep the membrane leafs apart a relatively thin spacer net is inserted. The feed spacer promotes flow instabilities to enhance mass transfer and reduce concentration polarization (Fimbres-Weihs *et al.*, 2006; Gao *et al.*, 2013). However, aside from a beneficial impact, the feed spacer also contributes to pressure drop increase along the flow channel. Furthermore, a major drawback in membrane technology is fouling, i.e., accumulation of unwanted material on the membrane and spacer surface (Antony *et al.*, 2011; Baker & Dudley, 1998b; Flemming, 2002; Ridgway *et al.*, 1983; Salvador Cob *et al.*, 2012). Micro-sized particles, colloids, organic macromolecules can deposit and microbial cells can attach, grow and form biofilms on the membrane and spacer surfaces and decrease the filtration performance (Flemming, 1997; Tang *et al.*, 2011; Yiantsios & Karabelas, 2003). Baker and Dudley (1998a) reported that initial deposition of fouling occurred on the feed spacer filaments and on the membrane alongside the spacer and, with time, intruded upon the remaining free membrane area.

It was shown that fouling accumulation can be controlled to some degree by suitable hydrodynamic conditions (Sablani *et al.*, 2001). Moreover, the same amount of accumulated fouling material can have a different impact on performance, depending on the hydrodynamic conditions (Araújo *et al.*, 2012a; Bucs *et al.*, 2014a; Valladares Linares *et al.*, 2014; Vrouwenvelder *et al.*, 2011a). Feed spacer geometry and operational conditions (e.g., flow velocity, transmembrane pressure, etc.) have

Author(s): Szilárd Bucs, Joop Kruithof, Mark van Loosdrecht and Johannes Vrouwenvelder
doi: 10.2166/9781780409597_0179

a profound influence on the flow pattern and foulant deposition in spacer-filled channels (Radu *et al.*, 2014b; Vrouwenvelder *et al.*, 2009c). To design new spacers with the least pressure drop, minimal fouling propensity, and maximum mass transfer, it is therefore important to both be able to determine experimentally and to numerically calculate the hydrodynamics (i.e., velocity and pressure profiles, shear at the walls, etc.) in these flow channels.

The impact of different spacer geometries on the flow pattern is not easily measurable experimentally due to the inherently small length-scales involved. Alternatively, advances in numerical performance have led to computational fluid dynamics (CFD) techniques being the primary means to understand the fluid flow in spacer-filled channels. Some studies have simulated fluid flow and mass transfer in simplified two-dimensional setups (Radu *et al.*, 2010; Schwinge *et al.*, 2002a). However, in recent years, three-dimensional (3-D) numerical models are becoming increasingly accessible, developed mainly with an emphasis on the effect of feed spacer on hydrodynamics and mass transfer in feed spacer channels (Koutsou *et al.*, 2007, 2009; Saeed *et al.*, 2012). Simplified cylindrical shapes have been used for representation of spacer filaments in most of the numerical studies. However, microscopic observations of the feed spacer revealed that the spacers used in commercially available spiral-wound membrane modules have more irregular geometry, with filaments varying in thickness (Fimbres-Weihs & Wiley, 2010; Picioreanu *et al.*, 2009; Vrouwenvelder *et al.*, 2010b). Simulations by Picioreanu *et al.* (2009) showed that the feed channel pressure drop for a simplified spacer with cylindrical filaments is significantly different from more realistic spacer geometry with variations in filament thickness. The importance of realistic spacer geometries in numerical studies was also revealed by a recent experimental and numerical study on particle deposition in spacer channels at various feed spacer orientations (Radu *et al.*, 2014b).

In contrast with the abundance of numerical studies, only a few experimental investigations were carried out to directly visualize and characterize the flow patterns in feed spacer-filled channels. This may be partly due to the geometrical complexity and small dimensions of the flow channels (thereby the need for microscopic techniques), but also due to the high temporal resolution required (i.e., very fast flows require use of fast cameras).

Nuclear magnetic resonance (NMR) or magnetic resonance imaging (MRI) has been used to locally resolve (in 3-D) the velocity and shear stress distribution in tube reactors containing biofilms (Manz *et al.*, 2003; Wagner *et al.*, 2010a). In spacer-filled channels with and without biofilm, NMR was used to visualize and measure two-dimensional velocity profiles (Vrouwenvelder *et al.*, 2010b). Formation of preferential flow patterns was observed as biofouling evolved in time. However, as a limitation of the method used, the flow cell was operated with 10 times lower linear flow velocity than normally used in practice, and the measured velocity had to be averaged over the flow channel height.

Doppler optical coherence tomography (DOCT) was also used to measure and visualize the local velocity profile in a spacer-filled channel (Gao *et al.*, 2013). The DOCT could reveal the flow profile normal to the main flow direction at different sections along the flow channel. Furthermore, the development of eddies were observed next to the spacer filaments.

Particle image velocimetry (PIV) was used to characterize the flow next to the membrane surface at different linear flow velocities by Gimmelshtein and Semiat (2005) and Willems *et al.* (2010). Gimmelshtein and Semiat (2005) found laminar flow at commonly used linear flow velocities (0.06–0.17 m · s^{-1}), with (expectedly) increasing mixing intensity at higher linear flow velocities. The measured flow fields were in good agreement with the two-dimensional numerical simulations developed by the authors (Gimmelshtein & Semiat, 2005). Willems *et al.* (2010) used PIV techniques to achieve a quantitative description of the two-phase flow in spacer-filled channel. Although the temporal resolution was low, the study showed how the flow direction changed along the channel height and also evaluated the impact of bubbly flow on local velocity. PIV techniques were also applied to investigate the flow patterns in flow channels using modified saw-tooth spacer geometries (Liu *et al.*, 2015). However, the flow channel thickness (4 mm) in these experiments was much higher than in reverse osmosis/nanofiltration systems applied in practice.

In summary, all of these experimental methods have some limitations in terms of the flow channel thicknesses used, the applied linear flow velocity, the spatial or temporal resolution. As such, the objectives of this study were: (i) to evaluate the micro-scale flow patterns in a feed spacer-filled channel under conditions representative for current practical applications (i.e., similar linear flow velocity and channel dimensions) using a PIV system, and (ii) to compare the measured flow fields with three-dimensional numerical simulations.

MATERIAL AND METHODS

Experimental

Micro-scale fluid flow patterns in spacer-filled channels were experimentally investigated by particle image velocimetry (PIV).

Flow cell

For PIV measurements, a fully transparent flow cell was constructed from two high quality glass slides (2.5 cm width and 7.5 cm length, VWR International, West Chester, PA) (Figure 10.1). Commercially available non-woven Toray feed spacer with 31 mil (0.787 mm) thickness was inserted between the glass slides. The height of the flow cell was adjusted to fit the feed spacer thickness and to avoid bypass flow below or above the spacer, then the whole ensemble was glued with epoxy resin on

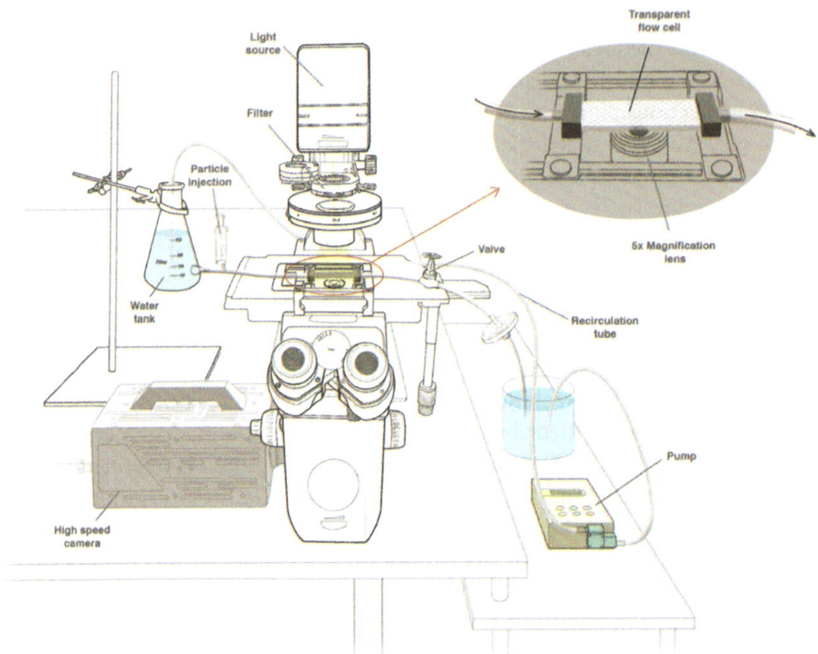

Figure 10.1 Experimental setup used for water flow measurements and visualization, using particle image velocimetry (PIV). The experimental setup consisted of an elevated water tank, a transparent flow cell with a feed spacer, a particle injection point, a microscope and a high speed camera.

the sides. Water connectors were fabricated from nylon tubing with 4 mm (for inlet) and 6 mm (for outlet) inner diameter (Figure 10.1).

In order to keep the flow cell transparent, no membrane sheets were inserted into the flow channel and the flow cell was operated without permeate production. The impact of permeation on the cross flow velocity in our lab scale flow cell (7.5 cm long) is negligible since the permeate velocity (typically in the order of 10 $\mu m \cdot s^{-1}$, (Radu *et al.*, 2010) is four orders of magnitude lower than the linear flow velocity of the feed water (in the order of 10 $cm \cdot s^{-1}$). Even without permeation, the presence of a membrane sheet instead of glass wall would not visibly influence the flow pattern, because the average roughness of an RO membrane is in the order of 50 nm (Ishigami *et al.*, 2012), three orders of magnitude lower than our depth of field for flow visualization in z direction.

To eliminate pulsations generated by pumps, hydrostatic pressure was used instead to drive the water flow through the flow cell. A 3 L reservoir was placed about 1 m above the flow cell and connected to the flow cell inlet. The flow rate was adjusted with a valve connected to the flow cell outlet. To maintain constant flow rate the water level was kept constant in the reservoir by recirculating the water (Figure 10.1). The Reynolds number was calculated according to Schock and Miquel (1987), equation 10.1.

$$\mathrm{Re} = \frac{u_m \cdot d_h \cdot \rho}{\eta} \tag{10.1}$$

with u_m being the imposed linear (i.e., average) flow velocity. ρ and η are the water density and viscosity at 20°C (998 kg · m^{-3} and 0.001 Pa · s, respectively), and d_h is the hydraulic diameter (≈1 mm), calculated from the total volume between the glass slides, V_{Tot}, the spacer volume, V_{Sp}, and the wetted surface, A_w including feed spacer and both glass slides (Schock & Miquel, 1987).

$$d_h = \frac{4(V_{\mathrm{Tot}} - V_{\mathrm{Sp}})}{A_w} \tag{10.2}$$

The flow rate was measured gravimetrically at the outlet. Measurements were performed at three linear inflow velocities: $u_m = 0.163$ m · s^{-1} (Re = 160), 0.074 m · s^{-1} (Re = 70) and 0.294 m · s^{-1} (Re = 300).

Particles

For flow visualization, silver coated hollow glass beads of 10 μm diameter (neutrally buoyant, $\rho_p = 1.05$ g · cm^{-3}, ±5% size distribution) were injected into the inlet tubing of the flow cell during the measurements (P/N 10089, TSI Incorporated, Shoreview, MN). A volume of 10 mL suspension of particles (≈0.5 mg · L^{-1} concentration) was injected right before the PIV measurements were made. Seeding density was kept above 8 particles for an imaged area of 32 pixels × 32 pixels. A 5 μm pore size filter was mounted to the outflow tube to prevent recirculation and eventual deposition of the injected particles.

Visualization

For typical PIV measurements a laser strobe is used as light source with a cylindrical lens to limit the illuminated region to the area of interest. To use conventional PIV setup for visualizing the flow in a spacer-filled channel at different heights the laser beam should have been placed perpendicular to the camera, thus allowing illumination of the flow channel at various heights (*z* direction). The dimension of the flow channel and the presence of the feed spacer limited the possibility of using conventional PIV. For the presented measurements a modified PIV setup was used which allowed capturing the flow pattern at various heights within the spacer-filled channel.

Particles were visualized using an inverted microscope (Axio Imager, Zeiss Microscopy GmbH, Germany) in combination with a high-speed camera (Fastcam SA5, Photron Inc. San Diego, CA) and a 5× objective lens with 0.13 numerical aperture (NA). To narrow the depth of field (DOF), a 550 nm filter (BP 550/25,

537–563 nm) was placed in front of the light source, so that a 42 to 44 μm focal depth was achieved, equation 10.3

$$\mathrm{DOF} = \frac{\lambda \cdot n}{\mathrm{NA}^2} \tag{10.3}$$

where $\lambda = 0.550$ μm is the light wavelength and $n = 1.33$ is the refractive index of water.

The camera was set to capture 7000 frames per second (fps) with an exposure time of 1/30,000 s, sufficient to eliminate particle motion blur. The obtained image size was 1024 × 1024 pixels. For each measurement 100 consecutive frames were processed. The angle of view and DOF allowed capturing an area of approximately one feed spacer mesh element (approximately 4 mm × 4 mm) from the flow cell (Figure 10.2). To obtain a complete overview of the flow pattern in the channel, measurements were repeated at different feed spacer mesh elements (x and y direction) and at different positions along the channel height (z direction), for each experimental condition.

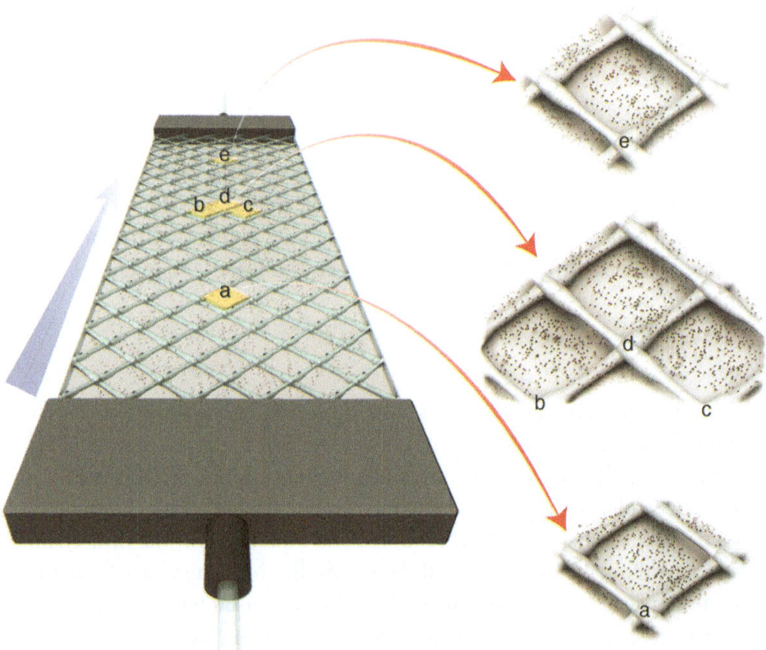

Figure 10.2 Schematic representation of the flow cell and the five feed spacer elements (a to e) used to assess the water flow reproducibility at different locations in the flow cell.

Calculation and visualization of velocity fields

The captured video clips were analyzed by a commercially available PIV software (DaVis, LaVision GmbH, Goettingen, Germany), using a sequential frame cross correlation method. For quantitative analysis the software was calibrated based on the pixel size (4 μm × 4 μm) on the captured video clips, which was the same for all PIV measurements. The analyses performed on the video data were the calculation of velocity fields and streamlines. The same velocity field calculation procedure was repeated for different z positions (i.e., at maximum 18 different heights in the flow channel). The measured vector fields were visualized and statistically analyzed by custom-written routines in MATLAB (MATLAB, 2014b, MathWorks, Natick, MA, www.mathworks.com). 100 PIV measurements (i.e., image frames over 0.14 milliseconds (ms)) were taken at each position and averaged in time to minimize the loss of tracked particles.

Numerical model

Experimental measurements were compared with results from three-dimensional CFD calculations. The model geometry (4.38 × 4.38 × 0.787 mm^3) included one feed spacer mesh element (Figure 10.3), with a realistic representation of the spacer geometry constructed with curved surfaces based on Scanning Electron Microscopy images (similar to the procedure from Radu *et al.*, 2014). The hydrodynamic calculations assumed stationary flow of incompressible fluid in laminar flow conditions, according to the Navier-Stokes and continuity equations 10.4:

$$\begin{aligned} &\rho(\mathbf{u}\cdot\nabla)\mathbf{u}+\nabla p=\nabla\cdot(\eta\nabla\mathbf{u}) \\ &\nabla\cdot\mathbf{u}=0 \end{aligned} \tag{10.4}$$

where $\mathbf{u} = (u_x, u_y, u_z)$ is the vector of local liquid velocity, p is the pressure, ρ and η are the density and dynamic viscosity of water (20°C). Similar to the approach used in Radu *et al.* (2014), periodic flow conditions were set between the inlet and outlet, as well as between the lateral boundaries (Figure 10.3b), implying identical flow velocity fields on pairs of boundaries (in with out and left with right). This approximates the flow profile corresponding to a spacer element situated within an array, sufficiently far from the flow cell walls not to experience entrance/exit or wall effects (Radu *et al.*, 2014b). The flow was driven by a pressure difference imposed between inlet and outlet, calculated from an additional constraint such that the desired linear flow velocity u_m is obtained. The water flow was calculated without permeate production, therefore no-slip boundary conditions were set to the top, bottom and spacer surfaces. The steady laminar flow equations were solved in COMSOL Multiphysics (v4.4, Comsol Inc., Burlington, MA) with finite element methods, on a tetrahedral mesh with maximum size of 50 μm.

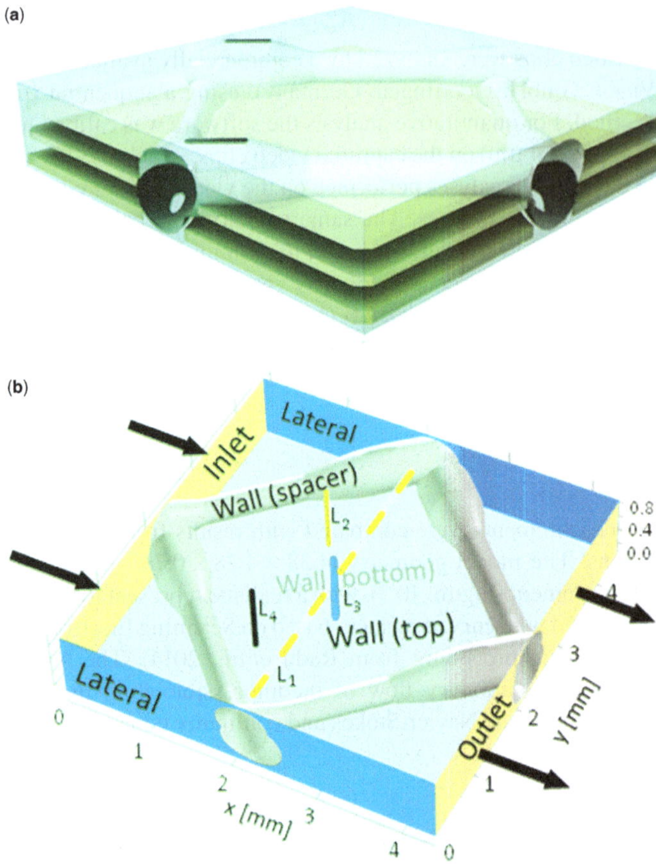

Figure 10.3 (a) Positions of *x-y* slices (*x* coordinate corresponds to the main flow direction and *y* coordinate is normal to the main flow direction) taken for comparing measured and calculated vector fields at different heights in the flow channel ($z = 200$, 400 and 650 μm). The three slices have a thickness of 40 μm each. (b) The computational domain for hydrodynamic simulations includes one spacer element, with spacer resembling the geometry used in practise and positions of cut lines for comparing measured (PIV) and calculated flow velocity as shown in Figure 10.9. Line L_1 is in the *x-y* plane at $x = 2$ mm, L_2–L_4 are in the *y-z* planes at $(x,y) = (1.48,3.12)$, $(2.06,2.22)$ and $(1.48,1.45)$ mm, respectively.

Calculation and visualization of velocity fields

The flow fields calculated with COMSOL were averaged over intervals of $\Delta z = 40$ μm before comparing simulation results with the PIV measurements. This averaging procedure was necessary because the measurements were recorded with a depth of field of approx. 40 μm. Therefore, the *x-y* velocity fields presented at a certain

z coordinate should be understood as averages between $z \pm 20$ µm (Figure 10.3a). The flow fields were plotted using a Python script with Matplotlib graphic library (Hunter, 2007), showing streamlines, velocity magnitude and flow direction.

Statistic similarity

The Bray-Curtis similarity was used to analyse the differences between measured and model-calculated 2-D velocity magnitude distributions, as well as the similarity between PIV measurements at different locations, and between velocity fields averaged over different time intervals. The Bray-Curtis dissimilarity or 'distance', d_{BC}, between two sets of data a and b, each containing n points to be compared, is a normalization method defined as (Krebs, 1999):

$$d_{BC} = \sum_{k=1}^{n} \left| a_k - b_k \right| / \sum_{k=1}^{n} (a_k + b_k) \qquad (10.5)$$

The similarity between the sets a and b was then defined as

$$S_{BC} = 1 - d_{BC} \qquad (10.6)$$

Since all velocity magnitudes are positive, the similarity S_{BC} is always between 0 and 1, with maximum value of 1 for perfectly similar velocity fields.

RESULTS

In this study the water flow patterns were investigated in spacer-filled channels using PIV in a flow cell and 3-D CFD techniques under various conditions: (i) varying location along the length and width of the channel, (ii) different measured time frames, and (iii) varying linear flow velocities.

Flow pattern change over channel height

The PIV measurements allowed the determination of flow velocity patterns in x-y planes, at different heights z in the channel, an example of which is presented in Figure 10.4. These planar velocity fields clearly show how the flow direction changes over the channel height, with two specific zones becoming apparent. Near the bottom (between $z = 0$ and 300 µm) and near the top walls (between $z = 550$ and 800 µm) the flow direction mostly follows the nearby located spacer fibre orientation, i.e., at 45 degrees with respect to the main flow direction (Figure 10.4). Thus, there is a 90 degrees change in the flow direction from bottom to top, so that in the middle of the channel ($z = 350$ to 500 µm) a convergence region develops, with the liquid following the main flow direction (left to right in x direction). Because the real spacer geometry includes various positions of filament thinning, the velocity magnitude changes creating zones with fast and slow flow, which

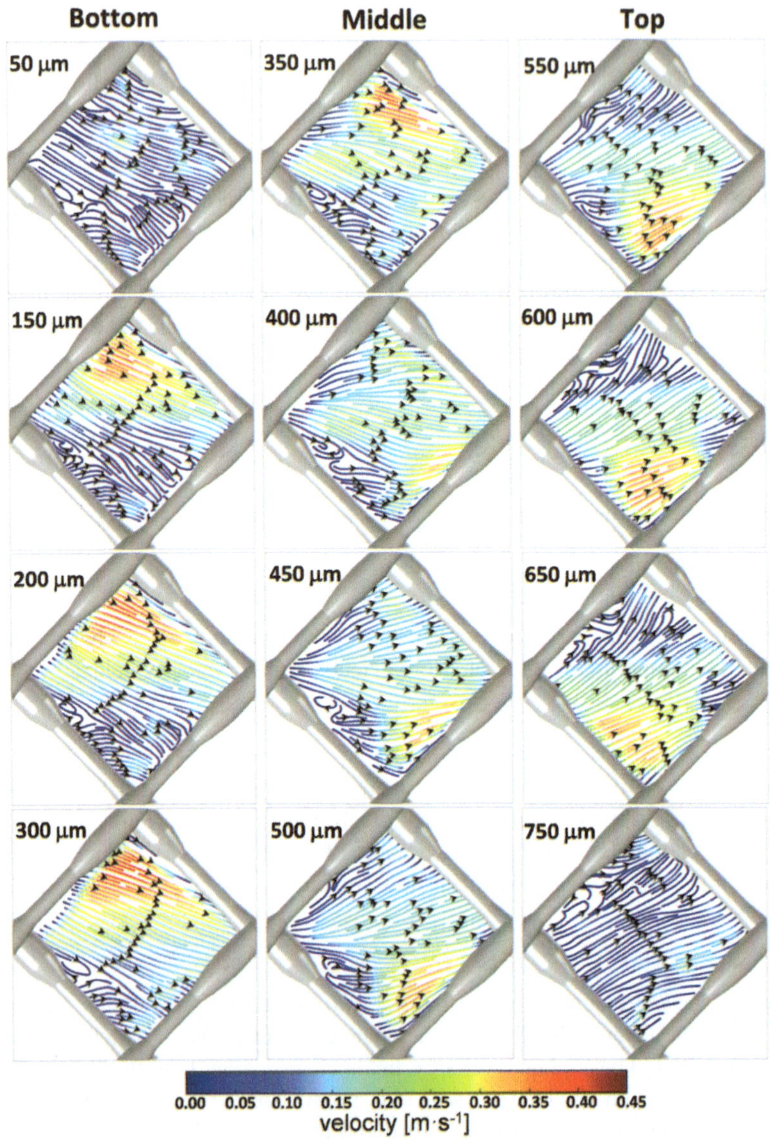

Figure 10.4 Measured flow pattern in *x-y* slices at different heights in the flow channel (*z* direction, the heights should be interpreted as the middle of the region where the measurements were taken, e.g., 50 μm represents the region between 30 and 70 μm in the channel) showing streamlines and flow direction (black arrows). The flow velocity magnitude is represented by the colour scale of a streamline. The main flow direction with an average flow velocity of 0.163 $m \cdot s^{-1}$ (Re = 160) was from left to right.

are not symmetric against the middle/centre plane at $z = 400$ μm. The maximum measured flow velocity within this experiment was 0.45 m · s^{-1} at $z \approx 300$ μm for the used average flow velocity of 0.163 m · s^{-1}.

Spatial reproducibility

To test the method reproducibility and the flow uniformity over the spacer-filled channel, the velocity distribution was measured at five locations (x,y) in the flow cell, as indicated in Figure 10.2. The flow patterns were similar at all measured locations, represented in Table 10.1 at three heights in the channel. The 2-D flow magnitude distribution at each location was compared by the Bray-Curtis similarity S_{BC} with the average over the five locations. The flow measured at the five locations proved to be very similar, with S_{BC} slightly better for the slices taken near the flow channel bottom and top ($S_{BC} = 0.96$–0.98) than for the middle slices ($S_{BC} = 0.94$–0.96), indicating that a measurement taken at any spacer cell could be considered representative of the entire flow channel. The rest of the measurements in this study were therefore done in the feed spacer mesh element *d*, located as shown in Figure 10.2.

Steady and unsteady flow – effect of different flow velocities

In order to study the effect of water velocity on the flow regime, measurements were done at three linear velocities: a reference velocity of 0.163 m · s^{-1}, which is commonly used in practical operation (Vrouwenvelder *et al.*, 2009), a lower velocity of 0.074 m · s^{-1} and a higher velocity of 0.294 m · s^{-1}. These correspond to Reynolds numbers 160, 70 and 300, respectively. The velocity magnitude for 100 consecutive measurements in time (all at the same point $x = 2$ mm, $y = 2$ mm) is plotted in Figure 10.5 for the three Re numbers. As the Re number increases, larger fluctuations of measured velocity can be observed. At Re = 70 the flow is

Table 10.1 Spatial reproducibility of measured flow field at five locations in the flow cell (*a* to *e*), at three different heights in the flow channel $z = 200$, 400 and 650 μm.

Height in the Channel \ Position*	Bray – Curtis Similarity (S_{BC}) a	b	c	d	e
$z = 200$ μm	0.97	0.98	0.96	0.98	0.97
$z = 400$ μm	0.95	0.94	0.94	0.96	0.95
$z = 650$ μm	0.96	0.98	0.97	0.98	0.97

*See Figure 10.2.

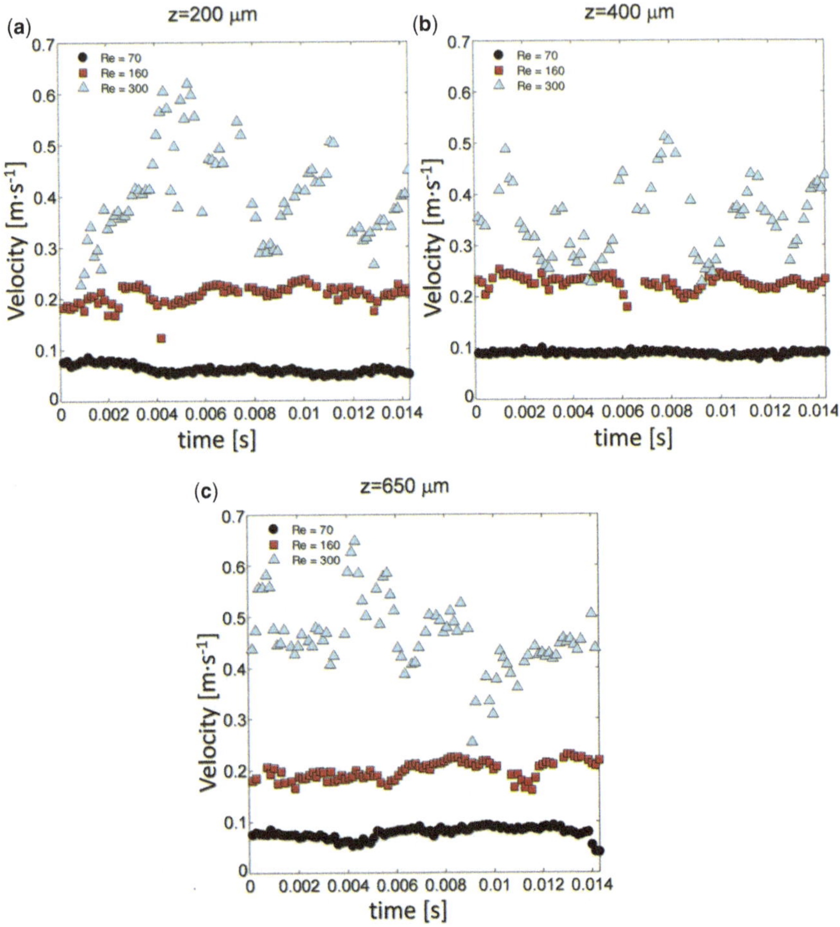

Figure 10.5 Measured local velocity magnitude over time at three flow velocities (corresponding to Re = 70, 160 and 300), in the middle of a feed spacer element ($x = 2$ mm, $y = 2$ mm) at different heights in the flow channel: (a) $z = 200$ μm, (b) $z = 400$ μm and (c) $z = 650$ μm. While the flow is mostly steady at Re = 70, it changes to unsteady (large velocity fluctuations) at Re = 300. The flow at Re = 160 is only slightly unsteady.

shown to be stable in time and it can be considered steady. At Re = 300 fluctuations between 20–50% of the linear flow velocity were recorded, which clearly indicate an unsteady flow regime. The flow at Re = 160 (as used in practice) shows only a mildly unsteady behaviour with fluctuations of 10% from average velocity. Most flow fluctuations seemed to occur near the top and bottom walls of the spacer-filled flow channel in the regions with the highest local flow velocities.

To quantitatively assess the error associated with time-averaging, the velocity field at Re = 160 was averaged over three time intervals Δt: 0.14 ms (2 frames or 1 measurement), 1.4 ms (10 measurements) and 14 ms (100 measurements), shown in Figure 10.6. The Bray-Curtis similarity was then calculated relative to the largest time scale (Δt = 14 ms), showing a 94% similarity for the velocity field averaged over 1.4 ms and 90% similarity for 1 measurement (0.14 ms).

The time averaged measured velocity distributions at three Re at different heights in the channel is presented in Figure 10.7. Similar flow regions as for the Re = 160 can also be observed for Re = 70 and 300: high and mostly one-directional velocities near the top and bottom walls and a flow convergence region with slower flow in the middle of the channel, where the flow changes direction.

For Re = 300, however, eddies (recirculation zones) were observed in the lower region of the channel (Figure 10.7, z = 200 μm, Re = 300). The maximum measured flow velocities were 0.20, 0.45 and 0.64 $m \cdot s^{-1}$ at Re = 70, 160 and 300,

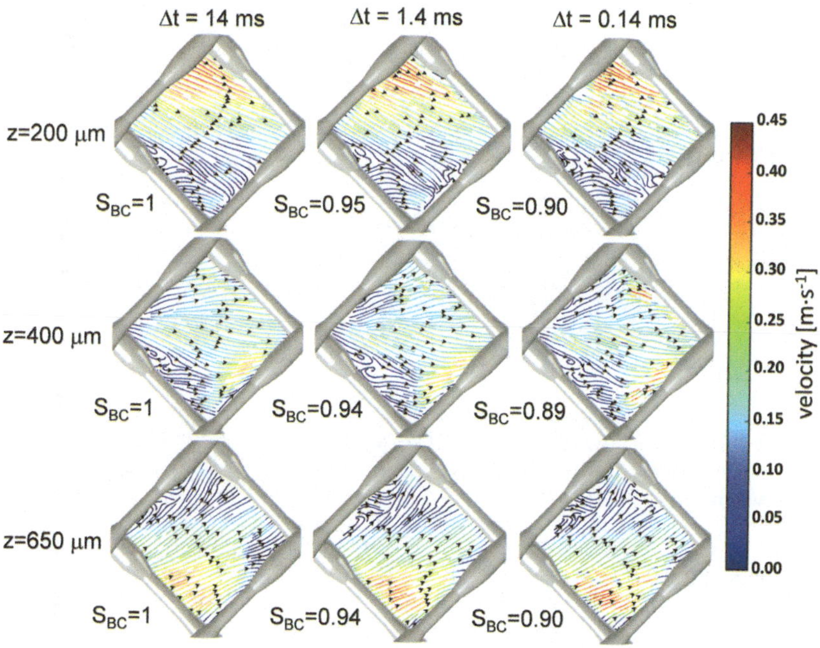

Figure 10.6 Time-dependence of the measured flow field at different heights in the channel z = 200, 400 and 650 μm. 2-D distributions of the flow velocity are shown averaged over periods of 0.14 ms, 1.4 ms and 14 ms. The Bray-Curtis similarity (S_{BC}) was calculated relative to the largest imaging time scale (Δt = 14 ms). The main flow direction with an average flow velocity of 0.163 $m \cdot s^{-1}$ (Re = 160) was from left to right. Streamlines and colour scale are as in Figure 10.4.

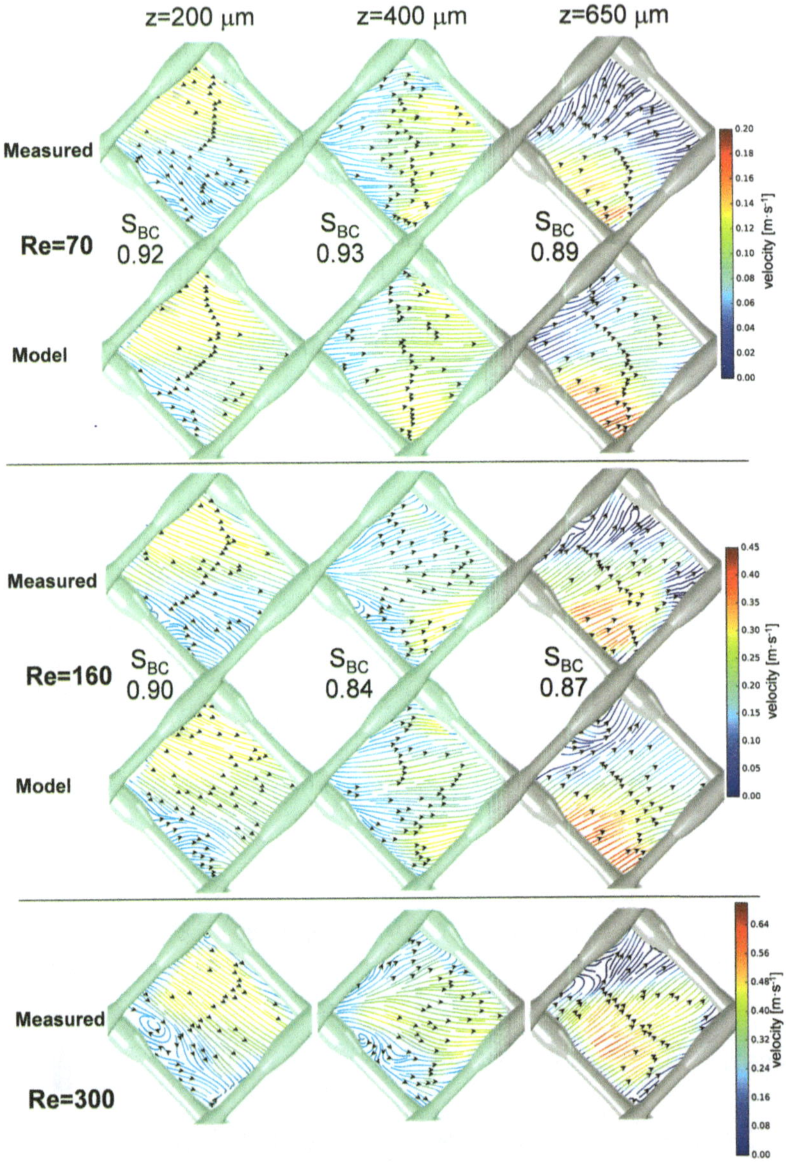

Figure 10.7 Comparison between measured and calculated flow fields at Re = 70 (top), Re = 160 (middle) and Re = 300 (bottom), at three channel heights $z = 200$, 400 and 650 μm. Streamlines show as colour scale the flow velocity magnitude. S_{BC} numbers represent the calculated similarity indices (Bray-Curtis) between measured and calculated 2-D velocity distributions.

respectively. The nonlinear increase of the measured maximum velocity with the increase of Re indicates a change in the flow regime from laminar to transitional or turbulent at Re = 300 (where at a linear increase the maximum velocity would have been 0.82 instead of the observed 0.64 $m \cdot s^{-1}$).

Measurements compared with CFD model

Flow fields determined with PIV at Re = 70, 160 and 300 were compared with those resulted from fluid dynamics calculations, using steady laminar incompressible Navier-Stokes equations. An example of computed flow represented by streamlines clearly shows the two main flow directions near the top and bottom membrane surfaces (Figure 10.8a). Velocity components u_x, u_y and u_z at three heights in the channel are presented in Figure 10.8b for a CFD simulation at Re = 160. The dominant velocity component is in the x direction (u_x between 0–0.4 $m \cdot s^{-1}$), whilst the u_y component is more important close to the top and bottom walls, with ranges between −0.3 and 0.3 $m \cdot s^{-1}$ and near-zero values in the middle of the channel (at $z = 400$ μm) where the flow changes direction. The vertical velocity component, u_z, has the largest values mostly close to the spacer fibres (between −0.2 and 0.2 $m \cdot s^{-1}$), in the regions where the flow passes from one feed spacer mesh element to another.

Comparisons between measured and calculated x-y velocity projections, $\sqrt{u_x^2 + u_y^2}$, in different planes z are presented in Figure 10.7. The Bray-Curtis similarity between measured and calculated velocity distributions are good overall ($S_{BC} > 0.84$). It appears that the model predicts better the velocity distributions at lower Re, where the laminar flow was shown to be steady (Figure 10.5). However, at Re = 160, where slight velocity fluctuations were measured, the model assuming steady laminar flow still compares favourably to the experimental data (Figure 10.7). At Re = 300, the steady laminar assumption does not hold anymore and the measurements cannot be compared with steady flow calculations. The measured maximum velocity (0.64 $m \cdot s^{-1}$) was ~30% less than the velocity calculated with steady laminar flow (0.82 $m \cdot s^{-1}$).

Measured velocity profiles can also be compared with numerical model results along cut lines in the z direction, as shown in Figure 10.3b. This comparison between measured and modelled flow data along a 1-D profile enables further (and more direct) evaluation of model accuracy. For example, calculated velocities on a line across the main flow direction in the middle of a feed spacer mesh element, shown in Figure 10.9a, exhibit good agreement with the measurements. Several steps and maxima can be observed in the measured velocity profiles near the bottom ($z = 200$ μm) and top ($z = 650$ μm) walls, which are also revealed by the simulations. Also, the velocity magnitude is correctly calculated, with a steep change from 0 $m \cdot s^{-1}$ next to the spacer to a maximum of 0.4 $m \cdot s^{-1}$ on the opposite side. In the middle section ($z = 400$ μm) the velocity changes within a narrower interval, showing 3–4 peaks.

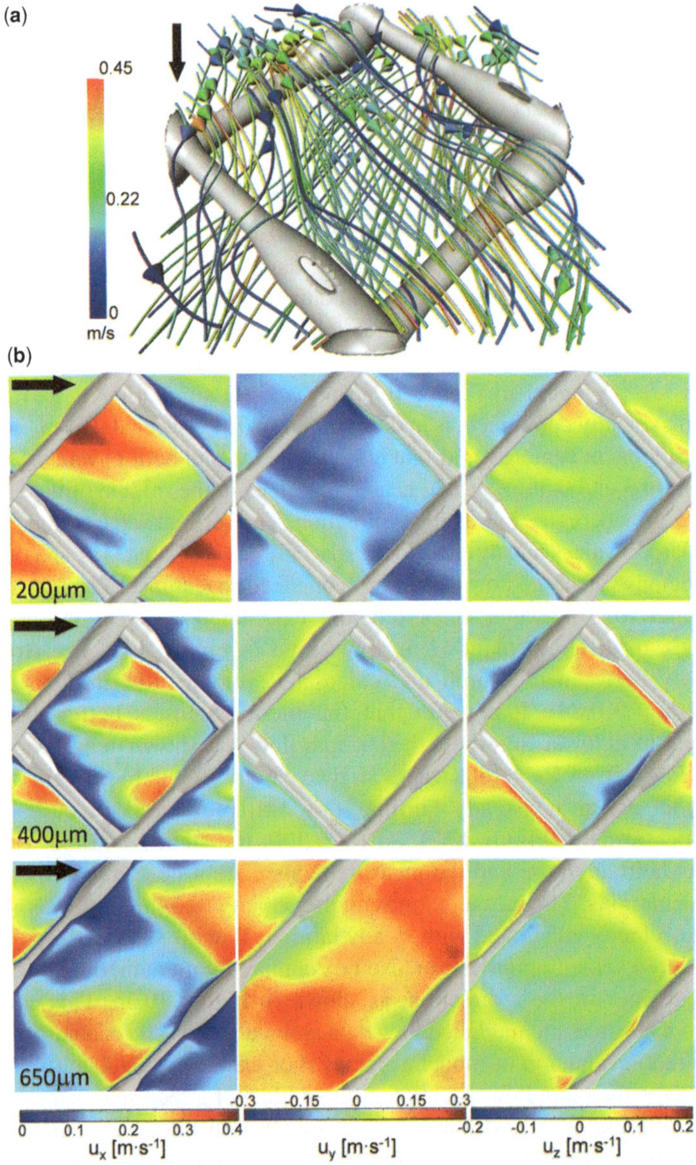

Figure 10.8 (a) Streamlines of simulated flow in a feed spacer element (flow direction from top to bottom). Colours show the velocity magnitude from 0 m · s^{-1} (blue) to 0.45 m · s^{-1} (orange). (b) Computed distribution of velocity components u_x, u_y and u_z at two heights in the flow channel (200, 400 and 650 μm). The CFD simulation was made at Re = 160, assuming steady laminar flow (direction from left to right).

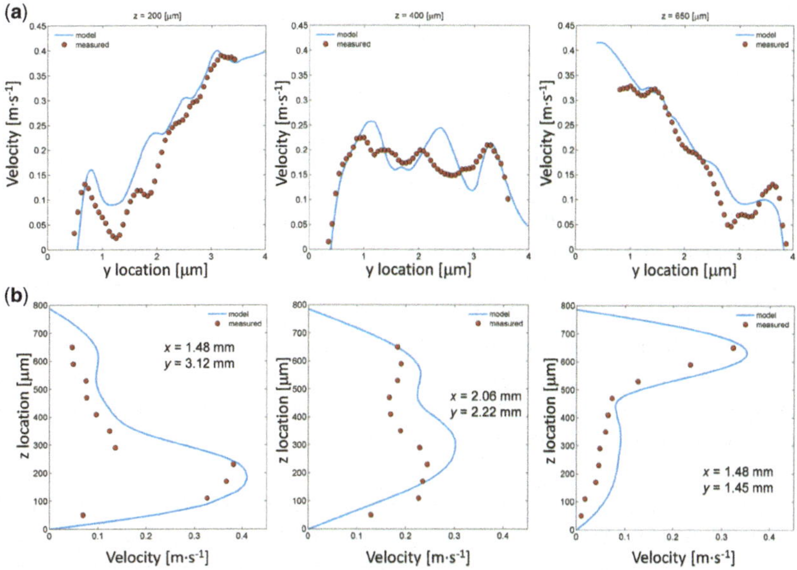

Figure 10.9 Comparison between measured (circle symbols) and calculated (lines) flow velocity magnitude along a line perpendicular to the main flow direction (a) line L_1 in the *x-y* plane at $x \approx 2$ mm and different channel heights ($z = 200$, 400 and 650 μm), and (b) lines L_2, L_3, L_4 in the *y-z* plane at different *x* and *y* positions (see Figure 10.3 for positions of L_1–L_4). The CFD simulation was made at Re = 160, assuming steady laminar flow.

The discrepancies in terms of local velocity peaks between the simulations and the experimental measurements can most likely be attributed to the simplified spacer geometry adopted in the model (the model geometry was constructed with parametric surfaces to approximate the SEM measurements). Measurements of the *x-y* projected velocities at different heights *z* in the channel, shown in Figure 10.9b, are more difficult to perform and with less resolution. Along line L_3 in the middle of the feed spacer mesh element two velocity peaks can be observed on the measured and calculated velocity profiles.

DISCUSSION

In this study water flow in a spacer-filled channel under hydrodynamic conditions used in practice was evaluated. PIV measurements at different locations along the spacer-filled channel of the flow cell at varying heights revealed steady and laminar flow conditions at linear flow velocities commonly used in practice ($u_m = 0.05$–$0.2\ m \cdot s^{-1}$), where also the 3-D CFD simulations were in good agreement with the measured data.

Flow regimes

An important result of this study is that the possibility of laminar and steady flow in the velocity range used in practice ($u_m = 0.07–0.2\ m \cdot s^{-1}$) was confirmed. The flow appears to be laminar with only slight unsteadiness in the upper velocity limit. Although a velocity of $0.2\ m \cdot s^{-1}$ can be achieved in the lead module of a membrane plant, as permeate is produced along the modules, the velocity drops towards the end and it can reach $0.07\ m \cdot s^{-1}$ (Malek *et al.*, 1996). Higher velocities, where the flow was observed to be unsteady (e.g., $0.3\ m \cdot s^{-1}$), are not commonly used in practice for spiral wound reverse osmosis or nanofiltration modules, as these would result in large axial pressure drops.

However, the presence of laminar flow in our experimental conditions should be cautiously interpreted, because in these flow-cell measurements a constant inlet velocity was achieved. It is still possible that in practice the unsteady flow starts to develop at lower velocities if the inlet flow rate is not very well controlled and fluctuates.

A previous study on PIV measurements for Re up to 4000 (Gimmelshtein & Semiat, 2005) reported laminar flow up to Re = 1200 with small fluctuations using feed spacer with a thickness of 800 µm (≈31 mil). These results are in contrast with the findings of the current study, where for Re > 200 unsteady transient flow was measured. However, one should note that the equation used to calculate Re in Gimmelshtein and Semiat (2005) was based on the channel height and did not include any correction to account for the spacer presence.

$$\mathrm{Re} = \frac{2 \cdot u \cdot h \cdot \rho}{\eta} \tag{10.7}$$

where u is the magnitude of local liquid velocity, ρ and η are the density and dynamic viscosity of water and h is the channel height.

Based on the equation used in Gimmelshtein and Semiat (2005), the Re at linear flow velocities used in this study are 110, 256 and 463. Mojab *et al.* (2010) evaluated experimentally and with a numerical model flow characteristics for a spacer geometry scaled-up by a factor 10 compared to the commonly used spacer thickness for Re ranging from 100 to 1000 and reported that transition from steady to unsteady flow occurred above Re = 250.

Flow field measurements using Doppler OCT indicated the presence of swirling flows next to the spacer filaments in the vertical (*y-z*) plane at a linear flow velocity of $0.03\ m \cdot s^{-1}$ (Gao *et al.*, 2013). It was expected to capture the impact of the swirling flows (in the *y-z* plane) on the flow captured with PIV in *x-y* planes, taking in consideration that the Doppler OCT measurements were performed at a much lower linear flow velocity than the PIV measurements, however, no eddies or disturbed flow were found until an inflow velocity of ~$0.3\ m \cdot s^{-1}$. These differences might point to the need for an improved interpretation of Doppler OCT images, as also suggested by Gao *et al.* (2013).

The water flow velocities over the spacer-filled channel height were measured in this study with a much higher spatial resolution ($\Delta z = 40$ µm, Figures 10.4 and 10.9b) than in a previous report of spacer-filled channels for membrane filtration processes ($\Delta z = 300$–600 µm in Willems *et al.*, 2010). This allowed capturing the gradual change in flow direction over the flow channel height (z direction). Focusing on only one feed spacer mesh element provided a more accurate x-y resolution than in other studies, revealing more details of the flow structure and the existence of small recirculation zones behind the spacers at Re = 160 (Figure 10.4), and even larger recirculation zones at Re = 300 (Figure 10.7).

Repeated measurements in different feed spacer mesh elements along the flow cell showed that the flow pattern is very similar at all monitored locations. Similar results were obtained in particle deposition experiments (Radu *et al.*, 2014b). This indicates that flow observations performed in small flow cells are representative, at least for clean (i.e., not fouled) modules.

Solute transfer

Understanding the detailed (micro-scale) flow dynamics is relevant for understanding solute transport (e.g., salts) and for the design of feed spacers that promote the best mixing possible with the least energy input. Our results clearly demonstrate the existence of faster flow near the top and bottom membrane surfaces and a slower flow in the middle of the flow channel. The fast flow will result in better solute transfer near the membranes, where the concentration polarization layer forms. However, the existence of recirculation zones at the linear velocities used in practice seems to be limited. Moreover, the slow laminar flow in the last module (at the end of the pressure vessel with the lowest velocity) would favour salt precipitation and scaling. Knowledge of the detailed flow features could allow further investigations of mass transfer coefficients in the feed channel, using the constant wall concentration approach as suggested in Koutsou *et al.* (2009) and Fimbres-Weihs and Wiley (2010).

CFD model

A key result of this study is the good agreement between the PIV measured and the computed water flow fields. It reveals that the steady laminar flow assumption in the CFD simulations allows a good description of the actual flow in the spacer-filled channel. Furthermore, the obtained results also imply that for clean flow channels the computational domain can be reduced to one feed spacer mesh element with periodic boundary conditions for hydrodynamic calculations, if permeation flux contribution to velocity change is negligible. This approach gives confidence in pursuing model-based studies for designing other feed spacer geometries with reduced pressure drop and fouling, while still providing good solute transfer. Flow pattern has been shown (Radu *et al.*, 2014b) to be of crucial importance in the deposition and development of foulant layers, both for scaling and for biofouling.

The velocity field resulted from numerical simulations is sensitive to the accuracy (or realism) of used feed spacer geometry, which influences the comparison with experimental measurements. It is therefore imperative that the 3-D spacer geometry is represented in the model as closely as possible to the real shape of spacers used in practice. This has been pointed out also by other numerical and experimental studies (Picioreanu *et al.*, 2009; Radu *et al.*, 2014b). In future studies we will use directly the solid geometries reconstructed from computerized tomography (CT) scans.

Further studies

The focal depth on the channel height (z) was set here with approximation within 40 μm. A narrower focal depth would allow taking more optical slices, thus more resolution over the channel height. For this, other lenses with higher numerical aperture and other light sources with lower wavelength (e.g., UV light) could be used.

Tomographic PIV would allow for capture of the flow velocity component in the z-direction (Casey *et al.*, 2013; Elsinga *et al.*, 2006). However, for such a narrow flow channel it would be very difficult to set up cameras to measure the flow from a lateral view. Especially, the presence of an opaque feed spacer will not allow the lateral observation of the flow channel.

For further model validation, other spacer geometries and spacer orientation should be tested. It is of interest to evaluate the impact of fouling on the local velocity field, since the water flow may become unsteady at lower linear flow velocities in the presence of fouling. It should also be tested whether the methods could be applied for forward osmosis (FO) and pressure-retarded osmosis (PRO) flow channels.

CONCLUSIONS

A particle image velocimetry (PIV) method was implemented to measure detailed water flow fields in a flow-cell simulating at small scale conditions encountered in the feed spacer-filled flow channels of spiral-wound membrane modules. Narrow focal depth allowed optical sectioning at multiple (maximum 18) levels along the channel height of this lab-scale spacer-filled flow channel.

At linear velocities between $u = 0.07\text{–}0.2\ \text{m} \cdot \text{s}^{-1}$ (as used in practice) the measured flow appears to be laminar with only slight unsteadiness in the upper velocity limit. At higher velocities ($0.3\ \text{m} \cdot \text{s}^{-1}$) the flow was observed to be unsteady with appearance of local recirculation zones.

Measurements taken at different locations in the flow cell (in several feed spacer mesh elements) showed very similar water flow patterns ($S_{BC} = 0.94\text{–}0.98$), indicating the same hydrodynamic conditions along and across a small flow channel without permeate production.

Computational fluid dynamics simulations based on steady laminar flow assumption were in good agreement with the measured flow fields (lowest $S_{BC} = 0.84$). This supports the use of model-based studies in the optimization of spacer geometries and operational conditions.

Chapter 11

Spacer geometry and particle deposition in spiral wound membrane feed channels

INTRODUCTION

Membrane processes play an important role in drinking and industrial water production. Membranes can be used to remove a wide range of contaminants, from colloids and suspended matter (with microfiltration, MF and ultrafiltration, UF) to ionic species (with nanofiltration, NF and reverse osmosis, RO). A common issue in membrane separation is the accumulation of fouling material at the feed side membrane surface.

Particulate and colloidal matter with size ranging from nm to μm, together with microbial cells and organic macromolecules can be encountered in the NF/RO feed water and deposit to the membrane surface (Tang *et al.*, 2011; Yiantsios *et al.*, 2005). While MF/UF pretreatment may be more effective than conventional granular media filtration (GMF) in preventing particles of different kinds from entering the NF/RO, microorganisms from the feed water have been reported to be present in the NF/RO even after UF pretreatment (Bereschenko *et al.*, 2007).

Bacteria are known to colonize most surfaces in membrane systems and form biofilms (Flemming, 1997) which can result in biofouling. Deposition of bacteria on a surface is a first step required for biofilm formation (Van Loosdrecht *et al.*, 1990) and involves bacterial transport in the liquid followed by adhesion to the surface (Bryers & Characklis, 1982).

To study bacterial deposition, abiotic particles have been used as surrogates for microbial cells in different porous media. Paramonova *et al.* (2006) reported comparable collision efficiency/deposition behavior for microspheres and bacteria in granular activated carbon filtration studies. Passmore *et al.* (2010) suggested that microspheres could be a good model for studying bacteria/virus transport

Author(s): Szilárd Bucs, Joop Kruithof, Mark van Loosdrecht and Johannes Vrouwenvelder
doi: 10.2166/9781780409597_0199

and removal in soil and subsurface environments. Moreover, studies in parallel plate flow cells showed deposition was similar for bacterial cells and polystyrene particles (Meinders *et al.*, 1995).

In-situ microscopic observations have been used in many experimental studies on deposition carried out for MF/UF systems, but only considering simple geometries, i.e., flat-sheet membrane channels (Kang *et al.*, 2004; Knutsen & Davis, 2006; Li *et al.*, 1998, 2003). Reports concerning particle deposition for NF/RO membrane modules are rare (Subramani & Hoek, 2008). In an early study of Li *et al.* (1998), direct observation through a microfiltration membrane was used to evaluate deposition of yeast cells and latex particles of various sizes, in relation to critical flux. Kang *et al.* (2004) studied in detail the importance of physico-chemical interactions for deposition of bacteria, yeast and latex particles for a variety of solution chemistries in a flat-sheet MF membrane flow cell. Their interaction force model indicated that permeation drag and electrostatic repulsion were highly important for the deposition. In a similar study for NF/RO, Subramani and Hoek (2008) found that membrane surface properties (i.e., nano-scale roughness and functionality) may impact the deposition of microorganisms. Moreover, the authors suggest that due to rejection of salts, resulting in concentration polarization layer at the membrane surface, complex effects related to destabilization of colloidal suspensions may play a role.

For NF and RO, the current industrial practice makes use of spiral wound elements, containing feed spacers that keep membrane sheets apart and create the flow channel (Schwinge *et al.*, 2004a). The pioneering work of Neal *et al.* (2003) studied deposition of latex particles for different feed spacer orientations (diamond and ladder) in relation to critical flux of an MF/UF membrane. Their observations indicated that the feed spacer orientation influences the deposition pattern. Recently, Ngene *et al.* (Ngene *et al.*, 2010b) analyzed particle deposition and biofilm formation in micro-structured membrane systems and feed spacer channels. In an attempt to distinguish deposition only from microbial growth, they observed various deposition patterns, function of microstructure geometry.

Besides some experimental investigations, several numerical models for particle deposition have been developed for parallel plate channels (Bowen *et al.*, 1976; Margalit *et al.*, 2013), porous media and membrane systems (Altena & Belfort, 1984; Chellam *et al.*, 1992; Kang *et al.*, 2004; Kim & Zydney, 2006; Ramon & Hoek, 2012; Song & Elimelech, 1995). A recent review (Henry *et al.*, 2012) summarizes the different approaches used for modeling particle transport and deposition. Two approaches were identified depending on the scale: continuum (Eulerian, 'volume-averaged', leads to Partial Differential Equations for convection-diffusion material balances) and discrete (Lagrangian, "particle-based", leads to Ordinary Differential Equations for motion). In the discrete approach, forces act on each particle determining their trajectory (particle-fluid interactions) and deposition (particle-surface interactions). Chellam and Wiesner (Chellam *et al.*, 1992) evaluated the relative importance of various forces on particle transport and concluded that inertia, gravity and drag play a role in particle migration in the far-field region. Altmann and Ripperger (1997) proposed the balance between lift and drag determined particles transport to the surface of a MF membrane. Kim and Zydney (2006) simulated particle trajectory and subsequent deposition in MF/UF membranes considering electrostatic, inertial lift, Van der Waals and Brownian forces. Their numerical model suggested the membrane could remain free of particles due to electrostatic repulsion.

Particle-fluid interactions require knowledge of the fluid flow pattern, involving thus Computational Fluid Dynamics (CFD) for complicated geometries. CFD studies have shown that spacers create a complex flow pattern in the feed channel (Fimbres-Weihs & Wiley, 2010; Koutsou *et al.*, 2009; Picioreanu *et al.*, 2009; Schwinge *et al.*, 2004a).

Recently, Chaumeil and Crapper (2013) simulated particle deposition in spacer feed channels, coupling CFD with discrete element method (DEM). The method involved calculation of several forces (drag, gravity, shear induced lift, surface interaction), resulting in high computational requirements. Moreover, their numerical simulations captured deposition only on very small areas around the joints of spacer filaments and could not provide any overall information on deposition on the membrane.

Despite extensive research on fouling in membrane channels, a systematic study on the effect of feed spacer configuration on particle deposition is still lacking. In-situ observation studies under controlled conditions representative for spiral wound systems are required for a better understanding of bacteria/particle deposition.

The objective of this study was to evaluate both experimentally and with a numerical model particle deposition in feed spacer channels for various spacer orientations under cross-flow conditions. The predictions of the numerical model combining fluid flow simulations in complex geometries with particle trajectory calculations were compared to experimentally observed deposition patterns. In addition, the impact of cross-flow velocity and permeate flux on particle deposition patterns was investigated.

MATERIAL AND METHODS

Experiments

Particle deposition in feed channels was investigated by in-situ microscopic observation within membrane fouling simulators (MFS, i.e., flow cells with hydrodynamic conditions representative for spiral wound modules (Vrouwenvelder *et al.*, 2006)).

Flow cells and cross-flow setup

Two different setups and flow cells were constructed for operation under cross-flow conditions: without permeate production (Figure 11.1a) and with permeate production (Figure 11.1b).

The flow cell for cross-flow studies without permeation (MFS1) consisted of a polyvinylchloride (PVC) bottom connected with screws to an aluminum clamp that

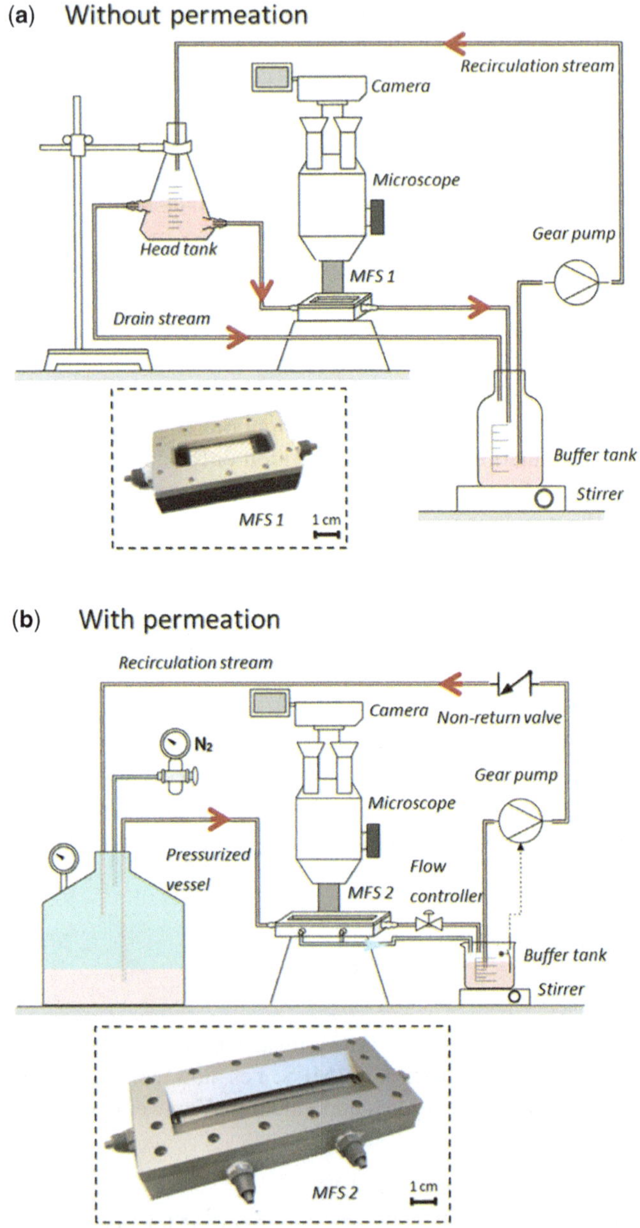

Figure 11.1 Schematic overview of the two setups for direct observation of particle deposition (a) without permeation, (b) with permeation. The inserts illustrate the two flow cells used in the experiments: MFS1 (without permeation) and MFS2 (with permeation).

embedded a 1 mm thick glass cover (Figure 11.1a). The glass cover allowed in-situ non-destructive observation of pattern development. The membrane and spacer coupons with a size of 35 mm × 15 mm were tightly packed within the flow channel.

In order to minimize flow pulsation, a gravity driven flow system was used: the feed (suspension of particles in water) passed from the head tank through the flow cell into a buffer tank (Figure 11.1a). The flow rate was set by adjusting the height of the head tank with respect to the MFS1 outlet and measuring the volumetric liquid flow. To ensure a constant flow rate, the liquid level in the head tank was kept constant by means of a drain stream into the buffer tank. A magnetic stirrer was placed under the buffer tank, keeping the suspension homogeneous. Recirculation was established by pumping the particle suspension from the buffer tank back into the head tank with a gear pump (Cole Parmer, IL, USA).

To investigate the impact of permeate flux on particle deposition patterns, both the flow cell and the setup needed several adjustments to allow operation under pressure. The flow cell for permeation studies (MFS2) was made of aluminum and included a transparent 2 mm thick hardened glass cover that can withstand up to 600 kPa of pressure (Figure 11.1b). In addition, a permeate compartment and two outlets for permeate collection were present. The membrane with feed spacer (120 mm × 40 mm) and permeate spacer (108 mm × 28 mm) were tightly packed within the flow channel.

A pressure vessel was used to deliver a smooth flow to the MFS2. A pressure of 500 kPa was set in the vessel with nitrogen atmosphere. Measured permeate flux at this operating pressure was ~35 L m^{-2} h^{-1}. A manual flow controller (FC8942, Brooks Instrument, PA, USA) situated downstream of MFS2 ensured a constant feed flow rate during the experiment. The suspension was collected in a beaker, equipped with a floating level sensor connected to a microcontroller (Applikon Biotechnology B.V., Schiedam, The Netherlands), which activated a gear pump (Cole Parmer, IL, USA) to send liquid back to the pressure vessel. In the recirculation stream a non-return valve prevented the suspension back-flow into the beaker. Permeate produced in MFS2 was collected in the buffer tank.

At the end of each experiment, rhodamine B, a red dye was injected in the feed to check occurrence of bypass flow from the feed to the permeate side. The collected permeate remained transparent, while the feed solution was colored, indicating the feed and permeate compartments were well sealed. In experiments performed with MFS2 without permeation, the permeate outlets were closed.

Membranes, spacer and particles

Flat sheet nanofiltration membranes (TS80, TriSep, CA, USA) together with 34 mil (863 μm) thick commercially available feed spacers (DOW Chemical Company, MI, USA) were placed in both flow cells. A detailed characterization of membrane properties (i.e., contact angle and surface charge) has been previously reported in Verliefde *et al.* (2009). Permeate spacers (DOW Chemical Company, MI, USA) were added in the corresponding compartment in MFS2.

Red dyed, mono-disperse polystyrene microspheres (Polybead, Polysciences Inc., PA, USA) with a diameter of 3 μm were used. With no surface functionalization, the particles are approximately neutrally buoyant in water, having similar size and density (1.05 g cm^{-3}) to bacterial cells. The feed suspension for each experiment contained approx. 3.6×10^9 particles L^{-1}. Since the focus of this work was to study deposition in relation to hydrodynamic conditions and feed spacer geometry, particle properties and water chemistry were identical for all experiments.

Fouling protocol

New membrane and spacer samples were soaked in ultrapure water (MilliQ, Millipore Corporation, MA, USA) for 2 h prior to each experiment. After placing the flow cell on the microscope stage, the feed flow rate was set to obtain the desired average velocity in the feed channel. Unless mentioned otherwise, the average velocity was 0.14 m s^{-1}, within the range commonly encountered in NF/RO modules (Vrouwenvelder *et al.*, 2009b). To minimize the occurrence of air bubbles within the flow cell, ultrapure water was recirculated during 12 h. For MFS2, during these 12 h a stable permeate flux was obtained. Any bubbles still present in the feed channel were removed by syringe. The particle stock suspension was then injected into the buffer tank that contained MilliQ water (without additional electrolytes). The experiment duration was set to 8 h, long enough to allow the development of clear deposition patterns on the feed spacer, membrane and glass surfaces.

Imaging

The development of deposition patterns was monitored through a zoom stereomicroscope system. The flow cell (MFS1 or MFS2) was placed on top of an SZH-ILLD microscope stage (Olympus, Tokyo, Japan). A G12 PowerShot digital camera (Canon, Tokyo, Japan) was connected to the microscope through an adapter (Carl Zeiss, AG, Oberkochen, Germany), allowing high quality imaging. Two lamps illuminated the flow cell from the sides.

During the deposition experiment, the flow cell was kept in a fixed position so that the same spacer element could be monitored on the camera display. Separate photos of this spacer element were taken with focus on the membrane surface and on the glass surface, at one-hour intervals. At the end of the experiment (after 8 h of recirculation of the feed suspension), several pictures were taken at different positions within the feed channel, as well as an overview picture at lower magnification including multiple spacer elements.

Numerical model

A three-dimensional (3-d) numerical model was constructed to investigate particle deposition patterns in relation to hydrodynamics in membrane feed channels. The model included the following steps: (i) geometry construction, (ii) calculation of fluid flow field around a spacer element and (iii) determination of particle trajectories along the fluid streamlines followed by deposition on different surfaces.

Model geometry

In order to accurately reproduce the geometry used in the experiments, feed spacer samples were imaged by Scanning Electron Microscopy (SEM). The feed spacer samples were sputtered with gold with a JFC-1200 Fine Coater (JEOL Technicks Ltd., Tokyo, Japan) and a JSM 6480 LV SEM microscope (JEOL Technics Ltd., Tokyo, Japan) was used under high vacuum conditions at 10 kV accelerating potential. The micrograph revealed that the spacer consisted of two layers of filaments with different thicknesses. Moreover, the filaments did not have a constant radius but regions with thinning. Measurements of various filament features were performed on the image obtained with the software package ImageJ (National Institute of Health, MD, USA). The measured values were used as parameters for constructing the 3-d geometry in COMSOL Multiphysics (v4.2a Comsol Inc., Burlington, MA, www.comsol.com). The resulted computational domain consisted of a rectangular channel containing a representative spacer element as obstacle in the flow (Figure 11.2).

Fluid flow calculations

The fluid flow in the feed spacer channel was calculated using the Navier–Stokes momentum balance for steady state incompressible laminar flow:

$$\begin{aligned} &\rho(\mathbf{u}\cdot\nabla)\mathbf{u}+\nabla p=\nabla\cdot(\eta\nabla\mathbf{u}) \\ &\nabla\cdot\mathbf{u}=0 \end{aligned} \tag{11.1}$$

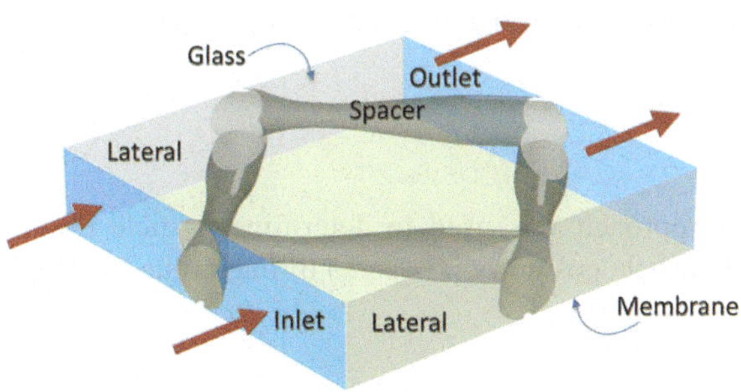

Figure 11.2 Computational domain and boundary conditions used for calculating the flow field. The red arrows indicate the main flow direction. The feed spacer is shown in grey. The top surface corresponds to the glass of the flow cell, while the bottom surface represents the membrane.

with $\mathbf{u} = (u_x, u_y, u_z)$ the vector of local liquid velocity, p liquid pressure, ρ liquid density and η liquid dynamic viscosity. It was assumed that the suspension is very diluted, such that the presence of particles does not affect the liquid flow (Henry *et al.*, 2012).

Periodic flow conditions were imposed between the inlet and outlet boundaries ($u(0,y,z) = u(L_x,y,z)$), as well as between the lateral boundaries ($u(x,0,z) = u(x,0,L_z)$). This approximates the profile corresponding to a spacer element situated within an array, sufficiently far from the flow cell walls not to experience entrance/exit or wall effects. The flow is driven by an imposed pressure difference between inlet and outlet ($\Delta p_{fc} = p(0,y,z) - p(L_x,y,z)$), while no pressure difference exists between the lateral boundaries ($p(x,0,z) = p(x,L_y,z)$). The pressure difference Δp_{fc} required to drive the flow at a certain average cross-flow velocity is obtained by including an additional constraint, $u_{\text{in,avg}}^{\text{set}} = u_{\text{in,avg}}(\Delta p_{fc})$. The calculated average inlet velocity $u_{\text{in,avg}} = \int_{A_{\text{inlet}}} u_x(0,y,z) \cdot A / A_{\text{inlet}}$ needs to match the experimental average velocity corresponding to a spacer element, $u_{\text{in,avg}}^{\text{set}} = Q/(WH\varepsilon)$, determined by the measured flow rate Q, flow cell width W and height H, as well as spacer porosity ε. Without permeation, no-slip boundary conditions were set to all other surfaces (membrane, glass and spacer).

To evaluate numerically the influence of permeate production on particle deposition patterns, the boundary conditions used for fluid flow calculations needed several adjustments. The membrane was treated as a permeable surface and the local permeate flux was defined as $u(x,y,0) = (0,0,L_P\Delta p)$, where L_P is the experimentally measured membrane permeability and Δp represents the local trans-membrane pressure. In order to ensure mass conservation under these conditions, the inlet-outlet periodic condition was replaced by a fully established laminar flow in the inlet with average velocity and a set pressure in the outlet, $p = p_{\text{out}}$. To minimize entrance-exit effects, the size of the computational domain was extended with three more spacer elements in axial direction (x) and the flow field obtained for the third element (out of four) was considered as representative for further particle deposition calculations.

Particle trajectory and deposition

A Lagrangian approach was used to determine particle trajectory, where the progression of particle position $r = (r_x,r_y,r_z)$ through time follows from its velocity $v = (v_x,v_y,v_z)$:

$$\frac{dr}{dt} = v \tag{11.2}$$

In the Newtonian approach, the change in particle velocity over time is its acceleration due to a net resultant force $F = (F_x,F_y,F_z)$ acting on the particle with mass m_p. Forces resulting from interactions of a particle with other surrounding bodies (Henry *et al.*, 2012), such as the liquid (e.g., drag force, shear-induced lift, etc.), other particles (e.g., electrostatic, etc.), a nearby surface (e.g., van der Waals, electrostatic, etc.) or the presence of force field (e.g., electric, gravitational, etc.) are not considered in the current model. Rather, it was assumed that massless particles follow the fluid streamlines (pure advection), such that the particle velocity **v** equals the calculated fluid velocity **u**. As highlighted in Adamczyk *et al.* (2000), hydrodynamics can influence particle deposition via two main mechanisms: macroscopic (convective) transport towards the interface and microscopic (force dominated) shearing close to a surface. Only near rigid interfaces ("walls") the particle velocity starts to differ from the local fluid velocity. It has been proposed in Adamczyk *et al.* (2000) that particle interactions with the wall become significant only at distances smaller than the particle diameter. A simple deposition criterion was therefore chosen: the particle attaches when situated within a threshold distance d_{att} from membrane or glass surfaces. Although considered in other models (Henry *et al.*, 2012), in the present approach the already deposited particles do not influence the next depositing particles.

Model solution

The fluid flow equations 11.1 are solved in COMSOL Multiphysics (v4.2a, Comsol Inc., Burlington, MA, www.comsol.com) with finite element methods on a tetrahedral mesh (maximum size 30 μm).

The resulted flow field **u** is exported on a 3-d Cartesian grid for particle trajectory calculations in MATLAB (MATLAB, 2012a, MathWorks, Natick, MA, www.mathworks.com). A fixed number of particles (N_P) are traveling at each time through the computational domain. Each followed particle is seeded at a certain position on the inlet boundary ($r(0,y,z)$ at $t = 0$ s). The position is defined through a random distribution proportional to $ux(0,y,z)$ (i.e., more particles enter the domain through areas with faster flow). Particle trajectories are calculated for a time interval of 2 s using a forward Euler discretization of equation of motion (11.2) with a time step of 10^{-5} s.

When a particle is situated within $d_{att} = 5$ μm from membrane or glass surfaces or when its velocity is approaching zero ($\|\mathbf{u}\| < 1$ μm s^{-1}), it is denoted as stuck and removed from the pool of moving particles. Deposited particles as well as particles exiting through the outlet are replaced by new particles seeded on the inlet boundary, so that the number of moving particles is always NP. For lateral boundaries, periodicity was imposed, i.e., particles leaving through one boundary re-enter at the corresponding opposite side. This approach allows an increased computational efficiency, however, it alters the real concentration of particles present in the system at a given time. Therefore the actual deposition rates cannot be predicted, but only the place of deposition.

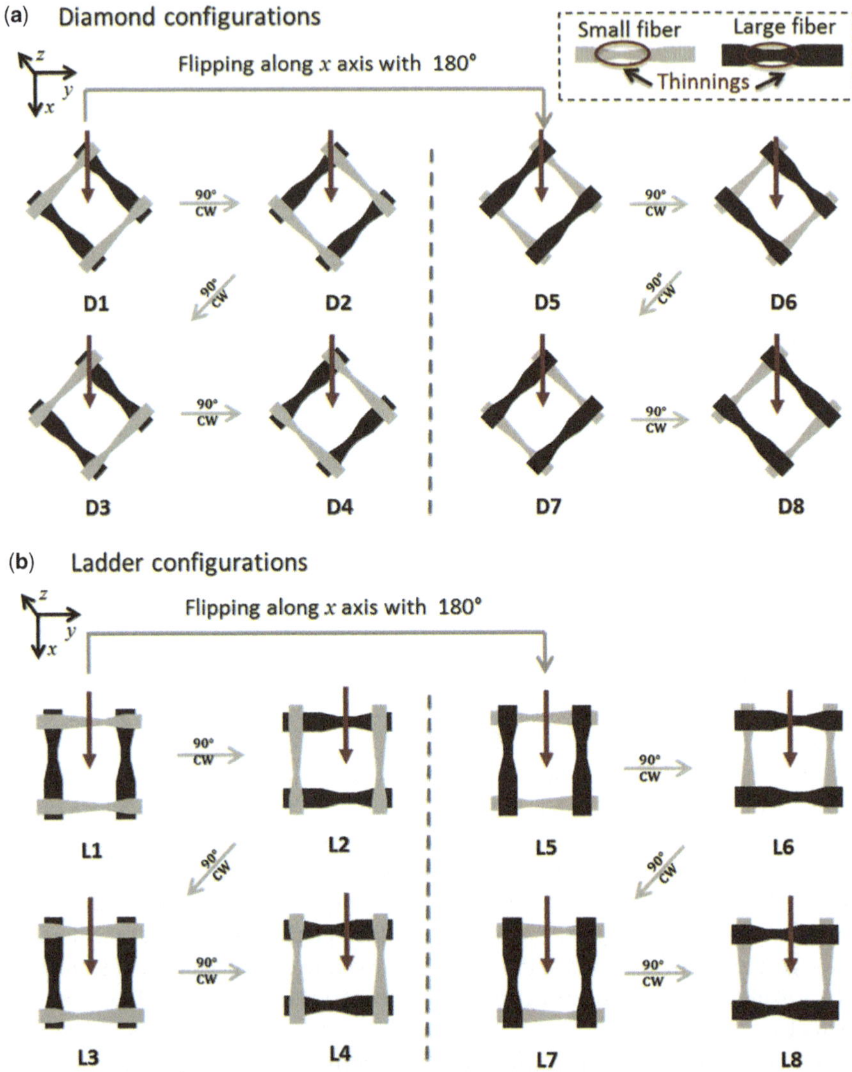

Figure 11.3 Schematic representation of different configurations that can be obtained from the same feed spacer mesh in (a) diamond and (b) ladder orientations. By rotating clockwise (CW) in plane a diamond spacer element, configurations D1–D4 can be obtained. By flipping these configurations, D5–D8 can be obtained. With similar operations on a ladder-oriented spacer element, configurations L1–L8 can be obtained. The red arrow indicates the flow direction. The top layer of spacer fibres is in contact with the glass, while the bottom layer is touching the membrane. The insert indicates specific fibre features.

At the end of a simulation the particles deposited are displayed in different colors according to the surface they are stuck to: particles on the membrane are shown in black, while those on the glass are red.

RESULTS

Two generic feed spacer orientations can be distinguished based on the flow attack angle (α) with respect to the feed spacer (Schwinge *et al.*, 2004a): diamond orientation (D, $\alpha = 45°$) and ladder orientation (L, $\alpha = 90°$). Due to the asymmetric nature of the spacer fibres (i.e., different thickness of the perpendicular layers and variations in fibre diameter) multiple spacer configurations can be derived. For example, in configuration D1 (Figure 11.3) the large fibre is in contact with the membrane, while the small fibre is in contact with the glass. Moreover, both fibre thinning are situated upstream with respect to the centre of the spacer element. Rotation of configuration D1 in the x–y plane in steps of 90° clockwise yields configurations D2–D4. In addition, rotation of configuration D1 along the main flow axis (i.e., flipping) results in configuration D5, from which D6–D8 can be derived. In a similar manner, eight distinct spacer configurations can be obtained for the ladder orientation (Figure 11.3).

Several experiments and numerical simulations were performed to investigate the effect of feed spacer orientation on deposition patterns in membrane systems. Additionally, the impact of cross-flow velocity and permeation were evaluated. An overview of all experiments performed is given in Table 11.1.

Table 11.1 Experimental conditions for particle deposition studies.

Configuration	Cross-flow Velocity	Permeate Flux	Flow Cell
Development of deposition pattern I time			
D5, D7, D8, L4	0.14 m s^{-1}	Without	MFS1
D1–D4	0.14 m s^{-1}	Without	MFS1
D5–D8	0.14 m s^{-1}	Without	MFS1
L1–L4	0.14 m s^{-1}	Without	MFS1
Effect of cross-flow velocity			
D1	0.07 m s^{-1}	Without	MFS1
D1	0.28 m s^{-1}	Without	MFS1
Effect of permeation			
D1	0.14 m s^{-1}	Without	MFS2
D1	0.14 m s^{-1}	With	MFS2

Development of deposition pattern in time

A set of images illustrating the development of the deposition pattern in time is shown in Figure 11.4 for configuration D8. At the start of the experiment, the spacer, membrane and glass surfaces are free of particles. After circulating the particle suspension through the flow cell for 1 h, it can be observed that particles deposit on the feed spacer filaments, particularly in the thinning zone of the filaments. No clear deposition areas are noticeable at this moment with the current microscopic technique on the membrane or glass surfaces. After 4 h, particles deposited on all surfaces are visible. Specific regions are covered with particles on the membrane and glass surfaces. The feed spacer filaments seem to attract more particles. This may be an indication of a higher affinity of particle for the feed spacer material (polypropylene) compared to the membrane material (polyamide), or it could be due to a specific hydrodynamic effect.

Interestingly, deposition areas observed on the membrane and glass after 8 h are not symmetric. On the membrane, a patch upstream the thinning of the small filament (region A) extends along the center of the spacer element (downstream, region B). Additionally, deposition also occurs upstream the crossing of the two layers of filaments (i.e., spacer nodes, region C). On the glass, a patch upstream the large filament is visible (region D), blurry because it is out of the image focus.

For all other configurations (D1–D7) the specific deposition patterns develop gradually in time, similar to the discussed configuration D8. The observed patterns for various feed spacer configurations show good reproducibility across the flow cell (Figure 11.5). The same clean and particle covered membrane areas can be clearly distinguished, repeating in neighboring spacer elements. At the feed spacer nodes, no deposition is observed because these nodes are in contact with both glass and membrane, leaving no space for particle transport. This is the first study to present such a reproducible fouling pattern for different feed spacer configurations and orientations.

Effect of feed spacer orientation on particle deposition

As shown in Figure 11.3, due to the asymmetric nature of spacer filaments, for a certain feed spacer geometry there can be eight distinct diamond and eight distinct ladder configurations. The importance of these geometry changes in relation to particle deposition was studied at constant cross-flow velocity (0.14 m s^{-1}).

Diamond orientation

Deposition patterns for diamond configurations that result from simple rotations in plane of the spacer element are shown in Figure 11.6. Most of the feed spacer surface was covered with particles in all cases. Deposition occurred on both the membrane and the glass, with specific patterns formed depending on the feed spacer orientation. The numerical model is able to describe most of the deposition patterns formed on the membrane and on the glass (Figure 11.6, middle column images).

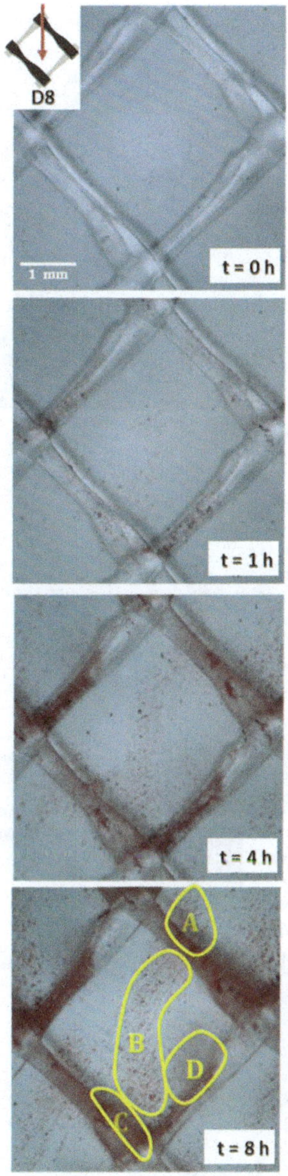

Figure 11.4 Particle deposition pattern development for configuration D8 (Figure 11.3). The focus of the microscope is on the membrane surface in all images. Flow direction is from top to bottom. At $t = 8$ h the yellow contours indicate the specific particle deposition regions on the membrane (a–c) and on the glass (d).

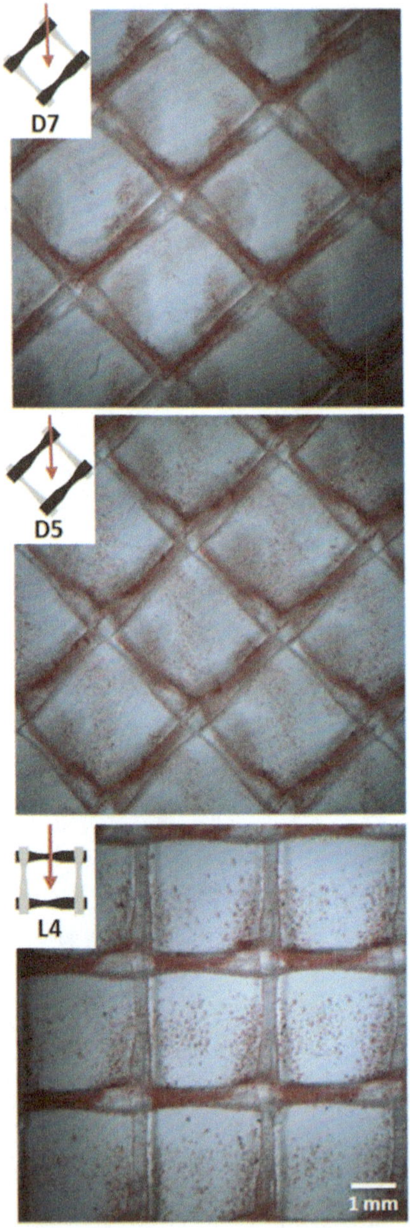

Figure 11.5 Reproducibility of the observed particle deposition patterns within the flow cell (MFS1) for different spacer orientations (D7, D5, L4) after 8 h of operation. Flow is from top to bottom as indicated by the red arrow.

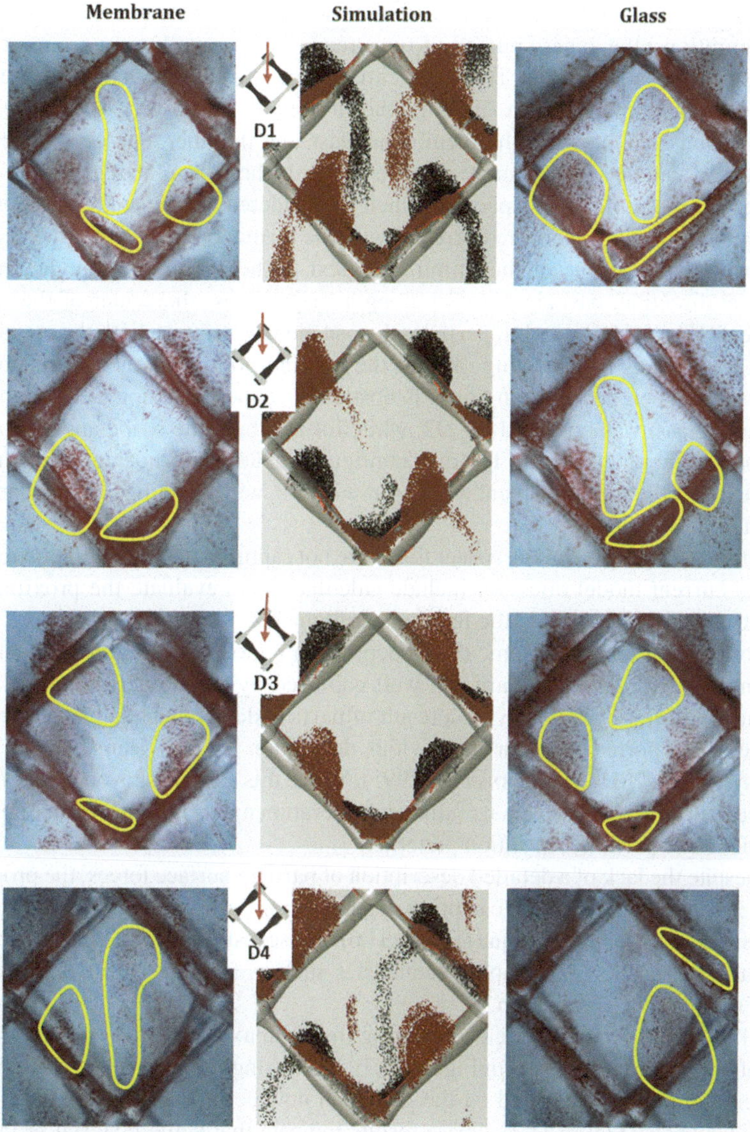

Figure 11.6 Experimental results and numerical model showing particle deposition for spacer configurations D1–D4 at cross-flow velocity 0.14 m s^{-1} without permeation. The experimental images show deposition patterns for microscope focus on the membrane (left panels) and on the glass (right panels). In simulation results (middle column panels) particles on the membrane are shown in black, while those on the glass are red.

Certain areas on the membrane and glass seem to be prone to deposition for all configurations: the zones situated just upstream the spacer filament are important deposition areas (region A). This is potentially due to a sudden change in direction of the liquid flow and distortions in the flow path. Moreover, within the flow constriction zone between the membrane/glass and the fiber, one can expect that the accelerated flow brings more particles towards these specific regions.

For the particular feed spacer geometry considered in this study, the position of filament thinning seems to correlate with the formation of a stripe-like deposit. Whenever at least one of the thinning is next to the spacer mesh node situated in axial direction, particles deposit along a stripe across the center of the spacer element (Figure 11.6). For D1, particles deposit along a stripe on both the membrane and the glass, as the thinning of both the top and bottom filaments are situated upstream. When one thinning only is upstream, stripe formation only on the top surface (glass) was observed for D2, while for D4 the same stripe appears on the bottom (membrane) surface. The only configuration for which no deposition though the center of the spacer element was observed is D3, with thinning of both filaments situated downstream.

Particles deposited on the spacer fibers are not captured by the simulation results, as the current model does not include calculations to evaluate the proximity of particles within 5 μm from the feed spacer filaments.

For configuration D3, some of the deposition regions observed experimentally downstream the filaments are not well captured by the numerical model. The formation of these areas may be a result of particle detachment and re-attachment processes, reported to occur in previous deposition studies (Henry *et al.*, 2012; Lecuyer *et al.*, 2011; Meinders *et al.*, 1995). While this aspect is beyond the scope of the current numerical model, its future consideration may contribute to an improved prediction of particle deposition patterns.

Despite the lack of a detailed description of particle-surface forces, the proposed numerical model is capable to capture the deposition patterns observed via the non-invasive microscopic technique (Figure 11.6). These results illustrate the importance of particle transport in complex geometries such as feed spacer channels for the deposition pattern prediction.

Flipping the feed spacer element along the main axis of the flow yields another set of 4 geometries (D5–D8 in Figure 11.3). Interestingly, for all configurations, the patterns previously obtained on the glass can now be observed on the membrane and vice-versa (Figure 11.7). These results indicate that while material properties may have an influence on a quantitative level on particle deposition rate, the main deposition areas are mostly affected by the flow pattern.

Like in the case of configurations D1–D4, the presence of a deposition area across the center of spacer filament correlates with the position of the filaments thinning.

The main difference between geometries D2 and D6 (for example) is that thick filaments are replaced by thin filaments and the other way around, while the

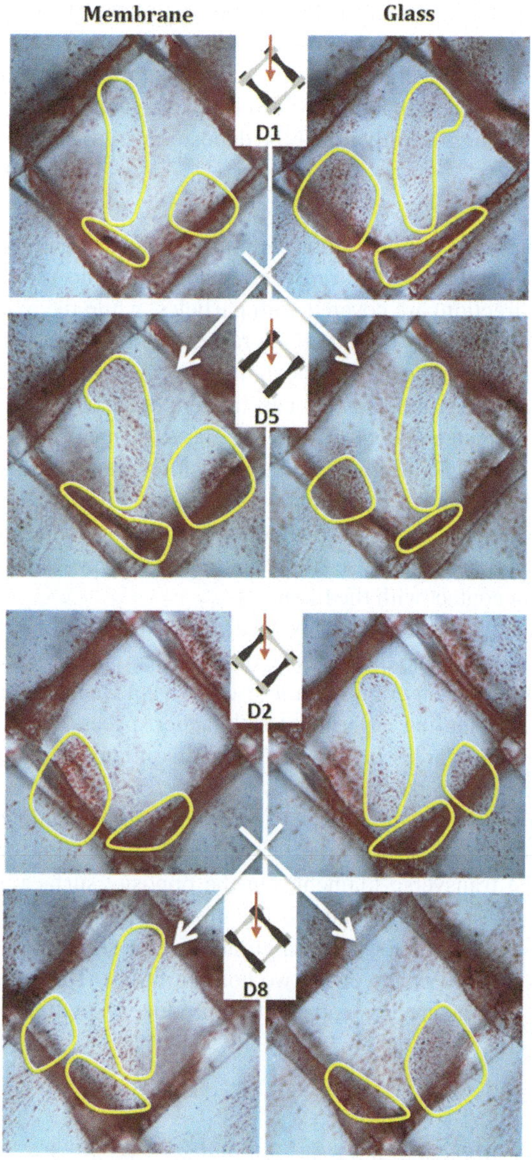

Figure 11.7 Observed particle deposition patterns (marked with yellow contours) for spacer configurations D1, D2 and their flipped analogues D5 and D8 respectively. For the pictures on the left, the microscope focus is on the membrane, while on the right the focus is on the glass. Patterns on the membrane/glass in D1 and D2 correspond to patterns on the glass/membrane in D5 and D8 respectively (see white arrows). Flow is from top to bottom as indicated by the red arrow.

position of the thinning remains the same. For both D2 and D6 configurations the same characteristic deposition regions are observed: an area across the center of the spacer element and patches on the glass and membrane in the filament thinning regions. This may suggest that for the current spacer geometry, the thickness of the filaments does not significantly affect the deposition areas.

Ladder orientation

The observed and simulated deposition patterns for 4 ladder configurations are shown in Figure 11.8. For all cases, the transverse spacer filaments (i.e., perpendicular to the main flow direction) are covered with particles, while the axial filaments (i.e., parallel with the main flow direction) are rather clean. The spacer mesh nodes are free of particles, like in the case of the diamond orientation.

The position of the transverse filaments influences the observed deposition areas. For transverse filaments in contact with the glass, the membrane seems to remain rather clean (L1 and L3), while the main deposition areas can be observed on the glass (Figure 11.8). Conversely, when the transverse filaments are situated in contact with the membrane (L2 and L4), deposition occurs mostly on the membrane (Figure 11.8). Downstream the transverse filaments there is always a deposition free area on the surface in contact with the filaments.

A similar result was reported by Neal *et al.* (2003) for latex microsphere deposition and attributed to the presence of a flow recirculation area. Furthermore, ex-situ imaging based on confocal microscopy has revealed similar patterns for initial stage of a biofouling experiment with ladder spacers (Suwarno *et al.*, 2012), with clean membrane areas behind the transverse filaments and biofilm-covered areas upfront the transverse filament.

Numerical results agree qualitatively with the patterns obtained experimentally (Figure 11.8). Simulations illustrate considerable deposition on the surface in contact with the transverse filament. On the opposite surface, some deposition occurs close to the spacer mesh nodes. The modeled patterns agree better with the experimental observations for the configurations with small filament in transverse position (L1 and L3). For the other two configurations (L2 and L4), the particle trajectory calculations are less capable to reproduce the rather uniformly spread particles across the surface.

Effect of cross-flow velocity

In full-scale NF/RO installations the cross-flow velocity changes from the feed towards the brine end due to permeate production. In addition, depending on feed water type and plant design, several cross-flow velocities can be used during operation. Most NF/RO installations have cross-flow velocities in the range 0.07–0.2 m s^{-1} (Vrouwenvelder *et al.*, 2009b). In this study, the effect of different cross-flow velocities on particle deposition was investigated for feed spacer configuration D1.

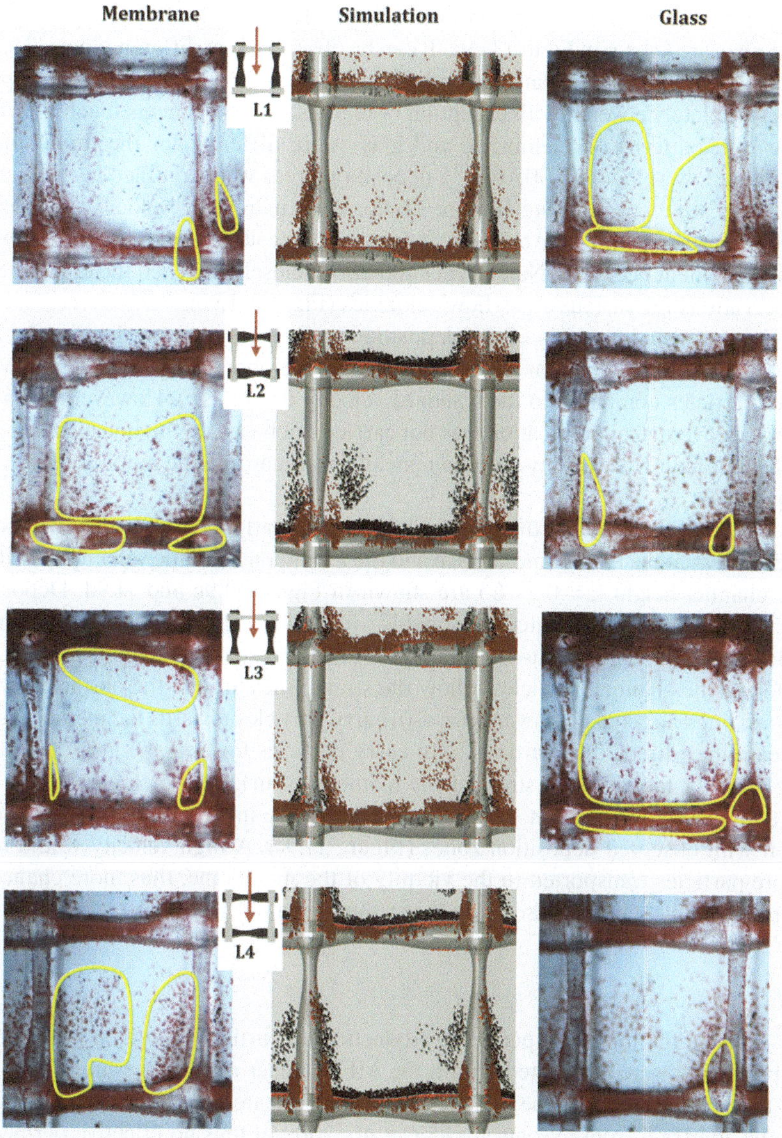

Figure 11.8 Experimental results and numerical model showing particle deposition for spacer configurations L1–L4 at cross-flow velocity 0.14 m s^{-1} without permeation. For the pictures on the left side the microscope focus is on the membrane, while on the right side the focus is on the glass. In simulation results particles on the membrane are shown in black, those on the glass in red.

Experimental deposition patterns obtained after 8 h for three cross-flow velocities are shown in Figure 11.9a. It can be observed that different flow velocities led to distinct deposition patterns. Indifferent of cross-flow velocity, the feed spacer was completely covered with exception of regions where filaments touch a surface. However, patterns on membrane and glass were different for the three studied cases. At low velocity (0.007 m s^{-1}), deposition areas were confined to the vicinity of spacer filament thinning (Figure 11.9a). With increasing velocity, besides the characteristic patches close to filament thinning, deposition along the diagonal of the spacer element occurred. Numerical simulations closely matched the experimental observations for various cross-flow velocities (Figure 11.9b).

From a visual analysis of the deposition patterns, it appears that for the high velocity (0.28 m s^{-1}) both on the glass and on the membrane larger areas are covered with particles compared to the standard velocity (0.14 m s^{-1}). However, given that surface coverage quantification was not carried out based on the images taken, it is rather difficult to draw any definite conclusions regarding the amount of particles deposited.

In an attempt to relate the observed deposition patterns to flow parameters, the shear stress at the membrane surface and the z component of the velocity at halfway the channel height ($z = 0.5 \cdot L_z$) are shown in Figure 11.9c and 11.9d. Deposition areas seem to be associated reasonably well with zones of high magnitude of velocity u_z, perpendicular to the membrane and glass surfaces (Figure 11.9d). Upon entering the channel, particles follow the streamlines around the obstacles, where streamlines directed downwards ($u_z \leq 0$) carry particles towards the membrane and streamlines directed upwards ($u_z \geq 0$) carry particles towards the glass. Similarly, the areas of higher shear stress at the membrane surface (which correspond also to a higher velocity within ~10 μm proximity of the membrane) seem to correlate well with observed deposition zones (Figure 11.9c). A high velocity would mean more particles transported in the vicinity of the membrane, thus more chance for deposition to occur, but also more shear, thus potentially more detachment.

Effect of permeate production

To evaluate the impact of permeate production on particle deposition patterns, two experiments were conducted within the MFS2 under a feed pressure of 500 kPa. In the first experiment, the control case, the permeate tubes were closed (similar to the approach from (Vrouwenvelder *et al.*, 2009d)) thus no effective permeation took place. In the second experiment, permeate was collected in the buffer tank and recirculated together with the feed solution back to the pressure vessel. The numerical model was also adjusted to reproduce similar conditions. Figure 11.10 compares the experimentally observed deposition patterns within MFS2, with and without permeate production. Our results indicate that the same particle deposition pattern was formed even when a permeate flux of ~35 L m^{-2} h^{-1} was produced. Supporting the experimental observations, simulations reveal the formation of identical deposition

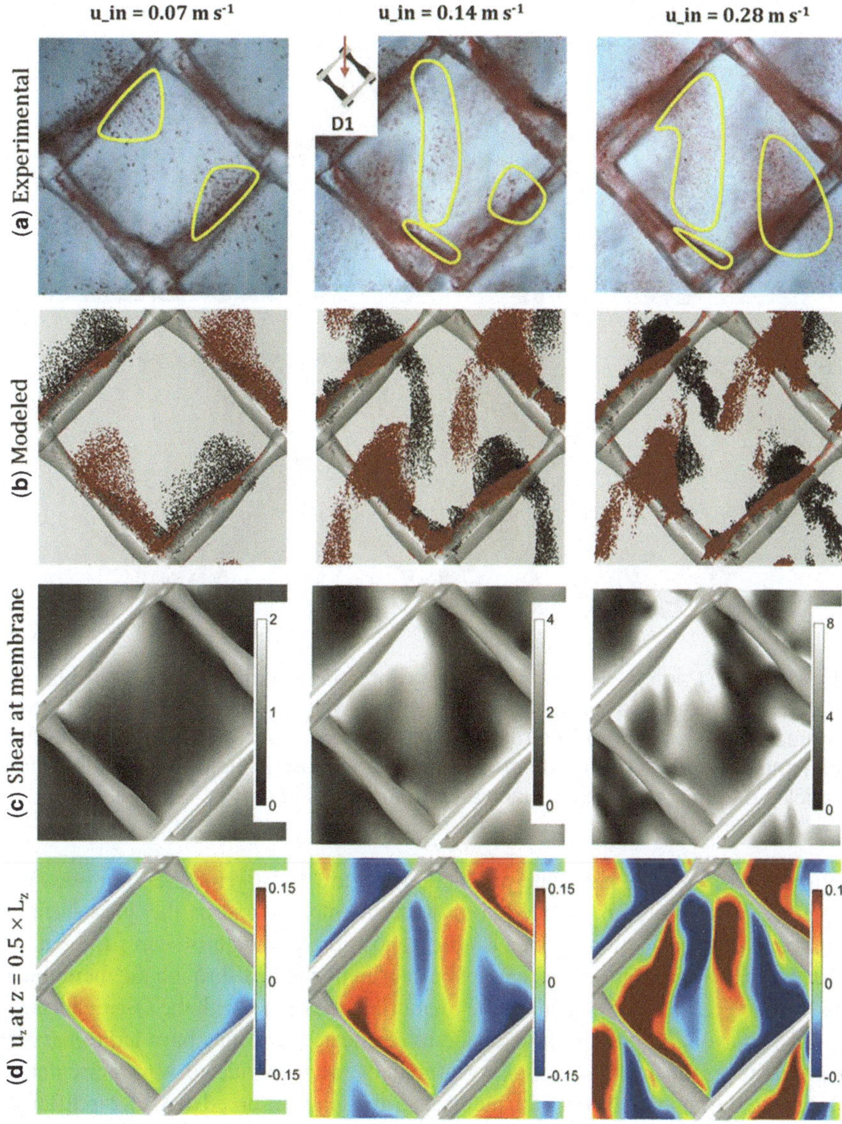

Figure 11.9 The impact of cross-flow velocity on particle deposition patterns for spacer configuration D1: (a) Experimentally observed deposition at the membrane surface; (b) Modelled deposition (particles in black are deposited on the membrane, those in red on the glass); (c) Calculated shear stress at the membrane surface (grey scale, in Pa); (d) Calculated z-velocity component halfway the channel height (colour scale represents the velocity in m s^{-1}). Flow is from top to bottom as indicated by the red arrow.

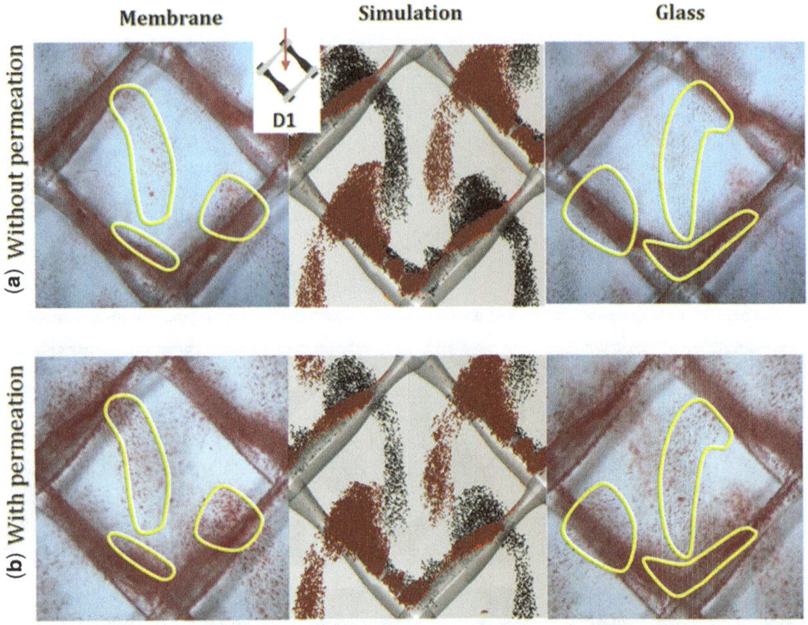

Figure 11.10 Particle deposition patterns for spacer configuration D1 obtained in MFS2 at reference cross-flow velocity: (a) without permeation and (b) with permeation. The pictures illustrate the patterns observed on the membrane (left side) and glass (right side). Numerical simulation results in the middle column show particles on the membrane in black and on the glass in red.

patterns indifferent of permeate production (Figure 11.10). However, the observed patterns appear more pronounced both on the membrane and on the glass surfaces with permeate production. Since also the deposition on the glass seems to be more enhanced with permeation, this cannot be caused by the flux through the membrane. Rather, the grease used for sealing the membrane in contact with the O-ring (needed in order to avoid by-pass from feed side to permeate side), may have led to particle aggregation or an increased sticking efficiency when permeate was recirculated. Moreover, the patterns on both the membrane and glass observed in MFS2 indifferent of permeate production are consistent with the ones obtained in the MFS1 for the same spacer configuration D1 (Figure 11.6) under the same cross-flow velocities. This shows that the two flow cells provide reproducible hydrodynamic conditions.

DISCUSSION

In this study, in-situ microscopic observations have revealed distinct deposition patterns depending on feed spacer orientation (Figures 11.6–11.8) and cross-flow velocity (Figure 11.9). Other not reported simulations with our model investigating variations in spacer geometry (e.g., different form of thinning in the threads) also showed a high sensitivity of the deposition pattern on the geometry. Characteristic deposition areas were not influenced by permeate production (Figure 11.10). The numerical model including hydrodynamics and simple deposition mechanism described well the experimentally observed patterns. This demonstrates the importance of hydrodynamic conditions and feed spacer geometry for particle deposition.

Importance of hydrodynamic conditions and spacer geometry

As previously discussed in several studies (Elimelech & O'Melia, 1990; Henry *et al.*, 2012; Meinders *et al.*, 1995), particle deposition is controlled by transport of particles and their subsequent fixation to surfaces. Before any interface forces can become significant, a particle needs to be in the vicinity of a wall (Adamczyk *et al.*, 2000). Within this work, we focused on transport of particles in the feed channel with spacers and assumed a very simple "rule" for deposition, based on capture distance.

Spacer geometry and orientation

The experimental results and numerical simulations presented here are in agreement with the deposition patterns visualized by Neal *et al.* (2003) for two generic spacer orientations, diamond and ladder. However, a more extensive analysis is carried out in this work, considering all possible spacer configurations that can be obtained due to asymmetry of fibers. Our results illustrate that simple rotations in plane of the feed spacer element have consequences on hydrodynamics and transport properties, leading to quite different fouling patterns (Figures 11.6–11.8). This is the first study to point towards such a detailed characterization of particle deposition in membrane feed channels. When flipping the feed spacer geometry along the x axis the main consequence is a mirroring of the flow pattern. The very good correspondence between deposition patterns from the membrane/glass from D1–D4 with the glass/membrane in D5–D8 supports the ideas that: (1) the spacer geometry itself creates an asymmetry in the flow pattern; (2) particle transport is more influential than surface properties (which would influence sticking efficiency) in our system (Figure 11.7). In addition, experiments performed with carboxylate-modified particles (Polybead, Polysciences Inc., PA, USA) gave identical deposition patterns with the non-functionalized ones (data not shown). This further suggests that not surface properties but hydrodynamics is in the studied conditions the main factor affecting the deposition patterns.

The feed spacer geometry (i.e., fiber diameter and thinning regions) is shown to determine the observed deposition patterns by affecting the hydrodynamics (Figures 11.6–11.9). Other computational studies (Picioreanu *et al.*, 2009) also pointed to the importance of considering a detailed spacer geometry, which can significantly influence the flow pattern, pressure drop, and mixing. Simply representing the spacer strands having variable diameter with cylinders is not an accurate approximation for membrane feed channel simulations.

Flow velocity

Cross-flow velocity can affect deposition in various ways: on one hand a faster cross-flow may result in particle removal (Belfort *et al.*, 1994) while on the other hand a higher velocity means a higher particle load, associated with more potential for deposition (Busscher & Van Der Mei, 2006). Previous experimental studies report increased deposition at low cross-flow assumingly due to lower shear (Chong *et al.*, 2008b; Subramani & Hoek, 2008), while in other works higher cross-flow and shear stress is associated with enhanced deposition (Li *et al.*, 2006; Paris *et al.*, 2007). Koutsou *et al.* (Koutsou *et al.*, 2009) also concluded that observed regions of more fouling with humic acids can be related to simulated areas of higher shear stress. However, feed spacer geometry (which our study shows can be important for deposition patterns) was not identical in the simulation and experiment of Koutsou *et al.* (Koutsou *et al.*, 2009). Other studies found no effect of cross-flow velocity on microbial deposition (Kang *et al.*, 2004). Higher flow leads to more particle transport to a surface, but also to a higher detachment (Lecuyer *et al.*, 2011). The combination of these opposing effects can result in an optimal velocity for particle accumulation. The numerical study of Margalit *et al.* (2013) shows indeed that a maximum deposition coefficient exists function of shear rate and such an effect is dependent on the particle size. Our experimental observations and numerical model indicate deposition of particles in the micrometer size range in areas of large shear, with apparently more covered area at higher cross-flow velocity (Figure 11.9). These findings can be attributed to more particles transported to the vicinity of the membrane at higher cross-flow velocity, without yet having crossed the maximum deposition shear stress.

Permeation effects

It has been previously reported that due to permeate drag, there may be an enhanced deposition of various foulants in NF/RO (Eshed *et al.*, 2008; Subramani & Hoek, 2008; Tang *et al.*, 2011). This is for particles often related to the critical flux concept (i.e., the flux below which no deposition occurs, (Howell, 1995); or the flux at which fouling becomes noticeable, (Bacchin *et al.*, 2006)). In our experiments and simulations there was no essential difference between deposition patterns obtained without permeation or with a permeate flux of ~35 L m^{-2} h^{-1} (which is higher than commonly applied in practice for NF/RO). The deposition in absence of permeation clearly shows the critical flux concept is not applicable for spiral wound NF/RO systems. Despite a common principle for all membrane systems (a cross-flow and a permeation flow, (Belfort *et al.*, 1994)), the different permeation flows relative to the cross-flow velocity and module design (e.g., hollow fiber, flat sheet or spiral wound) will result in rather different behavior in terms of fouling (Vrouwenvelder *et al.*, 2009d). For operational parameters in accordance with the experiments from our study (i.e., identical pressure, cross-flow, membrane permeability), the model predicts a local permeate velocity of ~10 $\mu m\ s^{-1}$, orders of magnitude lower than the average cross-flow velocity of ~0.14 $m\ s^{-1}$. Especially in NF and RO systems, the permeate flux is so slow that this convective transport due to permeation can hardly enhance the particle deposition. The extrapolation of various concepts applicable for MF/UF to spiral wound NF/RO should therefore be done with care.

Particles need to be transported close to a surface in order to have any particle-surface and particle-fluid interactions that may affect their deposition (Adamczyk *et al.*, 2000). In the current model we assumed a 100% sticking efficiency, which is not realistic (Meinders *et al.*, 1995; Paramonova *et al.*, 2006; Yiantsios & Karabelas, 2003). The experiments where a small amount of grease was present showed a faster accumulation of particles, likely caused by increased sticking efficiency due to more hydrophobic surfaces, indicates the importance of this factor. Models for predicting sticking efficiency are available in literature (Elimelech & O'Melia, 1990; Hahn & O'Melia, 2004) however they contain a large number of unknown parameters. Furthermore, as NF/RO membranes reject salts, high variations in local salt concentration due to concentration polarization can be found at the membrane surface in spacer filled channels (Radu *et al.*, 2014a) and may impact particle-surface interactions (Subramani & Hoek, 2008). Therefore, permeation might influence the sticking efficiency of particles approaching the membrane surface when electrostatic and van der Waals interactions (DLVO) are dominant, but it is not expected to change the deposition pattern significantly. The fact that the transport model proposed in this study already gave good qualitative results on deposition patterns and considering the uncertainty when estimating particles-surface interaction parameters led us to the decision not to pursue a more complicated model.

While more sophisticated force interactions would be required for a full quantitative evaluation (Henry *et al.*, 2012), the good agreement between our model and experiments emphasizes that bulk transport in spacer filled channel is of major importance for the particle deposition pattern in membrane systems. However, it must be noted that the current model does not predict deposition rates, but only the deposition pattern.

Implications for practice

The results presented in this study have practical implications for membrane fouling studies and evaluations by autopsy. For cases when an asymmetric feed spacer is present in the channel it is clear that considering only the bottom or only the top membrane sheet can result in significant errors in fouling quantification during autopsies. Moreover, the heterogeneous flow distribution corresponding to different spacer configurations may become a source of inaccuracy when comparing experimental findings from different experimental runs (as in (Suwarno *et al.*, 2012)). The way the spacer is introduced in laboratory membrane systems is usually not described in detail, however as indicated there are 8 different options to place the spacer at a 45° attack angle. When in parallel experiments the spacer is placed differently this is a cause of experimental variability.

If one assumes that the abiotic particles are representative for microbial cells (Meinders *et al.*, 1995; Paramonova *et al.*, 2006; Passmore *et al.*, 2010) that would develop into a mature biofilm, then the biofilm formation would start at the spacer filaments. Deposition and the resulted biofilm on the feed spacer filaments is a practical concern resulting in dramatic increase of feed channel pressure drop, as shown both experimentally (Vrouwenvelder *et al.*, 2009b, 2009d) and numerically (Picioreanu *et al.*, 2009). Although the actual contribution of attached cells to increase in biofilm amount is expected to be rather low (as the biofilm is mostly the result of cell growth, not deposition), the location of the initial deposits could impact performance indicators of the separation process (Bucs *et al.*, 2014a).

Further studies

It would be of great interest to develop a quantitative approach for assessing the particle deposition. Particle deposition patterns need to be observed in-situ, under flow conditions, which limits the range of usable methods. For example, details regarding the three-dimensional structure of the deposit could be obtained by coupling fluorescent abiotic particles or labeled microorganisms with in-situ Confocal Scanning Laser Microscopy (Beaufort *et al.*, 2011).

While the use of abiotic particles for deposition studies has several advantages by creating controlled conditions (e.g., well defined shape and density, no growth, no extracellular polymer production) and facilitate visualization, ultimately, various bacterial strains may behave differently. Given the number of microbial species that have been identified during membrane autopsies in membrane plants (Bereschenko *et al.*, 2007; Pang *et al.*, 2005) it would be rather difficult to select for a certain microorganism as representative. Other studies could examine deposition patterns in spacer-filled channels of NF/RO for a variety of colloids (ferric oxides, silica as well as natural organic matter, humic acids (Tang *et al.*, 2011), and transparent exopolymeric particles (Bar-Zeev *et al.*, 2012). It would be particularly important to further examine the impact of flow conditions and feed spacer geometry for colloidal particles of smaller size (i.e., sub-micron range) that can pass even through advanced pretreatment systems and deposit on NF/RO membranes.

For quantitative prediction, the numerical approach needs further refinement. The model should consider also deposition on the feed spacer filaments. Moreover, it would be of interest to develop a multi-scale model. In this way, the particle transport in the bulk fluid presented in this study could be combined with DEM simulations including various forces (e.g., drag, shear-induced lift, etc.) acting on a particle in the vicinity of a wall (Chaumeil & Crapper, 2013).

Anti-fouling developments in membrane processes have been mainly focused on improving material properties for higher fouling resistance (Kang & Cao, 2012; Rana & Matsuura, 2010). In the future, considerable interest should also be oriented towards optimization of the membrane module. As shown in this study, small changes in feed spacer geometry (e.g., the thinning in the spacer threads resulting from extrusion) can have an impact on particle deposition patterns. This opens the potential to design spacers that minimize deposition effects and thereby fouling related to particle and bacterial cell accumulation.

CONCLUSIONS

A detailed experimental and computational evaluation of particle deposition patterns in spiral wound NF/RO feed channels revealed that:

- Distinct and reproducible deposition patterns were formed as function of feed spacer orientation. Rotations or mirroring of the same spacer element resulted in deposition at different locations.
- Three-dimensional numerical simulations combining fluid flow and particle trajectory calculations could realistically describe the patterns observed experimentally.
- Cross-flow velocity affected specific deposition zones. High shear at the membrane surface roughly corresponded to areas covered with particles.
- Deposition occurred under cross-flow conditions on the spacer, membrane and glass surfaces, in the absence of any permeate production. Permeation did not significantly influence the deposition regions

Chapter 12

Characterization of feed channel spacer performance using geometries obtained by X-ray computed tomography

INTRODUCTION

Membrane operations are used increasingly for the removal of various contaminants from water. For instance, microfiltration (MF) and ultrafiltration (UF) separate particulate material, while nanofiltration (NF) and reverse osmosis (RO) retain charged solutes. Industrially, the removal of these contaminants in order to produce potable water is performed using spiral-wound membrane modules. Optimisation of the performance of spiral-wound modules has been focussed mainly on the development of membranes (Lee *et al.*, 2011) and the design of feed channel spacers (Fimbres-Weihs & Wiley, 2010).

Research into feed spacer design concentrates mainly on the effect of the spacer geometry on the hydrodynamics in the spacer-filled channel. Hydrodynamics, in turn, influences all other performance indicators. The spacer geometry determines the power input required to overcome the hydraulic resistance imposed on the flow, which is dominated by form drag of the spacer filaments and losses due to the change in direction of flow (Da Costa *et al.*, 1994). The dissipation of momentum and the resulting flow pattern are linked to the distribution of shear stress on the spacer filaments and membranes. Disruption of the hydrodynamic boundary layer by means of liquid recirculation regions influences the transport of solutes from and of particulate material towards the membranes ("The effect of turbulence promoters on mass transfer – numerical analysis and flow visualization," 1982), which in turn affects scaling (Radu *et al.*, 2014a), biofouling processes and the permeate flux (Bucs *et al.*, 2014a; Radu *et al.*, 2010). Moreover, these processes are coupled to one another to increasing extent with increasing time of operation (Vrouwenvelder

Author(s): Szilárd Bucs, Joop Kruithof, Mark van Loosdrecht and Johannes Vrouwenvelder
doi: 10.2166/9781780409597_0227

et al., 2009a). Therefore, all of these phenomena have to be taken into account when optimising the feed spacer design (Radu *et al.*, 2015).

Numerical modelling is a powerful tool to assess the impact of design parameters on the process performance, for a wide range of operational conditions encountered in spiral-wound modules. Computational fluid dynamics (CFD) coupled to transport of solutes has been employed in various studies to model flow and permeation, trying to find the optimal feed spacer design (Fimbres-Weihs & Wiley, 2010; Wiley & Fletcher, 2003). Ranade and Kumar (Ranade & Kumar, 2006) showed that model solutions obtained using a flat spacer-filled computational domain did not significantly differ from those using a curved domain when studying the hydrodynamics in a spiral-wound module, which had also been shown experimentally by Schock and Miquel (Schock & Miquel, 1987). Flat spacer-filled computational domains have been thus used extensively for studying the effect of alignment and shape of spacer strands in two-dimensional (2-D) as well as three-dimensional (3-D) models. Most of the work in 2-D has been conducted on a ladder-type spacer strand configuration, the spacer strands being aligned transversal and axial to the main flow direction. The effect of the ratios between the diameter of cylindrical strands, strand distance and channel height on the (unsteadiness of) fluid flow (Da Costa *et al.*, 1994; Schwinge *et al.*, 2002b) and solute transport (Cao *et al.*, 2001; Fimbres-Weihs *et al.*, 2006; Schwinge *et al.*, 2002a) has been studied extensively. Furthermore, the alignment of cylindrical strands along the channel height has been investigated using various configurations of the transversal strands, namely zig-zag (alternating on top and bottom membranes), fully submerged and all aligned with either top or bottom (Schwinge *et al.*, 2002a, b, 2003). Moreover, these models have been extended to assess the impact of strand size and alignment on scaling and (bio) fouling in reverse osmosis for desalination (Radu *et al.*, 2010, 2014a, 2015), and on nutrient removal performance of membrane biofilm reactors (MBfR) (Martin *et al.*, 2013). Although mostly cylindrical strands have been considered, other cross-sectional shapes like triangular, squared (Ahmad *et al.*, 2005), saw-tooth (Liu *et al.*, 2015) and elliptical (Sousa *et al.*, 2014) have also been evaluated.

Simulations in 2-D have allowed for fast computation and qualitative assessment of design parameters for ladder-type spacers. However, quantitative evaluations of spacer designs of which the strand geometries are not symmetrical in the direction lateral to the main flow cannot be performed in 2-D. The increasing availability of computational resources (i.e., powerful computers; stable, fast and accurate numerical methods; versatile and user-friendly CFD software) has allowed the study of various feed spacer designs using 3-D models without this limitation. Work in 3-D has been performed for multiple filtration technologies using spiral-wound membrane modules (NF, MF, FO), although most work has been centred around reverse osmosis and ultrafiltration (Fimbres-Weihs & Wiley, 2010). The increasing availability of 3-D modelling also triggered the design of new, non-conventional feed spacers with strand cross-sections other than circular (Ranade & Kumar, 2006; Santos *et al.*, 2007). Furthermore, multi-layer spacers with more than two layers of crossing spacer strands used conventionally (Li *et al.*, 2005; Schwinge *et al.*, 2004b) and spacer designs consisting of unconnected support structures, varying in shape, have been evaluated (Ngene *et al.*, 2010b). These efforts aside, the majority of studies on optimal spacer design remained focussed on the impact of alignment of cylindrical spacer filaments (flow attack angle, α, and internal strand angle, β) and the impact of their dimensions in relation to their mutual distance and channel height on hydrodynamics and solute transport (Fimbres-Weihs & Wiley, 2007; Karode & Kumar, 2001; Koutsou *et al.*, 2004, 2007; Lau *et al.*, 2009; Li *et al.*, 2002a, b; Ranade & Kumar, 2006; Shakaib *et al.*, 2007). Mainly, this is due to their presumed resemblance to designs already used in practice.

While efforts to accurately represent spacer geometries used in practice in 3-D modelling remained limited to the alignment and diameter of cylindrical strands, less work was directed towards the investigation of the shape and tilt of spacer filaments themselves. However, it has been shown using stereo light microscopy and scanning electron microscopy (SEM) (Bucs *et al.*, 2015; Radu *et al.*, 2014b) and optical coherence tomography (OCT) (West *et al.*, 2016) that the filaments from designs used in practice vary in thickness when looking from the top. These strands are thus far from cylindrical, most likely due to the polymer extrusion technique used in manufacturing. Geometries formed by filaments constructed from extruded circles with varying diameter and truncated cones have been used to better represent these physical characteristics (Picioreanu *et al.*, 2009; Radu *et al.*, 2014b). Furthermore, Picioreanu *et al.* (Picioreanu *et al.*, 2009) reported that, compared to these more realistic representations, model solutions obtained from geometries with cylindrical filaments may overestimate the pressure drop over the computational domain representing the feed channel. Moreover, this approach has been successfully used to compute axial pressure drop, which was in agreement with measurements using a flow cell for a new spacer design produced using 3-D printing (Siddiqui *et al.*, 2016).

More accurate geometry representations, however, remain crucial when one's goal is to compare results from modelling studies to experimental measurements on commercially available spacer designs. Furthermore, insight in the effect of spacer strand shape may lead to improved spacer designs. A recent study in which CFD results were compared to experiments using particle image velocimetry opted for more accurate 3-D spacer geometry representations (Bucs *et al.*, 2015). In addition, modelled biofouling on spacer filaments was shown to be more detrimental to the calculated pressure drop than fouling on the membrane (Bucs *et al.*, 2014a), underlining the importance of accurate geometry representation for the evaluation of a spacer design.

Compared to other available 3-D imaging methods (e.g., confocal laser scanning microscopy, CLSM, optical coherence tomography, OCT, magnetic resonance imaging, MRI), X-ray computed tomography (CT) provides a series of advantages: it can be used for non-transparent samples, requires little to no sample preparation, has a high spatial resolution, is unaffected by sample surface gloss often encountered with optical methods, allows the presence of metals and the scanning equipment is becoming increasingly available. The CT procedure uses 2-D images acquired at different angles on the sample to reconstruct the actual 3-D object (Herman, 2009).

The aim of this study was to develop a method to obtain accurate 3-D geometry representations for any given spacer design. First, a workflow was developed to construct precise 3-D geometry representations from X-ray computed tomography scans for several commercially available feed spacers. The pressure drop calculated from simulations using CFD was compared to experimental data measured using a flow cell for average effective flow velocities up to 0.2 m/s as used in industrial reverse osmosis membrane operations (Vrouwenvelder *et al.*, 2009a) Moreover, the feed spacers were compared based on shape-induced friction, efficiency of membrane shear stress generation compared to the power input and local shear and velocity distributions. Finally, CFD model solutions were compared to results obtained using conventional representations using strands with circular cross-section from literature.

MATERIAL AND METHODS

Spacer types

Five type of spacers, all 34 mil thick (~863 μm), were analysed in this study (Figure 12.1). The first two types are commercially available from (Conwed Plastics, Minneapolis USA), called here *CON-1* and *CON-2*). These have a similar filament shape and an internal strand angle (β) of 90 degrees, but different strand count (i.e., number of strands per unit of length) and a slight difference in strand thickness. Typically, the fibres have an quasi-elliptical section, tilted in respect to the main flow direction (Figure 12.1a, b). The third spacer was from Hydranautics, also with 90 degrees internal strand angle and elliptical in cross-section, but with profound thinner regions between the strand intersections (called *HYD*) (Figure 12.1d). A spacer from DOW Chemical with an internal strand angle of 70 degrees was also evaluated (called here DOW). This spacer presented strand cross-sections with a small eccentricity (i.e., a small deviation from circular) and almost uniform strand thickness (Figure 12.1e).

Finally, a new spacer design from Lanxess AG (*LXS-ASD*) had alternated strand thickness with the scope of increasing the porosity of the flow channel and therefore reducing the observed pressure drop (Figure 12.1c).

Spacer geometry acquisition and processing

The workflow used to convert the 3-D density information from thc CT scanning procedure to a solid shape suitable for calculations using computational fluid dynamics (CFD) is summarized in Figure 12.2.

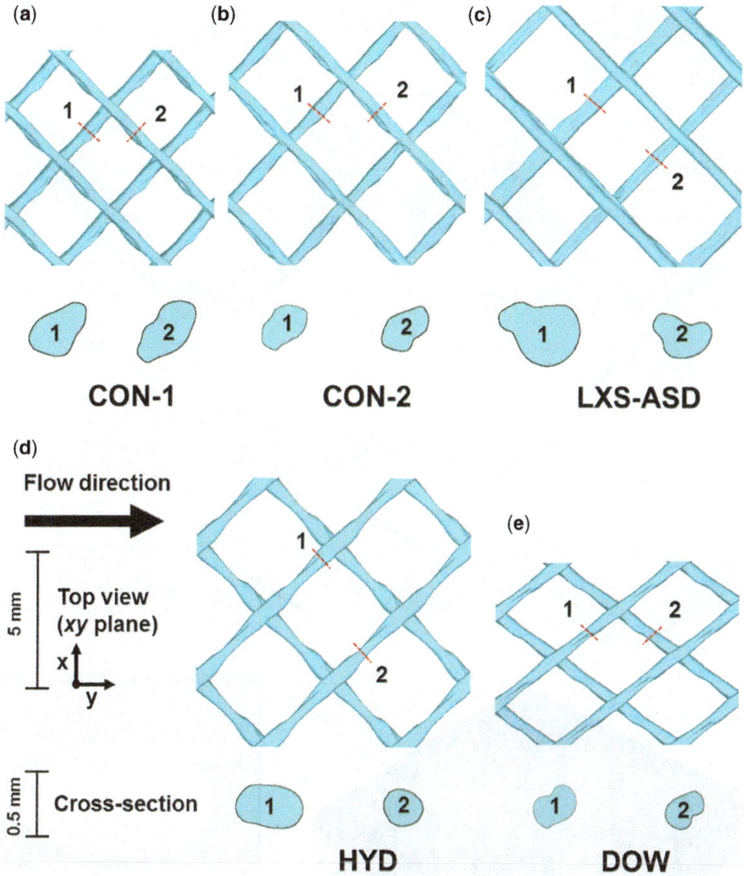

Figure 12.1 CT scans of the evaluated feed spacer designs. The top views show the spacers in the xy-plane. Two cross-sections normal to the strand axis (90°), at positions along the strands indicated by dashed red lines, are shown below each top view. All images are presented at the same scale (scale bar for top views, 5 mm; for cross-sections, 0.5 mm). The acronyms refer to the spacer manufacturer: Conwed (CON-1, CON-2), LANXESS (LXS-ASD), Hydranautics (HYD) and DOW Chemical (DOW).

X-ray computed tomography procedure and surface triangulation

Raw CT scans of all selected feed channel spacers were obtained from various sources. For the scan of CON-1, the spacer net was sandwiched between two pieces of expanded polystyrene (Styrofoam) to increase the flatness. However, the other spacers were glued on a support with the plane of the spacer net in vertical position since the rigidity of the small samples was sufficient to ensure the flatness. The scanning procedure used for each spacer design and the properties

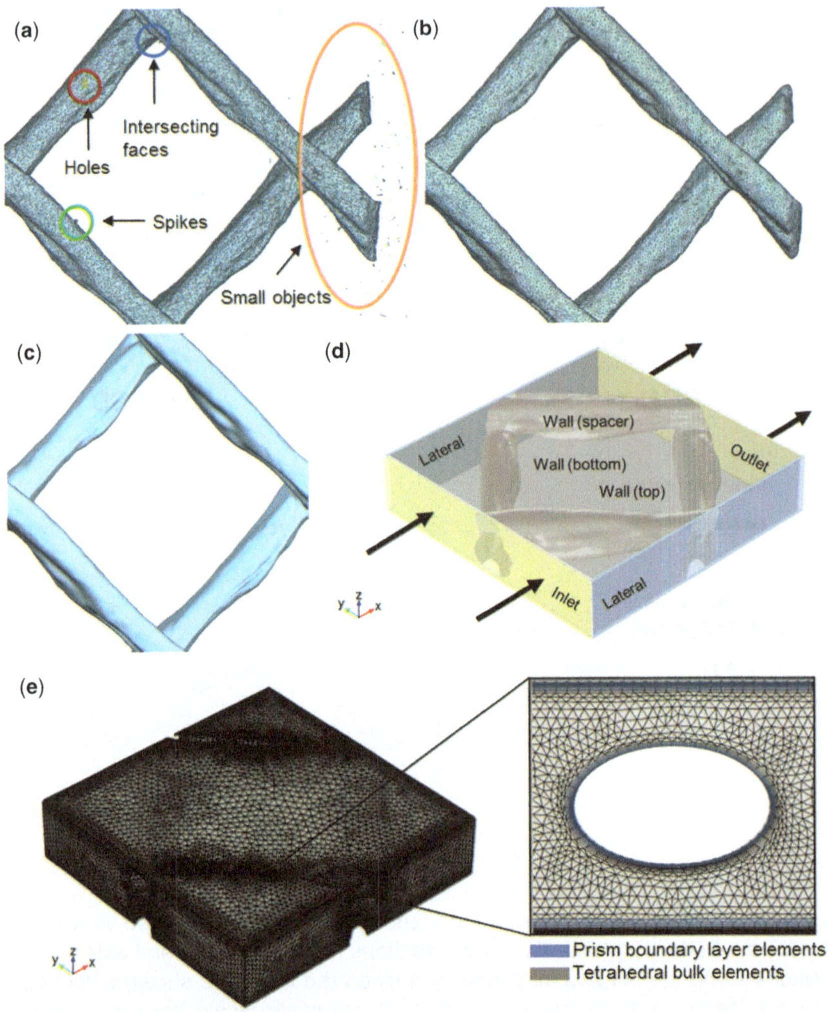

Figure 12.2 Overview of the workflow of CT scan processing for CFD modelling for spacer CON-1. (a) Image of an initial three-dimensional surface triangulation containing holes (non-manifoldness), spikes, intersecting surface tringle faces and other small objects (noise). (b) Image of a cleaned triangulated three-dimensional surface. (c) Solid spacer geometry created from surface fitting with non-uniform rational B-spline (NURBS) patches. (d) The computational domain (flow channel) and boundary conditions for CFD. (e) Mesh overview and detail showing boundary layer meshing.

Table 12.1 Overview of properties of CT scanned spacer samples and parameters which were used when the CT scan was performed.

Feed Channel Spacer	CON-1	CON-2	HYD	DOW	LXS-ASD
Sample properties					
Number of strand intersections	17	24	12	17	18
Number of repeating frames	6	9	5	6	3
CT scan procedure					
Voxel size (μm)	15–18	7.5	7.5	7.5	7.5
Image acquisitions per rotation	1200	1440	1440	1440	1440
Equipment used	Carl Zeiss Metrotom 800	Phoenix nanotom m, GE Life Sciences	Phoenix nanotom m, GE Life Sciences	Phoenix nanotom m, GE Life Sciences	Phoenix nanotom m, GE Life Sciences

of the samples that were scanned are summarized in Table 12.1. The 3-D density information obtained from these scans was subsequently processed to extract only the spacer material by applying a threshold (VGStudio MAX, Volume Graphics GmbH). The surface of the resulting object was then triangulated and the obtained mesh was stored in a common stereolithography file format (STL) [WR1] using the Iso2Mesh [WR2] toolbox for MATLAB (R2014a, MathWorks Inc.). This format contains the positions of vertices and the edges forming the surface triangles, including the normal vector for each triangle. An example of such a mesh is shown in Figure 12.2a.

Processing of triangulated surfaces

The surface triangulation introduces various defects in the 3-D representation of the spacer geometry resulting in holes, sharp peaks, intersecting faces and artefacts not on the spacer surface caused by noise in the CT scan (as shown, for example, in Figure 12.2b). These defects were removed using the open source application MeshLab [WR3], with the goal of creating a manifold (i.e., watertight) geometry. Prior to conversion of the surface mesh to a solid shape, geometry rotation operations were performed with a MATLAB routine in order to align the spacer with the xy plane of the used system of coordinates.

Surface fitting and conversion to a solid shape

The netfabb Basic application (3D Systems, Inc.) [WR4] was subsequently used to cut a repeating spacer frame from the manifold surface mesh. A piece slightly larger than required was cut in this operation to avoid alterations in the computational domain due to necessary smoothing of the cutting edges. The resulting piece was imported in GeoMagic Studio (v12.0, 3D Systems, Inc.) and prepared for surface fitting by using the automatic mesh repair functionality, which would smooth the sharp angles between edges introduced upon cutting the surface mesh in the previous step. Then, a surface was fitted onto the mesh using several non-uniform rational B-spline (NURBS) patches based on lines drawn along the important contours of the geometry. The patches were merged to form a single surface, which was subsequently converted to a solid shape (CAD-geometry) for use in the CFD code. An example of a resulting solid geometry is presented in Figure 12.2c.

Construction of computational domain

The computational domain was created by imposing a block with sizes in x- and y-directions equal to the dimensions of one (or multiple) repeating spacer frames and with a height of 863.6 μm (34-mil). For each geometry, the block was moved along the z-direction to minimize the fluid-filled volume of the channel. Finally, the liquid domain was constructed as the difference between the volumes of the block and spacer (Figure 12.2d).

Spacer geometry measurements

Distance measurements (i.e., strand intersection thickness, parallel strand distance, strand thickness, etc.) were conducted on the processed triangulated surface mesh. The matrix representation of the spatial position of vertices allowed for faster processing in MATLAB compared to the solid shape of the spacer geometry. Volume measurements were performed on the solid spacer geometries using COMSOL Multiphysics (v5.1, COMSOL Inc., Burlington, MA) and MATLAB through the LiveLink® interface. Thickness measurements were performed using digital callipers. Volumes of physical spacer nets for porosity were calculated from weight measurements based on the material density. The average thickness and porosity of designs with alternating strands were calculated based on the nodes of two strands of the thickest type, assuming that these nodes support the membranes the most and thus determine the feed channel height in spiral-wound membrane modules.

Construction of simplified spacer geometry

Simplified spacer geometries were constructed to mimic the conventional approach of spacer geometry construction for CFD modelling in 2-D as well as 3-D. Measurements on the spacer dimensions have been obtained by light microscopy or scanning electron microscopy (Radu *et al.*, 2014b), always using views from the top of the spacer mesh. Assuming that spacer strands are rotationally symmetric along a central axis, a cylindrical strand with varying diameter could be constructed. The same procedure was applied here, but using instead the top view of the CT scan. A set of Bèzier curves was used to approximate the surface curvature based on these measurements. The simplified solid strand shape was finally created by revolving the Bèzier curves around a central axis, followed by the construction of the channel geometry by multiplying such obtained strand geometry units.

Numerical model

The computational domains constructed from the CT scans were used for three-dimensional CFD calculations. Stationary, laminar flow of an incompressible fluid was assumed (Navier-Stokes with continuity equations):

$$\begin{aligned} \rho(\mathbf{u}\cdot\nabla)\mathbf{u}+\nabla p &= \nabla\cdot(\eta\nabla\mathbf{u}) \\ \nabla\mathbf{u} &= 0 \end{aligned} \tag{12.1}$$

where $\mathbf{u} = (u_x, u_y, u_z)$ denotes the vector of local fluid velocity, p is the pressure, η is the dynamic viscosity of water and ρ is the density of water (30°C). No-slip boundary conditions were specified for the spacer surface and for top and bottom walls, thus performing calculations without permeate production.

To reduce the computational requirements, while still having a representative computational volume, periodic boundary conditions were applied on one typical spacer frame. The periodic flow conditions were set between the opposite boundaries lateral and axial to main flow direction. Considering that the CT scans contain small deviations in geometry from one repeating spacer frame to the next, the geometry of the opposite faces of the computational domain does not match exactly. To allow the use of periodic boundaries, the computational domain was slightly altered by adding very thin blocks (50 μm thick) on all four faces. Consequently, the periodic boundaries were actually applied on the perfectly matching rectangular faces of opposite blocks. The validity of this approach was tested by additional calculations using a simplified spacer geometry containing identical periodic faces with and without the blocks. Differences in the calculated pressure drop and velocity fields were negligible (less than 0.1%), thereby justifying this approach. A pressure difference between the axial pair of periodic boundaries was driving the flow. The needed pressure drop to obtain the desired average velocity ($\bar{u}_{\text{set}}$) resulted from an additional constraint:

$$\bar{u}_{\text{set}} = \bar{u}_x = \int_\Omega u_x dV \Big/ \int_\Omega dV \tag{12.2}$$

where $\bar{u}_x$ is the velocity in the main flow direction, averaged over the whole computational domain Ω.

Model solution and mesh convergence studies

COMSOL Multiphysics 5.1 was used to solve the stationary flow equations by finite element methods. A schematic overview of a computational domain in depicted in Figure 12.2d. For quantitative comparison with experimental data, it is essential the numerical results are accurately obtained on converged grids (or meshes). A triangular mesh was generated on one face of a pair of periodic boundaries and subsequently copied to the other face of the pair. Tetrahedral elements were then used to mesh the liquid volume of the computational domain with a maximum element size varying per spacer geometry (CON-1: 70 μm, CON-2: 75 μm, HYD: 78 μm, DOW: 66 μm, LXS-ASD: 108 μm). From this tetrahedral mesh, a number of prism shaped boundary layer elements were generated near the no-slip walls (membranes and spacer). The solution was calculated on successively refined meshes until a converged grid was obtained. The degree of convergence was assessed using the Grid Convergence Index (GCI) proposed by Roache (1994). The friction factor f was used as the quantity to assess the convergence. This is equivalent with using the pressure drop as mesh convergence criterion. In the simulated conditions, the GCI was below 10% for all spacers when a number of ten boundary layer elements was generated on the no-slip boundaries. This is considered as sufficient mesh accuracy in CFD (Fimbres-Weihs & Wiley, 2010). Further refinement of the tetrahedral mesh did not result in an increase in accuracy.

Experimental methods

Experimental pressure drop measurements on the spacers under evaluation were performed at cross flow velocities representative for membrane modules in practice (Vrouwenvelder *et al.*, 2009b), using membrane fouling simulators (Vrouwenvelder *et al.*, 2006) and sensitive differential pressure transmitters (Vrouwenvelder *et al.*, 2009c). The complete experimental set-up has been described in recent work (Bucs *et al.*, 2016a; Siddiqui *et al.*, 2016). Feed spacers (thickness of 34 mil) were selected based on earlier studies (e.g., Araújo *et al.*, 2012a).

Measurements on each spacer type were conducted using the same flow cell with a fixed channel height of 34 mil (863 μm) and inner width and length of 4.0 cm and 20.0 cm, respectively. A mould was used to cut small sheets of spacer to the aforementioned size. Filtered (10 μm pore size) tap water was used for this study with a controlled temperature of 30°C. Pressure drop measurements were performed in duplicate for a range liquid flow rates between 2 kg/h and 20 kg/h. Flow rates were converted to average velocities $\bar{u}_{eff}$ using the flow channel porosities determined from CT-scans.

Measures for evaluation of different spacer geometries

Pressure drop

Objective comparison of pressure drop resulted from different spacer designs is crucial for the evaluation of their performance. Experimental measurements and flow simulations on different spacer geometries must be compared based on the same average liquid velocity in the main flow direction $\bar{u}$ corrected for the porosity of the spacer-filled feed channel, ε, such that $\bar{u}_{\text{eff}} = \bar{u}/\varepsilon$. The channel porosity (i.e., the volume fraction available for fluid flow) is calculated from $\varepsilon = 1 - V_{sp}/V_{\text{tot}}$, where V_{sp} and V_{tot} are the volume of the spacer and the total volume of the channel including the spacer and the liquid.

Friction factor

When comparing the drag characteristics of different spacers resulting from the shape of the spacer strands and the orientation towards the liquid flow, the Reynolds number, $Re = \rho\bar{u}_{\text{eff}} d_h/\eta$, and the Darcy friction factor, $f = 2\Delta p\, d_h/(\rho\, l\, \bar{u}_{\text{eff}}^2)$, were used as proposed by Schock and Miquel (Schock & Miquel, 1987). The pressure drop Δp is related to a unit length of flow channel, l. Various definitions for the characteristic length scale of the flow in spacer-filled channels and 3-D simulations have been proposed, which either use only the channel height, h, ("The effect of turbulence promoters on mass transfer – numerical analysis and flow visualization," 1982) or the strand diameter d_S (Koutsou *et al.*, 2004), thus assuming on average cylindrical strands. However, the wetted area A_w (spacer and all other walls) and the shape of the spacer strands would also influence the friction factor. Therefore, the definition of a hydraulic diameter d_h proposed by Schock and Miquel (Schock & Miquel, 1987) is the most widely used (Fimbres-Weihs & Wiley, 2010) and allows for analysis of spacer designs without *a priori* assumptions on the strand shape.

$$d_h = \frac{4(V_{\text{tot}} - V_{sp})}{A_w} = \frac{4V_{\text{tot}}\varepsilon}{A_m + A_{sp}} \tag{12.3}$$

This hydraulic diameter takes into account the porosity of the spacer-filled channel, ε, the spacer surface area, A_{sp} and the membrane surface area, A_m. The advantage of CT scans is that it allows an accurate calculation of the spacer surface area and volume without assuming a particular shape of the strands.

Membrane shear stress generated per power input

Besides the pressure drop, the solute (e.g., salt) transport and (bio)fouling are important in the operation of industrial reverse osmosis modules and these are impacted by shear (Fimbres-Weihs & Wiley, 2007; Li *et al.*, 2002b; Radu *et al.*, 2014b). Therefore, feed spacer performance was also assessed in terms of the energy loss resulting from the form drag through flow around the spacer strands in a feed channel and shear stress on the membrane surface. The influence of the strand shape (i.e., regardless of the dimensions) was assessed by getting a modified friction factor (f_{mod}) proposed by Santos *et al.* (Santos *et al.*, 2007) as a function of the Power number, which allowed comparison of different spacer designs for the same power input. The Power number, Pn, was defined as $Pn = Re^3 f$. The modified friction factor is a function of shear at the membrane, $f_{\text{mod}} = \tau_m (d_h^2 \rho / \mu^2)$, where τ_m is the shear averaged over the whole membrane area. The f_{mod} should ideally be maximized for a given power input.

Membrane shear stress efficiency for a constant flow rate

Another operational mode of spiral-wound membrane modules is operation using a constant feed flow rate. Therefore, the shear stress on the membrane per unit of pressure difference for a spacer design as a whole was analysed for a fixed average velocity $\bar{u}_{\text{eff}}$ to assess the level of hydrodynamic boundary layer disruption on the membrane per amount of momentum dissipated. To be able to compare correctly different spacer designs, the dimensionless quantity $f_p = (\tau_m / \Delta p)(A_m / A_{\text{axial}})$ was used.

A_{axial} is the cross-sectional area of the computational domain perpendicular to the main flow direction.

Strand eccentricity

The strand eccentricity for a spacer of a certain cross-section was calculated by $e = \sqrt{1 - b^2 / a^2}$, where e is the eccentricity, b the shortest semi-axis and a the longest semi-axis. A circular cross-section yields an eccentricity of 0, while e increases asymptotically towards 1 for increasing eccentricity. Lengths a and b define the ellipse circumscribing a cross-section of spacer strand with the smallest circumference. Average eccentricity of the strand was determined by averaging the eccentricities of 20 cross-sections.

RESULTS

Geometric characteristics of feed channel spacers from CT scans

Visual inspection of the CT scans of the spacers under study reveals a variety of geometric features, hardly observable by previously used microscopic methods. Ridges, valleys and varying strand thickness in all three dimensions are apparent for all five feed channel spacer designs, features sometimes generated by twisted strands between fibres intersections. The CT scans clearly revealed that the cross sections of the fibres are not rotationally symmetric (see Figure 12.1). This means that the spacer strand orientation against the flow becomes important, unlike in previous spacer representations where the cross-sections were circular (Bucs *et al.*, 2015; Picioreanu *et al.*, 2009; Radu *et al.*, 2014b).

Quantitative geometry measures were extracted from the CT scans for each spacer design. In the following, we annotate with an asterisk (*) the measured quantities that were changed in the model due to the fixed height of the experimental flow channel ($h = 863$ µm). The average thickness of strand nodes, h^*, and the average parallel strand distance, L_{sp}, were measured to validate the dimensions of the CT scan. The volumes, $V_{sp}*$, and surface areas, $A_{sp}*$, of several repeating elements in each spacer geometry were calculated from the CT scans. From these, the porosity of the flow channel corresponding to each spacer, $\varepsilon*$, and the specific surface area, A^*_{sp}/V^*_{sp}, could be calculated. Calculated and physically measured values are listed in Table 12.2 for comparison.

The deviation of the average spacer thickness calculated based on the CT scans from the thickness measured using digital callipers varied per spacer design. The differences may occur either due to the small force exerted by callipers on the spacer net (which may slightly deform, e.g., LXS-ASD). Additional sources of inaccuracy could be the variation in spacer sheet thickness between different production batches and that different sheets had been used for CT scanning than for the physical measurements. Despite these possible sources of error, the calculated average spacer thickness was in good agreement with the measured thickness for spacers CON-1, CON-2, HYD and DOW.

Moreover, parallel strand distance calculations based on the CT scan matched well the measured values for each spacer geometry (Table 12.2). Based on these findings, the dimensions of the CT scans were concluded to be valid.

Flow channel porosities calculated using the CT scan compare well to those based on weight measurements. Because of higher accuracy, the CT scan porosities were further used in this study. CON-1, CON-2 and DOW have the lowest porosity (0.856, 0.859 and 0.862), while HYD and LXS-ASD have the largest (0.893 and 0.897). Accurate channel porosity determination is important for calculating the average liquid velocity and, subsequently, for comparing hydraulic characteristics (e.g., pressure drop) of different spacer designs.

Impact of resolution on geometry characteristics

The effect of the resolution at which a CT scan was acquired on the geometry characteristics of the 3-D representations was investigated by comparing the results of different resolutions. Characteristics for spacers CON-2 and DOW extracted from CT scans acquired with a resolution of 7.5 μm and 60–80 μm, respectively, are listed in Table 12.3. Comparison with measurements conducted using digital callipers clearly show that a scanning resolution of 60–80 μm results in a too coarse geometry representation, while the scan acquired with a resolution of 7.5 μm was in close agreement with the measurements. The specific spacer surface area (A^*_{sp}/V^*_{sp}) was found to be smaller using a higher resolution of 7.5 μm compared to 60–80 μm, indicating that the spacer volume is underestimated using the coarse CT scan. The intersection thickness was considered to be the most useful quality measure for CT scans of feed spacers, since too coarse CT scans yielded geometries with significantly smaller thickness compared to the measured data. Therefore, the resolution of 7.5 μm was preferred for obtaining digital geometry representations of feed channel spacers.

Table 12.2 Overview of geometric characteristics of spacer designs.

Spacer Characteristics	CON-1	CON-2	HYD	DOW[a]	LXS-ASD[b]
Measurements					
*Spacer thickness, h**					
From CT scan (μm)	852 ± 11	822 ± 6	816 ± 6	803 ± 9	903 ± 7
From calliper measurement (μm)	847 ± 24	825 ± 9	820 ± 22	806 ± 14	830 ± 15
Average parallel strand distance, L_{sp}					
From CT scan (mm)	2.28	2.65	2.91	2.34	3.11/2.47
From calliper measurements (mm)	2.26	2.67	2.85	2.43	2.95/2.63
Average strand thickness, d_{sp} (μm)	50	59	50	43/48	65/50
Maximum (μm)	56	67	62	48/52	74/53
Minimum (μm)	42	47	35	38/42	58/46
Average strand eccentricity[c]	0.45	0.62	0.59	0.52/0.43	0.67/0.68
Maximum	0.63	0.79	0.75	0.64/0.45	0.76/0.74
Minimum	0.20	0.47	0.42	0.20/0.41	0.46/0.57

Inner strand angle, β (°)	90	90	90	70	90
Specific spacer surface area, A^*_{sp}/V^*_{sp} (m^2/m^3)	8828 ± 87	8413 ± 44	8927 ± 2	9083 ± 27	9490
Flow channel porosity, ε^{*d}					
From CT scan	0.856 ± 0.002	0.859 ± 0.001	0.893 ± 0.001	0.862 ± 0.002	0.897 ± 0.001
From density based calculation	0.861 ± 0.012	0.863 ± 0.007	0.892 ± 0.005	0.878 ± 0.009	0.879 ± 0.002
Numerical model					
Dimensions					
flow channel height, h (μm)	863	863	863	863	863
spacer frame axial length, L_x (mm)	3.90	4.47	4.93	4.70	9.63
spacer frame lateral length, L_y (mm)	3.90	4.56	4.73	3.42	9.53
Mesh length over channel height, L_m/h	4.5	5.2	5.6	5.4	5.6
Specific surface area A_{sp}/V_{sp} (m^2/m^3)	8811	8422	8973	9052	9537
Flow channel porosity, ε^d	0.855	0.866	0.893	0.873	0.893
Hydraulic diameter, d_h (mm)	0.97	1.01	1.09	1.01	1.07

Table 12.3 Geometry characteristics of spacer CON-2 and DOW extracted from CT scans acquired with different resolutions compared to measurements using digital callipers.

Spacer	**CON-2**			**DOW[a]**		
Method	**Measured**	**CT Scan (7.5 μm)**	**CT Scan (60–80 μm)**	**Measured**	**CT Scan (7.5 μm)**	**CT Scan (60–80 μm)**
Intersection thickness (μm)	825 ± 9	822 ± 6	776 ± 13	806 ± 14	803 ± 9	749 ± 16
Parallel strand distance (mm)	2.67	2.65	2.71	2.43	2.34	2.31
Porosity	0.863 ± 0.007	0.859 ± 0.001	0.869 ± 0.002	0.878 ± 0.009	0.862 ± 0.002	0.877 ± 0.001
Strand thickness (μm)	–	59	49	–	43/48	46
Specific spacer surface area (m^2/m^3)	–	8413 ± 44	9117 ± 118	–	9083 ± 27	9897 ± 51

[a]The CT scan with a resolution of 7.5 μm revealed a slight difference in thickness between the bottom and top strand of the sheet of spacer DOW (or vice-versa).

Hydraulic characterization

Pressure drop measurements compared to model solutions using CT scans

Measured pressure drop data at a velocity commonly used in practice, $\bar{u}_{\text{eff}} = 0.163$ m/s (Vrouwenvelder *et al.*, 2009a), are shown in Figure 12.3. A similar pressure drop was measured for spacer designs HYD (159 ± 9 mbar/m), DOW (144 ± 7 mbar/m) and LXS-ASD (173 ± 9 mbar/m), performing the best of the evaluated spacers, while higher pressure drop was measured for CON-1 (260 ± 14 mbar/m) and CON-2 (270 ± 15 mbar/m). These differences may be attributed to several factors. First, spacers CON-1 and CON-2 both have a lower channel porosity compared to HYD, DOW and LXS-ASD. Second, the quasi-elliptical strand section of CON-1 and CON-2 (see average eccentricity of spacer strands in Table 12.2) has a flow attack angle of 20° due to the strand tilt, leading to more drag, compared with the elliptical section of HYD, which is aligned with the flow (Figure 12.1). Although strands of DOW are relatively circular in cross-sections, this spacer design has a smaller angle between strands in flow direction ($\beta = 70°$), which lowers the hydraulic resistance and results in a lower pressure drop, comparable to that from spacers with more eccentric

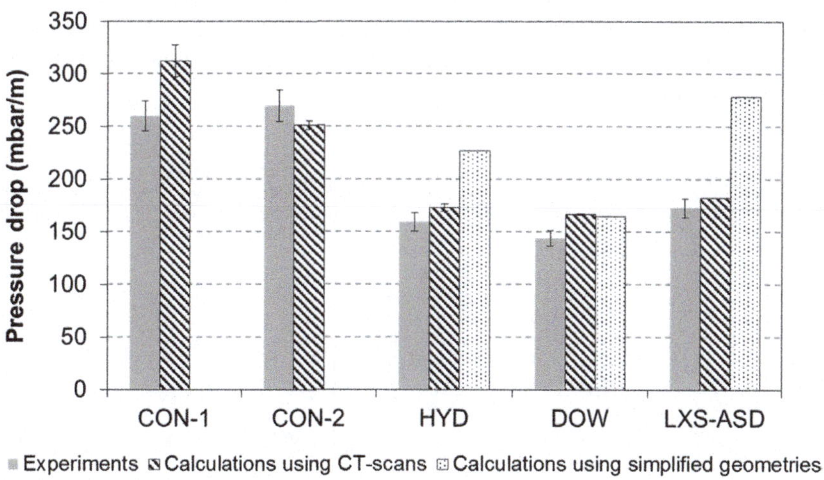

Figure 12.3 Comparison of measured and calculated pressure drop per meter, for all spacer designs, at = 0.163 m/s. Grey bars: experimental measurements; error bars represent the standard deviation between determinations ($n = 3$). Hatched bars: model results with spacer geometry from CT scans; errors bars represent standard deviation between calculations using different spacer frames from the CT scan ($n = 5$). Dotted bars: model results for simplified spacer geometries.

strands like HYD and LXS-ASD ($\beta = 90°$). Even though both strands of LXS-ASD are on average thicker those of HYD, the alternating strand configuration and the more eccentric strand shape of LXS-ASD yields a similar pressure drop compared to HYD. The non-rotationally symmetric strand sections and strand tilt with respect to the flow are thus important characteristics influencing the pressure drop.

Numerical calculations using CFD were performed on spacer-filled channels with a fixed channel height of 34 mil ($h = 863.6$ μm). A representative repeating frame from the CT scan of each spacer design was used to construct the computational domain. For statistical treatment, the calculations were performed on several single frames from the same spacer scan. The pressure drop calculated for spacers CON-2, HYD and LXS-ASD at $\bar{u}_{\text{eff}} = 0.163$ m/s was in agreement with the physical measurements (average differences of −7%, +10% and +6% respectively), while the pressure drop for spacers CON-1 and DOW were overestimated with respectively 17% and 16% (Figure 12.3).

Pressure drop measurements were performed for average effective velocities up to $\bar{u}_{\text{eff}} = 0.2$ m/s (Figure 12.4). Quadratic dependencies of the measured pressure drop on the average effective velocity were found, following Bernoulli's law. The level of agreement between the model solutions and the experimental measurements differs per spacer. The model solutions of the CT scanned spacer geometries yielded a slight over- (CON-1 and DOW) or underestimation in the upper region of the velocity range (CON-2) of the pressure drop compared to the experimental measurements, while a good agreement or a slight underestimation was obtained for spacers (HYD and LXS-ASD).

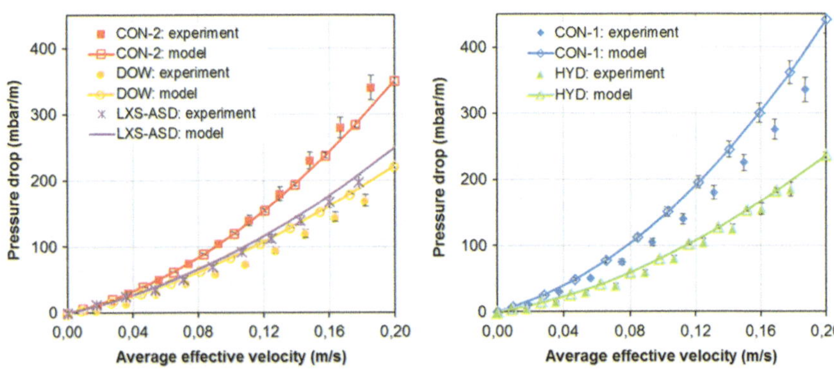

Figure 12.4 Pressure drop per meter at different average effective velocities up to 0.2 m/s, for all spacer designs. Symbols: experimental measurements; lines: average model data computed using CT scans. Error bars for experimental data represent standard deviation of triplicate measurements for each spacer. Error bars in model results originate from model solutions for different frames of each spacer design ($n = 5$).

The observed differences between measured and calculated pressure drop can be attributed to different reasons. Deviations in slope between experimental and model results in the upper region of effective velocity range (up to 0.2 m/s), most evident in the graph representing spacer CON-2 in Figure 12.4, could indicate the transition from steady to unsteady flow, which would invalidate the assumption of steady laminar flow for these velocities. Furthermore, the average thickness measured for spacers CON-1 and LXS-ASD (h^*) was very close to the channel height of the flow cell and filaments of all spacers are irregularly shaped. Therefore, part of the spacer exceeding the imposed channel height (h) was cut off when constructing the computational domain. If the spacer sheet is too thick at some locations, when used in a flow cell, it would deform instead, resulting in a mismatch between the scanned geometry and the actual geometry in the flow cell and thereby different flow patterns.

Furthermore, the average thickness measured for spacers CON-1 and LXS-ASD (h^*) was very close to the channel height of the flow cell and filaments of all spacers are irregularly shaped. Therefore, part of the spacer exceeding the imposed channel height (h) was cut off when constructing the computational domain. If the spacer sheet is too thick at some locations, when used in a flow cell, it would deform instead, resulting in a mismatch between the scanned geometry and the actual geometry in the flow cell and thereby different flow patterns.

Friction from strand shape and orientation

The friction factor f was used to assess the impact of the strand shape and orientation on the hydraulic performance of a particular spacer design. For comparison of different spacers, the friction factor was expressed as a function of the Reynolds number for an average effective velocity up to 0.2 m/s ($50 < Re < 200$). Results for both experimental measurements and model solutions are shown in Figure 12.5, where they are also compared with the correlation $f = 6{\cdot}23 \cdot Re^{-0.3}$ ($100 < Re < 1000$) proposed by Schock and Miquel (Schock & Miquel, 1987).

Both pressure drop and friction factor are providing the same quantitative basis for comparing the model with experiments by using the same d_h and ε based on the CT scans (see Table 12.2). However, the friction factor, being a dimensionless measure, does provide insight on the drag due to the shape of a certain spacer design regardless of its physical dimensions. A more detailed comparison of experimental and model results seems to indicate that the decrease in f with Re in the interval $125 < Re < 200$ levels off to a larger extent for the experimental data than for the numerical results obtained with steady laminar flow (this is particularly visible for spacer CON-2 in Figure 12.5a). The onset of unsteady flow phenomena could provide an explanation for this systematic observation, which has been reported in other modelling studies for the Reynolds numbers studied here (Fimbres-Weihs & Wiley, 2010).

The shapes of spacers CON-1 and CON-2 generate more drag compared with the spacers HYD and LXS-ASD (all having the same inner strand angle). In spite

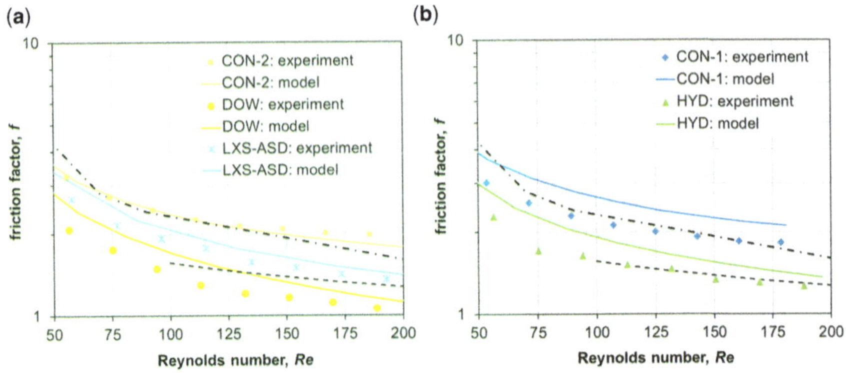

Figure 12.5 (a and b) Darcy friction factor function of Reynolds number, calculated from pressure drop measurements and model solutions, for all spacer designs. Symbols: experimental measurements; Solid lines: average model data computed using CT scans; Dashed lines: correlation proposed by Schock and Miquel (1987) for $100 < Re < 1000$; Dash-dot lines: numerical data from Fimbres-Weihs and Wiley (2007) for geometry with cylindrical strands ($Lm/h = 4$, $dsp/h = 0.5$, $\alpha = 45°$, $\beta = 90°$).

of having similar excentricity, the higher *f* for CON-1 and CON-2 can be attributed to smaller hydraulic diameters and, moreover, the strands are tilted with respect to the direction of the flow (the largest ellipse semiaxis is at an angle of approximately 20° with the main flow direction). The DOW spacer has the lowest friction factor of all spacer designs. Although the strands of DOW are close to cylindrical and the hydraulic diameter is smaller compared to HYD and LXS-ASD, the smaller friction factor results from the smaller inner strand angle (70° compared to 90°). LXS-ASD, even though it has the largest specific surface area of all spacers under study, it also increases the feed channel porosity due to the relatively thin alternating strand leading to a relatively low pressure drop and friction factor.

Shear at the membrane and power input

Although pressure drop is important to evaluate the performance of a particular spacer design, the mass transfer and the fouling potential are also critical characteristics, both depending on shear. Contour maps of the bottom membrane shear stress from model solutions for an average effective velocity of 0.163 m/s are presented in Figure 12.6. Comparison of the shear distribution of spacers CON-1 (Figure 12.6a) and CON-2 (Figure 12.6b) both yielded a maximum shear stress of 8.3 N/m² and a similar pattern consisting of high shear stress in the regions below the top filaments. CON-1 presents a curved region of relatively high shear along the middle of the mesh frame, while this pattern is less pronounced for CON-2 probably due to its larger mesh length.

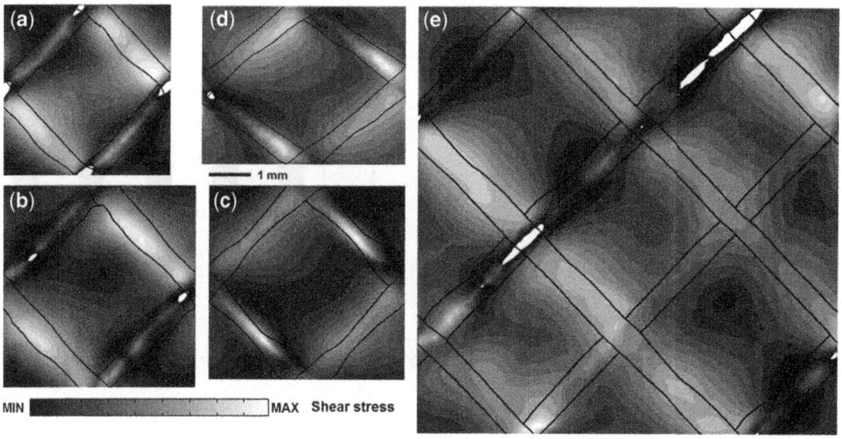

Figure 12.6 Contours of shear stress on membranes for all spacers at an average effective velocity of 0.163 m/s (a: CON-1, b: CON-2, c: HYD, d: DOW, e: LXS-ASD. The grey scale is relative to the maximum membrane shear stress of each spacer (a: 8.3 N/m^2, b: 8.3 N/m^2, c: 8.6 N/m^2, d: 6.9 N/m^2, e: 6.3 N/m^2). Overlaid black lines represent the outlines of the spacer geometry.

The maximum shear stress for spacer HYD (Figure 12.6c) was similar, 8.6 N/m^2, but the highest shear stress was localized below the thinnest part of the bottom filament. A distinct region of low shear was located on a line through the middle of the frame. The same high shear regions below the bottom filaments were observed for spacer DOW (Figure 12.6d), although the maximum shear stress was lower (6.7 N/m^2). Spacers CON-2, HYD and DOW all present a region of low shear downstream of strand intersections, while for CON-1 this is less pronounced. Finally, the model solution for LXS-ASD (Figure 12.6e) yielded a maximum shear stress of 6.3 N/m^2.

Spacer designs can be optimized on their ability to disrupt (with a minimum power input) the hydrodynamic boundary layer and increase the shear on the membrane, which would be beneficial for the mass transfer. To this aim, dimensionless relations were used, which correlate the Power number ($Re^3 f$) to a modified friction factor incorporating the average shear on the membrane (f_{mod}) introduced by Santos *et al.* (Santos *et al.*, 2007). The CFD calculations revealed that the spread in performance between the spacers is small over the whole range of Reynolds numbers evaluated (Figure 12.7a). A similar analysis was performed for the industrially-relevant average effective velocity of 0.16 m/s to estimate the shear generated on the membrane per unit of pressure drop, f_p (Figure 12.7c). The simulations revealed in more detail that the best performance in terms of shear obtained per pressure drop at this average effective velocity is delivered by spacer DOW, followed by LXS-ASD and HYD.

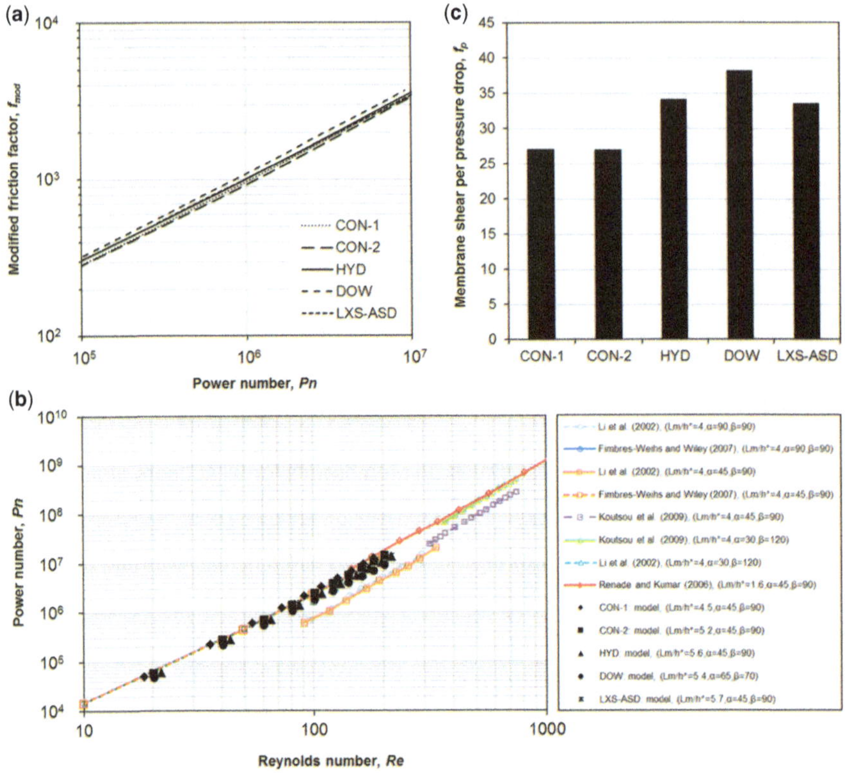

Figure 12.7 (a) Modified friction factor (incorporating the average shear stress on both membranes) function of the Power number (Re^3f) in CFD simulations for all spacer designs. (b) The Power number as function of the Reynolds number for all spacers compared to results obtained using cylindrical strands in previous work (figure adapted from Fimbres-Weihs and Wiley (2010). (c) Calculated average shear rate on the membrane surfaces (top and bottom) per unit of pressure drop for an average effective velocity of 0.163 m/s. The average membrane shear rates were 3556, 2919, 2579, 2745 and 2664 1/s from left to right.

Therefore, they exhibited the highest efficiency in converting the pumping power into membrane shear.

Spatial velocity distribution

Velocity fields computed for spacers CON-1, DOW and LXS-ASD at three heights in the flow channel are depicted in Figure 12.8, showing one full repeating frame for all designs for an average effective velocity of 0.163 m/s. The flow pattern near the membranes (at $z = 216$ and 648 μm) for spacers CON-1 and DOW is mostly guided by the spacer strand orientation in the bottom and top slice, and aligns with

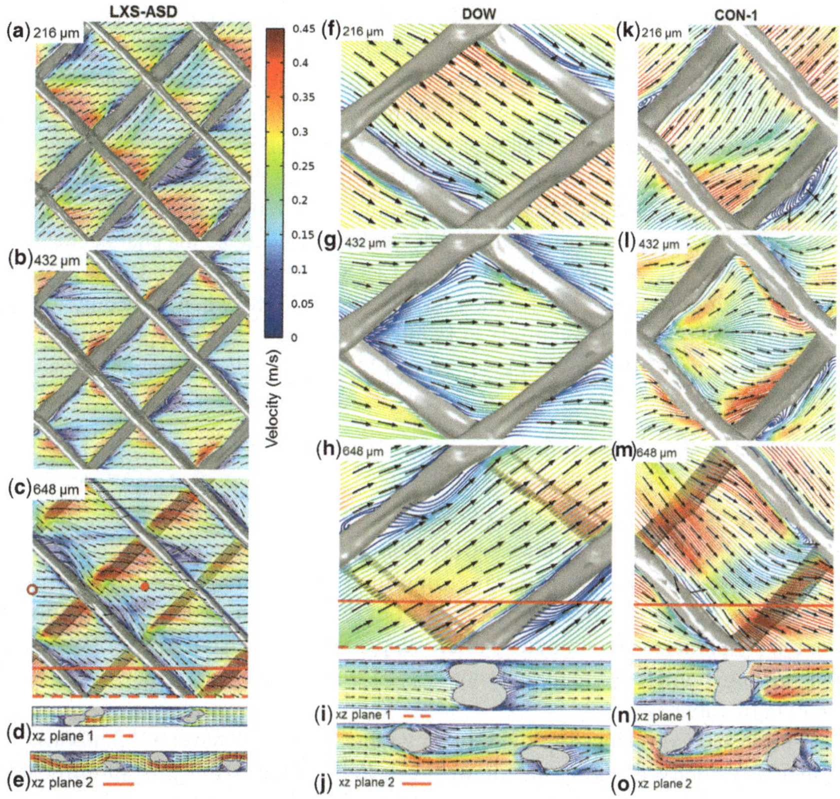

Figure 12.8 Flow fields calculated using spacer geometries from CT scan for LXS-ASD at average effective velocity 0.163 m/s. (a–c) Three top views (xy planes) for each spacer at different heights in the channel (distance from bottom): $z = 216$, 432 and 648 µm. (d, e) Two cross-sections (*xz* planes) are shown for each spacer, at positions indicated by lines on (c). Filled and open red circles in (c) indicate the position at which the average ratios $|ux|^2/|u|^2$ were calculated along the channel height for spacer LXS-ASD. Flow fields calculated using spacer geometries from CT scans for DOW and CON-1 at average effective velocity 0.163 m/s. (f–h, k–m) Three top views (*xy* planes) for each spacer at different heights in the channel (distance from bottom): $z = 216$, 432 and 648 µm. (i, j, n, o) Two cross-sections (*xz* planes) are shown for each spacer, at positions indicated by lines on (h) and (m).

the main flow direction in the middle of the channel ($z = 432$ µm). The change in flow direction occurs when the liquid passes above or below spacer strands, which can be concluded from the two axial slices shown in Figure 12.8. Noticeably, spacer CON-1 leads to vortex formation and higher velocities, probably due to the tilted strands, which may also explain the larger pressure drop obtained in this case. Qualitatively similar flow patterns were observed for spacers CON-2 and HYD.

The extent to which the direction of flow is determined by the strand orientations differs for the spacer with alternating strand thickness (LXS-ASD, Figure 12.8a–e). The thickest strands – supporting and thereby also imprinting the membranes – occupy a larger fraction of the available channel height and direct the flow to the largest extent. The thinnest strands and their intersections are mostly submerged in the channel, thus imposing less obstruction to liquid flow and channelling the flow to a smaller extent. Furthermore, the magnitude of the velocity component in the main flow direction $|u_x|^2$ and the total velocity vector magnitude $|u|^2$ were averaged along the channel height in the middle of two consecutive frames (Figure 12.8 shows with red circles the positions where velocities were averaged). The ratio $|u_x|^2/|u|^2$ was 0.80 when preceding a thick-strand intersection (open red circle in Figure 12.8c) and 0.96 when preceding a thin strand intersection (filled red circle in Figure 12.8c). This difference demonstrates that the thickest strands guide the flow to the largest extent.

Analysis of simplified model geometries

In order to assess the importance of using for CFD an accurate geometry obtained from the CT scans, calculations were compared for spacers HYD, DOW and LXS-ASD with those obtained using simplified spacer geometries. The simplified geometries were constructed using data from the CT scans, but only viewed from the top (*xy* plane) as if obtained by microscopy. Strands with circular cross-sections but varying diameter were in this way created, with geometric characteristics based on data in Table 12.2.

The pressure drop for simplified spacer geometries calculated at 0.163 m/s average effective velocity is shown in Figure 12.3. When comparing this simplification strategy with cylindrical strands to calculations using accurate CT scans, the difference in the calculated pressure drop depends on the amount of eccentricity of the spacer strands. For the DOW spacer the error is relatively small (14%) since the spacer strand cross-sections are rather circular, while for HYD and LXS-ASD the deviations are much larger (43% and 61%, respectively) due to the greater strand eccentricities. Based on these results, it is clear that assuming circular strand cross-sections (even of variable diameter) results in a too large simplification. The CT scan thus allows for a more accurate quantitative match to experimental data in terms of pressure drop, while decreasing the additional human bias introduced upon spacer geometry simplification.

DISCUSSION

Feed channel spacer geometry

The aim of this study was too find a methodology for obtaining an accurate three-dimensional representation of feed channel spacer geometries and to evaluate its effects on the numerical modelling of a spacer's hydraulic performance. Various spacer designs have been proposed and evaluated numerically in 3-D CFD simulations

(Dendukuri *et al.*, 2005; Koutsou & Karabelas, 2015; Li *et al.*, 2002b, 2005; Liu *et al.*, 2015; Ranade & Kumar, 2006; Santos *et al.*, 2007). However, the most studied spacer designs *in silico* were constructed from cylindrical strands (Fimbres-Weihs & Wiley, 2010) or strands with a circular cross-section with varying diameter (Bucs *et al.*, 2015; Picioreanu *et al.*, 2009; Vrouwenvelder *et al.*, 2010b) as resulted from SEM images (Radu *et al.*, 2014b). X-ray computed tomography used in the present study revealed much more detailed feed spacer geometries than the simplified ones frequently used to model hydrodynamics and mass transfer. In general, feed spacer strands assessed in this study turned out to be rather irregularly shaped, with cross-sections more or less elliptical and also differing in orientation with respect to the main flow direction. Since the X-ray imaging is nowadays a well-established procedure, future 3-D modelling studies should be performed using geometries obtained from CT scans, or from other 3-D imaging modalities able to capture the required spatial detail.

Pressure drop measurements compared to model solutions using CT scans

The calculated values for the pressure drop using the CT scan spacer geometries were in agreement with the values measured using a flow cell to a varying extent depending on the spacer. Errors in the geometry representation yielded a systematic overestimation of the pressure drop for spacers CON-1 and DOW, while the same holds for spacers HYD and LXS-ASD to a lesser extent (Figure 12.4). Comparison of experimental and model results for spacer CON-2 revealed an increasing deviation for an increasing effective velocity above 0.14 m/s. Moreover, differences in slopes between the graphs of experimental and model results depicting the friction factor as a function of the Reynolds number (Figure 12.5) could indicate the onset of unsteady flow phenomena for all spacers. Santos *et al.* (Santos *et al.*, 2007) reported the transition of steady to unsteady flow using ladder shaped spacers using a flow attack angle of 90° for $100 < Re < 300$. Furthermore, Shakaib *et al.* (Shakaib *et al.*, 2007) calculated that onset of unsteady flow takes place between $150 < Re < 175$ for a flow attack angle of 45° and an L_m/h ratio of 4.25 using cylindrical filaments. The latter is comparable to the spacers evaluated in this study (ranging from 4.5–5.6, see Table 12.2). A recent experimental study by Bucs *et al.* (Bucs *et al.*, 2015) employing particle image velocimetry measured a deviation of 10% from the mean pressure drop for a spacer geometry similar to HYD for $Re = 160$, confirming these modelling results. Based on these studies, the onset of unsteady flow most likely caused the observed differences in experimental and model results for $125 < Re < 200$ reported in this study.

Friction due to strand shape and orientation

Schock and Miquel (Schock & Miquel, 1987) derived an empirical correlation for the Darcy friction factor f as a function of Reynolds number, from experiments performed on several spacer types used in industrial spiral-wound membrane elements for reverse osmosis. We compared the experimental and numerical results from the present study with the correlation of Schock and Miquel, $f = 6.23 \cdot Re^{-0.3}$ (Figure 12.5). All the spacers tested in this study provided a friction factor equal or higher than the one from this correlation. Notably, spacers CON-1 and CON-2 do not line up with the Schock and Miquel correlation, yielding friction factors as high as double (for $Re > 100$). A likely explanation for the observed differences is that the proposed relation was based on friction factor calculations using cylindrical spacer strand approximation and less accurate geometry dimensions than those obtained from the CT scans. Furthermore, solutions obtained from a 3-D model using a simplified geometry with cylindrical strands ($L_m/h = 4$, $d_{sp}/h = 0.5$, $\alpha = 45°$, $\beta = 90°$) for $10 < Re < 200$ (Figure 12.5) showed good agreement with the trend of the model solutions obtained in this work, although the actual *f-Re* relation differs for each spacer geometry. One can therefore conclude that the established *f-Re* correlation from Schock and Miquel (Schock & Miquel, 1987) cannot describe all mesh-type feed spacer geometries, which is in line with conclusions from (Da Costa *et al.*, 1991) based on strand angle. Therefore, new spacer designs should be assessed individually based on accurate geometry measurements, which is in agreement with Picioreanu *et al.* (Picioreanu *et al.*, 2009).

Local membrane shear distribution

Shear stress has been studied using 3-D models utilizing CFD for the investigation of solute transport and deposition of particulate and bacterial fouling in spacer-filled channels. It has been reported that local membrane shear stress distribution and local velocity vectors are important factors determining the pattern of initial particle deposition on membranes (Fimbres-Weihs & Wiley, 2007; Li *et al.*, 2012). Radu *et al.* (2014b) reported that the alignment of spacer strands with slightly varying thickness with respect to the flow has a large influence on the local shear stress and the resulting deposition pattern on the membrane, based on particle deposition experiments and numerical modelling. X-ray computed tomography, as shown in this work, provides the opportunity to investigate numerically the fouling patterns in a more accurate way.

Power input with respect to friction

Numerical modelling results on the hydraulic performance using simplified spacer geometries with cylindrical strands were compiled by Fimbres-Weihs and Wiley (Fimbres-Weihs & Wiley, 2010). The power input in the form of *Pn* function of *Re* (such that $Pn = C \cdot Re^n$) was compared with results from the current study (Figure 12.7b). All the spacers with an internal strand angle of 90° and a flow attack angle of 45° evaluated by CT scans presented, for the same *Re*, a *Pn* well in agreement for *Re* up to 200 (Fimbres-Weihs & Wiley, 2007) or less than one order of magnitude larger for $100 < Re < 200$ (Li *et al.*, 2002b) to that obtained with simplified cylindrical geometries (CON-1, CON-2, HYD and LXS-ASD). Furthermore, spacer DOW ($\beta = 70°$) showed an almost identical *Pn-Re* relation. However, the dimensionless ratio of the mesh length over the channel height, L_m/h, was 4 for the simplified cylindrical filaments while geometry characteristics extracted from CT scans in this work yielded ratios larger for all spacers (ranging from 4.5 to 5.6, see Table 12.2), resulting in a larger void space in a feed channel. Despite these differences in strand alignment, the comparison shows that the use of cylindrical strands yields a *Pn-Re* relation that is in good agreement with the results obtained in this work using accurate geometries. Still, the geometry representations obtained from CT scans allows one to study the local hydrodynamic effects as well without the bias introduced when constructing simplified geometries.

Additionally, the variation in exponent n of the *Re* number in the power relation is, although in similar range) less for the CT scanned spacer geometries (2.3–2.43) compared to the one reported for the simplified geometries (2.25–2.83), indicating a smaller increase in average velocity encountered upon increasing power input. The differences in n found for the five CT scanned spacer geometries further underline the conclusion by Fimbres-Weihs and Wiley (Fimbres-Weihs & Wiley, 2010) that the choice of a spacer depends on the flow conditions expected to be encountered and no spacer design is optimal for all possible operating conditions.

Spatial velocity distribution

The zig-zag flow pattern calculated for spacers HYD, CON-1, CON-2 and DOW at $\bar{u}_{\text{eff}} = 0.163$ m/s (shown in Figure 12.8) has been experimentally observed by Da Costa *et al.* (Da Costa *et al.*, 1994) and in earlier numerical work (Fimbres-Weihs & Wiley, 2007; Karode & Kumar, 2001; Picioreanu *et al.*, 2009; Shakaib *et al.*, 2009). Recently, this was also experimentally measured using particle image velocimetry (Bucs *et al.*, 2015). In other work it was found that, as the parallel strand distance L_{sp} decreases for the same channel height, flow increasingly aligns with the strands until there is very little transition between the top and bottom layers (Shakaib *et al.*, 2007). When all spacers with an internal strand angle of 90° are ranked based on their average parallel strand distance (see Table 12.2), the obtained ranking matches the order obtained when ranking the same spacers based on the average of the ratio $|u_x|^2/|u|^2$ (ASD = 0.80, HYD = 0.76, CON-2 = 0.70, CON-1 = 0.53) for the whole computational domain. Moreover, this tendency was shown to decrease for a smaller inner strand angle, such as for spacer DOW (inner strand angle of 70°), were $|u_x|^2/|u|^2$ was 0.82 in the middle of a spacer frame element. In conclusion, flow pattern characteristics found in modelling studies using simplified spacer geometries are in good agreement with the model solutions reported in this work for commercially available feed spacers used in practice.

Mechanical deformation and membrane imprinting

Matching the modelling parameters to the exact physical properties of the object under study is crucial when model solutions are translated into strategies for improving the design of that object (here, the industrial feed channel spacers). It is clear that using CT scanned spacer geometries greatly improves the quality of CFD model solutions when striving towards more quantitatively accurate results. Although it was shown in earlier work that a flat flow cell could be used to study the hydrodynamics of feed channels in spiral-wound membrane modules (Schock & Miquel, 1987), it is difficult to reproduce the physical force exerted on spacer sheets in such a module, the subsequent deformation of the spacer and the level of membrane imprint. Furthermore, it is not yet established how to model the spacer deformation when constructing the computational domain for CFD. The area surrounding the place where the spacer is pressed into the membrane influences the flow pattern and thus affects the concentration polarisation and permeation (Radu *et al.*, 2010). These, in turn, influence scaling and (bio)fouling processes (Radu *et al.*, 2015). Thus, we believe that further studies should evaluate the effect of mechanical deformations and spacer-membrane imprints on the accuracy of model predictions.

Further studies

CFD modelling was successful in matching experimental pressure drop measurements using CT scans without the bias introduced when using simplified geometries constructed from 2-D measurements. In addition, the micro-velocimetry measurements (Bucs *et al.*, 2015) could be better matched with the velocity field calculated with CT scanned geometries. Based on the calculated flow pattern, modelling particle deposition in conjunction with experimental work (as described by Radu *et al.* (2014b)), would harness the potential of using CT scanning to model (bio)fouling initiation for a given spacer design. This could improve the particle deposition model predictions presented in Radu *et al.* (2014b). The spacer not only determines the flow pattern but also serves as a surface for initiation and subsequent fouling development. Since the position of foulant deposits affects the local rate of solute transport, this would allow a better estimation of the decline in process performance due to fouling. Therefore, simulations using the 3-D representation of a spacer geometry obtained from a CT scan should be used for accurate modelling of the solute mass transport in spacer filled channels.

Finally, more attention should be paid to the statistical treatment of the scanned geometries. In this work we extracted several (at least 5) spacer frames from each scanned spacer, performed CFD simulations for all these frames and then reported the average results with corresponding deviations. However, other possible approaches can be taken. A statistically adequate number of 3-D scans could be obtained (eventually from different fabrication batches), from which a "representative average shape" could be created. Even better, with an increased computational power, simulations over larger spacer patches should be performed, leading in principle to more representative results.

CONCLUSIONS

This study developed a method to obtain accurate three-dimensional (3-D) geometry representations for any given spacer design from X-ray computed tomography (CT) scans. Model solutions obtained using these geometries and experimental measurements led to the following conclusions:

- The CT scanning revealed that the filaments of several industrial spacers have a highly variable cross-section size and shape, which has not been addressed in fluid dynamics models until now.
- The advantage of using CT scanned spacer geometries in hydrodynamic calculations is that less bias is introduced compared to the use of conventional circular cross-section strand representations.
- Computational fluid dynamics (CFD) on four commercially available feed spacers and one new spacer with geometries from CT scans were in close agreement with experimental pressure drop measurements, for flow velocities up to 0.2 m/s as used in industrial reverse osmosis membrane operations.
- The simple cylindrical description of spacer filaments is still a good approximation for overall hydrodynamic design of spiral-wound membrane modules. However, by providing detailed insight on the spacer filament shape, CT scans allow better quantification of local distribution of velocity and shear, possibly leading to more accurate estimations of fouling and concentration polarization.
- A newly proposed spacer geometry with alternating strand thickness was tested (LXS-ASD), which led to lower pressure drop while being highly efficient in converting the pumping power into membrane shear. This shows that although the form drag is inevitable, the design of spacers should focus on directing flow toward the membrane while minimizing drag and generating high shear at the membrane surface.

Chapter 13

Development and characterization of 3D-printed feed spacers for spiral wound membrane systems

INTRODUCTION

Fresh water scarcity is one of the biggest problems faced globally. The most cost efficient technique to obtain fresh water from saline water is by application of spiral-wound nanofiltration (NF) and reverse osmosis (RO) (Shannon *et al.*, 2008; Van der Bruggen, 2003). When compared to conventional desalination processes such as thermal distillation, NF and RO provide a more selective separation, continuous and automatic operation, easy scale-up, and effective space utilization (Sonune & Ghate, 2004). However, NF and RO systems face a significant amount of energy loss due to operational problems due to fouling and in particular biofouling (Shannon *et al.*, 2008). Most types of fouling like particulate fouling, scaling and organic fouling can be controlled to a great extent by various pretreatment methods, whereas biofouling persists even after extensive water pretreatment (Vrouwenvelder *et al.*, 2008b). Biofouling is the amount of accumulated biofilm (biomass) causing unacceptable membrane performance loss (Flemming, 2002). Numerous studies have been carried out to control biofouling in spiral wound membrane systems (Al Ashhab *et al.*, 2014; Baker & Dudley, 1998b; Ben-Sasson *et al.*, 2014; Habimana *et al.*, 2014; Ridgway *et al.*, 1983; Schneider *et al.*, 2005; Schwinge *et al.*, 2004a; Tasaka *et al.*, 1994; Vrouwenvelder *et al.*, 2008b; Vrouwenvelder & Van Der Kooij, 2001; Ying *et al.*, 2013). Biofouling is influenced by many factors in the filtration process e.g. feed water substrate concentration, chemical dosage (e.g. anti-scalants) and bacteria count, linear flow velocity, feed spacers, etc. (Van der Hoek *et al.*, 2000; Van Loosdrecht *et al.*, 2012; Vrouwenvelder *et al.*, 2000, 2010a).

Author(s): Szilárd Bucs, Joop Kruithof, Mark van Loosdrecht and Johannes Vrouwenvelder
doi: 10.2166/9781780409597_0257

Feed spacers are used in spiral-wound membrane modules to create inter-membrane space and enhance water mixing to avoid concentration polarization. As reported by Baker *et al.* (1995) feed spacers provide surface for initial deposition of fouling that accumulates and eventually spreads to the free membrane area. Van Paassen *et al.* (1998) observed that biofouling accumulated on feed spacers causing an exponential increase in the pressure drop in membrane modules. Tran *et al.* (Tran *et al.*, 2007) indicated that the membrane area most affected by fouling was the area along the feed spacer strands. Vrouwenvelder *et al.* (Vrouwenvelder *et al.*, 2009a) indicated that the impact of biofouling on pressure drop was higher in the presence of a feed spacer. Strategies have been addressed focusing on the reduction of biofouling on feed spacers by periodic air/water flushing (Cornelissen *et al.*, 2007b) and by applying thicker feed spacers (Majamaa *et al.*, 2010). Therefore, feed spacers are important for membrane performance and play an important part in biofouling of membrane systems. Modifying feed spacers could be a strategy to reduce the impact of biofouling on the performance of membrane filtration based water treatment systems.

Recently, an increasing number of studies have been focused on modifying the surface chemistry of feed spacers to reduce biofilm attachment and growth in spiral wound NF and RO systems (Araújo *et al.*, 2012a, 2012b, Ronen *et al.*, 2015, 2016; Wibisono *et al.*, 2015b). One approach to reduce biofouling impact is to cover membranes and spacers with antifouling coatings to prevent microbial attachment and growth (Miller *et al.*, 2012; Ronen *et al.*, 2015, 2016; Wibisono *et al.*, 2015b). However, anti-fouling coatings wear out eventually, reducing with time the effect of the coating on biofilm accumulation (Araújo *et al.*, 2012a). Another approach is to modify the design of the feed spacers in such a way that the impact of accumulated biomass on membrane performance is reduced and biofouling removal from the membrane modules is enhanced during advanced cleaning strategies (Vrouwenvelder *et al.*, 2011b). Numerical studies have been done to evaluate biofouling development on feed spacers from practice with different thickness (Bucs *et al.*, 2014a). Studies have also been done on the hydrodynamics and the deposition behavior of particles, simulating bacteria, related to the geometry of the spacers (ladder-shaped and diamond-shaped geometry) (Radu *et al.*, 2014b). The locations of particle deposition predicted by the numerical modeling were in agreement with the locations where initial biofilm accumulation was observed in practice. A recent study using optical coherence tomography on feed spacer channels in RO membrane modules also showed that feed spacer geometry affects the structure and distribution of biofilm (West *et al.*, 2016).

Modifying the spacer geometry may (i) reduce the impact of biofouling, (ii) reduce energy requirements and (iii) avoid early membrane replacement. Commonly, feed spacers from practice applied in spiral-wound membrane modules have a diamond-shaped geometry with varying thickness ranging from 26 mil to 34 mil (1 mil is 25.4 μm) (Figure 13.1). Feed spacer geometries can be modified by computer

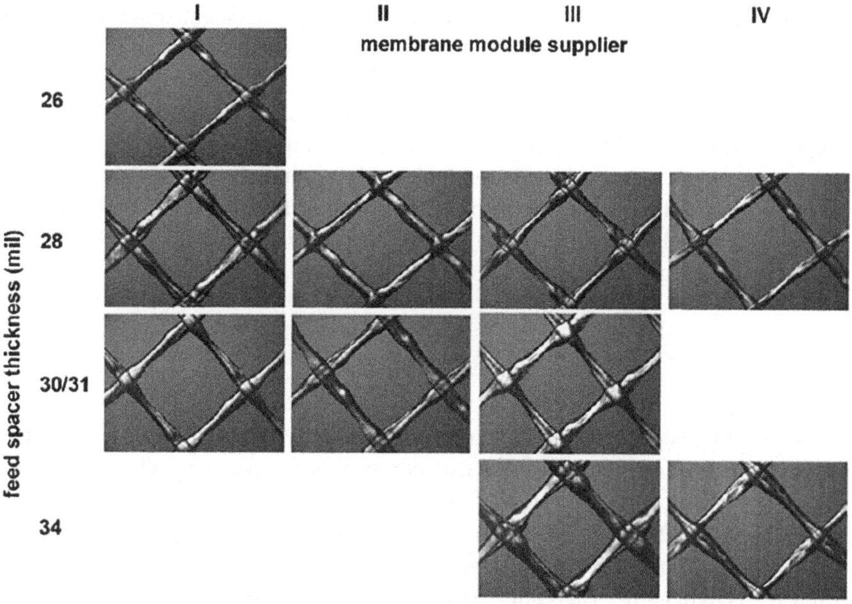

Figure 13.1 Inventory of feed spacers used in practice in spiral-wound reverse osmosis and nanofiltration membrane elements for drinking water production, supplied by four global membrane manufacturers coded I-IV. All feed spacers had the same geometry differing in thickness between 26 and 34 mil (1 mil equals 25.40 μm).

aided designing (CAD) and tested theoretically by numerical modeling/simulation (Fimbres-Weihs & Wiley, 2010; Madireddi *et al.*, 1999; Picioreanu *et al.*, 2009; Wiley & Fletcher, 2002). To test modified feed spacer designs experimentally, the modified spacers need to be manufactured. The start-up costs of the conventional process of manufacturing feed spacers is more expensive than the cost of bulk spacer manufacture.

The spacer production process requires special extrusion nozzles to create the mesh-like structure (diamond shape formed due to intertwining of spacer filaments) in bulk. Therefore, a new technique is needed to rapidly produce modified feed spacers at low cost in quantities sufficient for lab-scale studies.

Potentially, three-dimensional (3D) printing technology is a tool to produce complex geometry feed spacers. 3D printing technology has been used for over 25 years to develop prototypes for toys, tools, prosthetics etc. Several 3D printing techniques are available at present namely Fused Deposition Modeling (FDM), Stereolithography and Polyjet/Multijet printing. A suitable 3D printing technique and material must be selected for a feasible method to produce 3D printed feed spacers.

The objective of the study was to develop a strategy to design and evaluate feed spacers with a modified geometry for spiral-wound NF and RO membranes based on (i) numerical modeling for developing and selecting spacers with a modified geometry, (ii) selecting and applying a 3D printing technique to produce the modified spacers, and (iii) membrane fouling simulator (MFS) study to investigate the impact of feed spacers with a modified geometry on biofouling and pressure drop development. A feed spacer from practice was used as a reference and compared with a 3D printed feed spacer with the same geometry and a 3D printed feed spacer with a modified geometry.

MATERIAL AND METHODS

Numerical modelling

A three-dimensional mathematical model was developed to study the impact of different spacer geometries on hydrodynamics in spacer filled channels. For the hydrodynamic calculations, commercially available software COMSOL Multiphysics (COMSOL 5.0, Comsol Inc., Burlington, MA, www.comsol.com) was used. The feed spacer geometries were replicated and designed in SOLIDWORKS (SOLIDWORKS 2014, Dassault Systèmes SolidWorks Corporation, Waltham, MA, www.solidworks.com), used in the developed model and printed for experimental studies using 3D printing techniques. The model geometry included one spacer element (Figure 13.2), representative of the spacer geometry. The hydrodynamic calculations assumed stationary flow of an incompressible fluid in laminar conditions, according to the Navier–Stokes and continuity equation 13.1:

$$\begin{aligned} &\rho(\mathrm{u}\cdot\nabla)\mathbf{u}+\nabla p=\nabla\cdot(\eta\nabla\mathbf{u}) \\ &\nabla\cdot\mathbf{u}=0 \end{aligned} \tag{13.1}$$

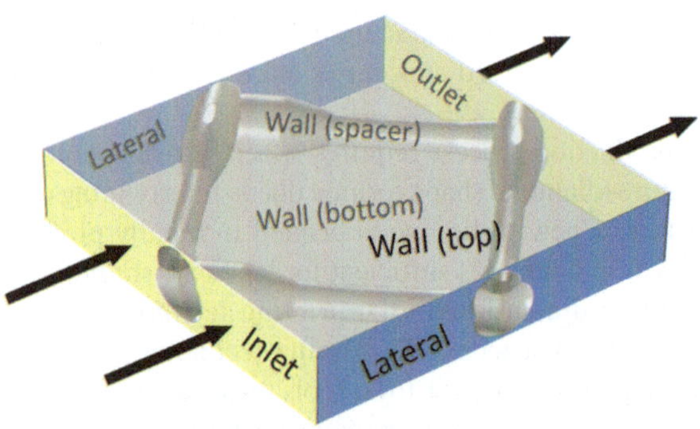

Figure 13.2 The computational domain and applied boundary conditions for hydrodynamic simulations including one feed spacer element. Arrows indicate flow direction.

Table 13.1 Numerical model parameters.

Parameter	Symbol	Value	Unit
Model geometry			
Channel length	L_X	2.6–6.8	mm
Channel width	L_Y	3.3–6.8	mm
Channel height	L_Z	863	μm
Model parameters			
Average inlet flow velocity	u_m	8.6	cm s^{-1}
Feed water density	ρ	1000	kg m^{-3}
Water viscosity	η	10^{-3}	Pa s
Temperature	T	293	K
Model geometry			

where $\mathbf{u} = (u_x, u_y, u_z)$ is the vector of local liquid velocity, p is the pressure, ρ and η are the density and dynamic viscosity of water (20°C) (Bucs *et al.*, 2014a; Picioreanu *et al.*, 2009; Radu *et al.*, 2014a). Periodic flow conditions were imposed between the inlet and outlet boundaries ($u(0,y,z) = u(L_X,y,z)$), as well as between the lateral boundaries ($u(x,0,z) = u(x,L_Y,z)$). This approximates the profile corresponding to a spacer element situated within an array (Table 13.1). The flow is driven by an imposed pressure difference between inlet and outlet ($\Delta p = p(0,y,z)\ p(L_X,y,z)$), while no pressure difference exists between the lateral boundaries ($p(x,0,z) = p(x,L_Y,z)$).

The pressure difference Δp required to drive the flow at a certain average linear flow velocity is obtained by including an additional constraint (Eq. 13.2).

$$u_{m,\mathrm{avg}}^{\mathrm{set}} = u_{m,\mathrm{avg}}(\Delta p) \tag{13.2}$$

$$u_{m,\mathrm{avg}} = \int_{A_{\mathrm{inlet}}} \frac{u_x(0,y,z)\,dA}{A_{\mathrm{inlet}}} \tag{13.3}$$

$$u_{m,\mathrm{avg}}^{\mathrm{set}} = \frac{Q}{w \cdot h} \tag{13.4}$$

No-slip boundary conditions were set to all other surfaces (top, bottom and spacer) thus the flow was calculated without permeate production. The steady laminar flow equations were solved with finite element methods, on a tetrahedral mesh with a maximum size of 50 μm.

3D printed feed spacers

Fused Deposition Modeling (FDM) was the first technique tested for producing 3D printed spacers. FDM proved to be unsuitable for producing feed spacers, since the resulting spacers and spacer strands were thicker than the feed spacers from

Table 13.2 Code used for the feed spacers studies.

Feed Spacer Code	Details	Thickness Mil*
SSP	Standard spacer as used in practice	34
3DS	3D printed spacer with similar dimensions as SSP	34
3DM	3D printed spacer with modified geometry from SSP	34

*1 mil equals 25.40 μm.

practice. Also, the FDM produced feed spacers were brittle. Stereolithography (SLA) was also tested as a possible option for printing feed spacers. SLA is an additive method of printing using UV light to cure a liquid resin. SLA spacers were found not to be suitable as they showed limited material rigidity and mechanical strength in water with time. Polyjet printing also known as Multijet printing (3D Printing Industry, 2015) was the next technique tested, allowing a higher resolution at lower cost compared to other current 3D printing technologies. A Polyjet printer utilizes multiple jet heads that deposit a polymer onto a platform, where the polymer is cured with UV light. The Polyjet printed end product had a smoother surface finish compared to the FDM spacer (Wang *et al.*, 1999).

Two sets of 3D printed spacers were produced using Polyjet printing: the first was designed with a similar geometry as the 34 mil (863 μm) thick feed spacer from practice (referred to as 3DS) and the second was a modified geometry feed spacer (based on numerical modeling) with the same thickness of 34 mil but differing in mesh-size (number of spacer filaments per unit area), contact angle between the filaments, and slightly in bottle-necking structure (referred to as 3DM, Table 13.2). The 3D printed feed spacer is manufactured from a urethane acrylate polymer, which is non-biodegradable.

A trial and error approach was applied to develop a modified geometry spacer. The best spacers from a pressure drop point of view were selected after testing 7 spacer geometries: three spacers with decreasing angles and three with increasing mesh sizes. The best performing angle was combined with the best performing mesh size to create the 7th spacer (modified spacer), which was tested experimentally in the biofouling study (3D printed modified geometry spacer, 3DM). In other words, both the feed spacer mesh size and contact angle were modified, to explore whether feed spacer modifications provide possibilities to reduce the impact of biofouling on membrane performance. The porosity of the two feed spacers 3DS and 3DM, calculated by using the numerical model were 0.85 and 0.90, respectively.

Microscopic observation

A Leica M205 FA Stereomicroscope was used to compare the geometry of the 3D printed spacers with the feed spacer from practice. The characterization was based on spacer filament shape, contact angle and mesh-size (Li *et al.*, 2002a). On a typical feed spacer from practice, a bottle-neck like structure is present, formed due to the intertwining of filaments during the manufacturing process. The 3D printed spacers were also designed with a bottle-neck structure on the spacer filaments.

Experimental set up

The experimental set up (Figure 13.3) consisted of a tank, gear pump (PRIMIX), temperature and differential pressure sensor (Deltabar, Endress + Hauser PMD75, Germany), back pressure valve (Hydra cell, Wanner Engineering Inc., USA), substrate dosing diaphragm metering pump (Stepdos 03, KNF Lab, Germany), flow sensor (Ultra flow sensor ULF01.H.0, POM/FPM) and a membrane fouling simulator (MFS). The accuracy of the differential pressure transmitter is ±0.05% (Vrouwenvelder *et al.*, 2009c). The tank was filled by tap water filtered by cartridge (pore size of 5 μm) and carbon filters (UPS BB3, Bluefilters Group, Netherlands) to remove particulate matter and residual chlorine respectively.

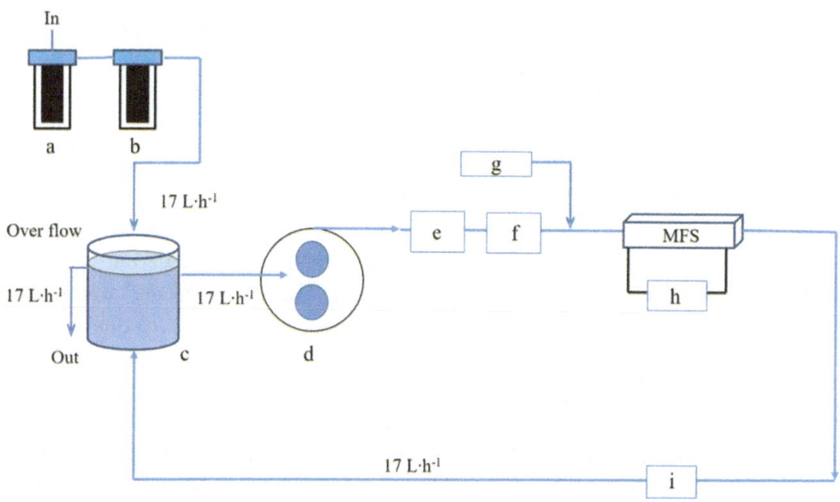

Figure 13.3 Scheme of laboratory test installation. The residual chlorine in the feed water (In) was removed using a carbon filter (a) and particulate matter was removed using a cartridge filter (b). The water entered a reservoir tank (c) at a constant flow and was pumped to the MFS by a rotary gear pump (d). The temperature sensor (e) and flow controller (f) were installed before the MFS to monitor the temperature and regulate the flow respectively. The pressure drop over the MFS was measured with a sensitive accurate differential pressure transmitter (h). Chemicals can be dosed using a substrate dosing pump (g) to the feed water of the MFS. A back pressure valve was used (i) to prevent back flow. The water was recirculated to the tank.

The MFS with external dimensions of 0.07 m × 0.30 m × 0.04 m was used in laboratory controlled stable conditions to study hydrodynamics and biofouling development (Bucs *et al.*, 2016a). The MFS has shown to be representative for the hydrodynamics and biofouling development (pressure drop development and biomass accumulation) of spiral-wound membrane modules (Vrouwenvelder *et al.*, 2006, 2007b). A coupon of virgin membrane and feed spacer of 0.04 m × 0.20 m was placed in the simulator in the same orientation as in spiral-wound membrane modules in practice. The RO membranes were provided by Trisep Corporation, USA. The feed spacer from practice was 34 mil (863 μm) thick, diamond-shaped, made of polypropylene, with a porosity of 0.85, and was provided by Conwed plastics, USA. No water permeates through the membranes in our studies; previous experiments have shown that permeate production has no effect on the measured feed channel pressure drop during biofouling experiments (Vrouwenvelder *et al.*, 2009a, 2009d).

Operating conditions

The hydrodynamic characterization of 3D printed feed spacers was carried out in MFSs by measuring the pressure drop at different flow rates, ranging from 10 L h^{-1} to 20 L h^{-1} (equivalent to a linear flow velocity of 0.09 m s^{-1} to 0.18 m s^{-1}). The relation between feed flow and pressure drop for the feed spacers are shown in the supplementary material.

To study biofilm development in the MFS, tap water was pumped to the MFS at a flow rate of 17.0 L h^{-1} equivalent to a linear flow velocity of 0.16 m s^{-1}, representative of practice (Vrouwenvelder *et al.*, 2009b). The outflow from the MFS was recirculated to the tank where fresh tap water was also added at a constant flow rate making a partially recirculated system. The tap (drinking) water fed to the monitors contained bacteria. No additional bacteria were fed to the tap water. A nutrient solution containing acetate, nitrate and phosphate in a mass ratio C:N:P of 100:20:10 was dosed to enhance the growth of bacteria (originating from the tap water) in the monitor (Vrouwenvelder *et al.*, 2010a). The concentration of substrate added to the MFS feed water was 1 mg C L^{-1}. Biofouling development was monitored by measuring the pressure drop increase over time along the length of the MFS (Vrouwenvelder *et al.*, 2009b) and analyzing the accumulated biomass in the MFS at the end of the study.

Sampling and biomass analyses

To analyze the accumulated fouling, the membrane and spacer coupons were removed from the MFS at the end of the biofouling study and placed in tubes containing 30 mL of autoclaved milliQ water. The tubes with the coupons were placed in an ultrasonic water bath for 2 min followed by mixing on a Vortex for 1 min to remove fouling from the membrane and spacer coupons.

The procedure was repeated three times and the solution left after removing the coupons was used to determine the active biomass and total organic carbon content. To determine the active biomass (live bacterial cells) and total organic carbon content, the methods used were Adenosine Tri Phosphate (ATP) and Total Organic Carbon (TOC) using an ATP Celsis Luminometer and a Shimadzu TOC analyzer respectively. Details of the methods have been described elsewhere (Vrouwenvelder *et al.*, 2009a).

RESULTS

A feed spacer from practice was recreated and modified in CAD software and the hydrodynamic behaviour from the practical and modified spacer was compared using numerical modelling. Subsequently, a feed spacer with the geometry from practice and a feed spacer with the modified geometry, selected from the numerical modelling study were 3D printed. An experimental comparison was made between a spacer from practice, the 3D printed spacer with the same geometry and the 3D printed spacer with a modified geometry. The spacers were compared with respect to (i) geometry (microscopic), (ii) hydrodynamic behaviour in MFSs (pressure drop vs. linear flow velocity) and (iii) pressure drop development during biomass accumulation in time in MFSs.

Numerical modeling and 3D printing of feed spacers

Feed spacers were modified by varying mesh-size and contact angle (Figure 13.4). The hydrodynamic characterization of the resulting spacer designs was modeled to assess possible improvement over the feed spacer from practice (Figure 13.5). On the basis of a more restricted pressure drop development, one spacer design with a larger mesh-size and reduced contact angle was selected for experimental studies (Figures 13.4d and 13.5d) and 3D printed. This modified spacer was used to study the combined effect of the larger feed spacer mesh-size and the smaller contact angle on biofouling development.

Comparison of feed spacer from practice and 3D printed spacer with same geometry

Visual comparison of the practical feed spacer (SSP) and the Polyjet produced spacer (3DS) was based on spacer filament shape, angle between filaments, and mesh-size (Figure 13.6). The two spacers were found to be similar in relation to spacer thickness (34 mil) and contact angle (90°), but were slightly different in bottle-necking. The hydrodynamic behavior was characterized in MFSs by determining the pressure drop at different flow rates. The feed spacers SSP and 3DS showed a similar relationship between pressure drop and linear flow velocity, in agreement with the numerical modeling data (Figure 13.7).

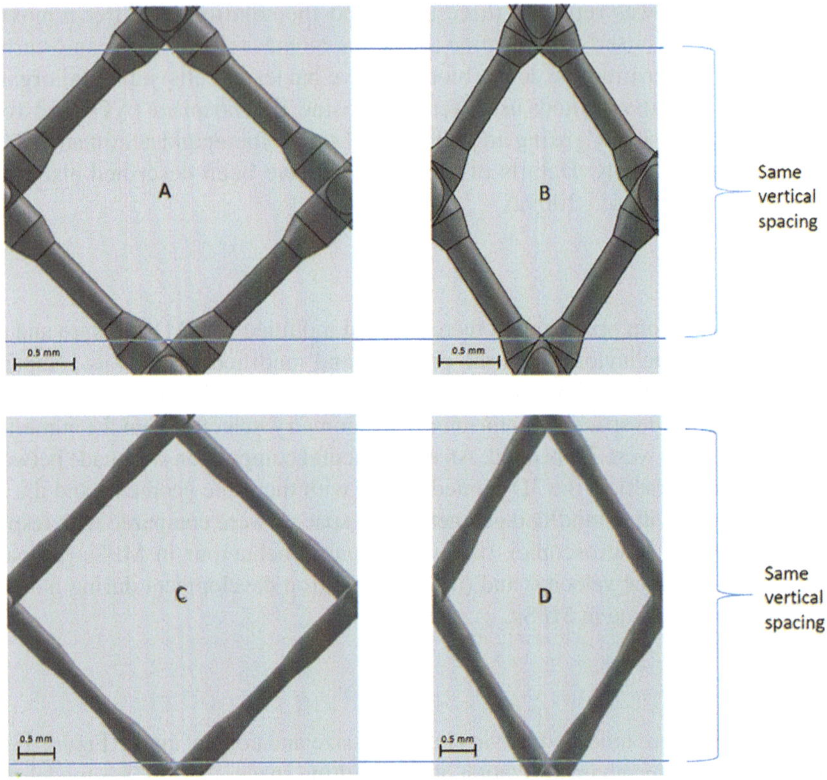

Figure 13.4 Mathematically modelled (a) standard feed spacer, (b) spacer with modified filament angle, (c) spacer with modified mesh size and (d) spacer with modified filament angle and mesh size.

During the biofilm development study at constant nutrient dosage, the pressure drop increased with time in MFSs containing SSP and 3DS (Figure 13.8a, b) and the accumulated biomass was quantified after 7 day MFS operation (Figure 13.8c, d). The initial pressure drop, pressure drop increase with time, and amount of accumulated biomass at the end of the study indicated similar performance of SSP and 3DS (Figure 13.8).

Based on the visual, hydrodynamic and biofouling characterization, 3D printing proved to be a suitable technique to develop and test feed spacers for lab scale studies.

Comparison of feed spacer from practice and 3D printed spacer with modified geometry

A feed spacer with modified geometry (3DM) selected based on the results of the numerical modeling (Figure 13.5d) was printed and experimentally compared with

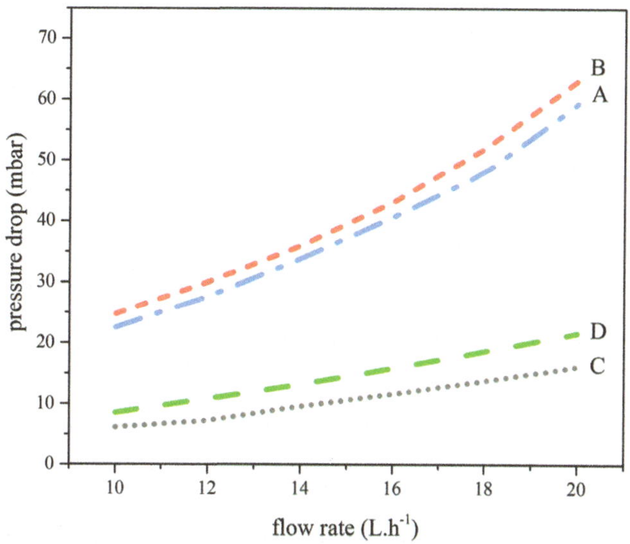

Figure 13.5 Mathematically modelled feed channel pressure drop (mbar) as a function of feed water flow ($L \cdot h^{-1}$) for (a) standard feed spacer, (b) spacer with modified filament angle, (c) spacer with increased mesh size and (d) spacer with modified filament angle and increased mesh size.

SSP. Visual observations of SSP and 3DM confirmed differences in mesh-size and contact angle between the spacer filaments, while the spacer thickness was the same (34 mil, Figure 13.9). 3DM had a larger mesh-size and smaller contact angle compared to SSP. Comparison of the hydrodynamic behavior of 3DM and SSP showed a much lower initial pressure drop and pressure drop development for 3DM,

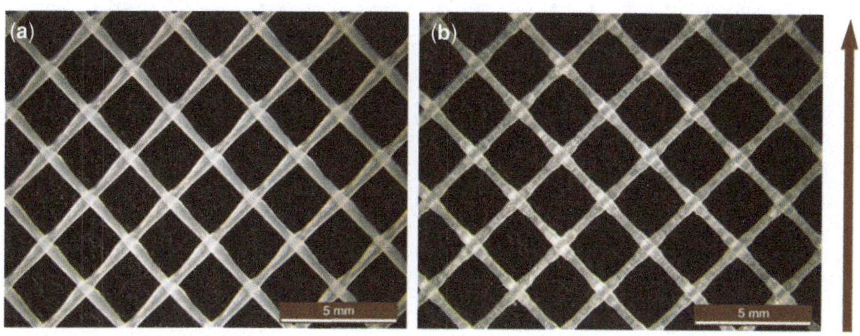

Figure 13.6 Microscopic top-view images of a 34 mil thick (a) standard feed spacer (SSP) and (b) 3D printed similar geometry feed spacer (3DS). Arrow indicates flow direction of water in the MFS.

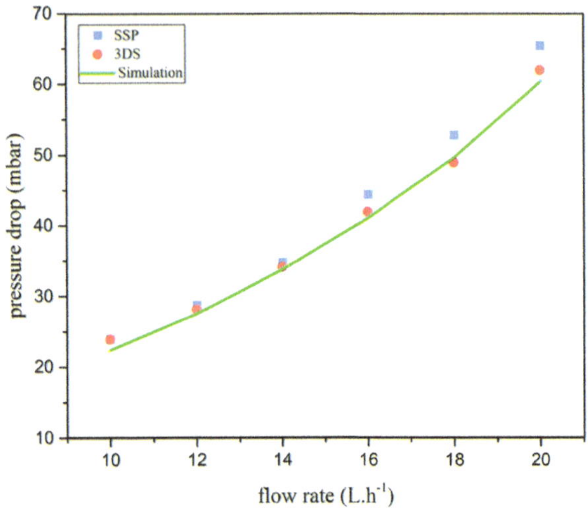

Figure 13.7 Feed channel pressure drop (mbar) as a function of the feed water flow ($L \cdot h^{-1}$) simulated for feed spacer 3DS using numerical modeling and measured for the feed spacers SSP and 3DS using MFS.

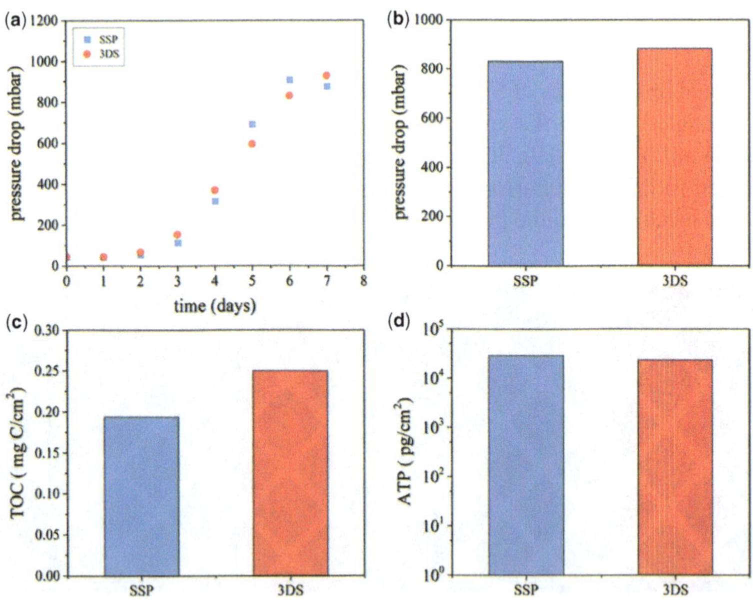

Figure 13.8 Pressure drop in time (a), pressure drop increase (b) and amount of accumulated biomass in the MFS (c and d) after 7 days of operation at constant feed flow.

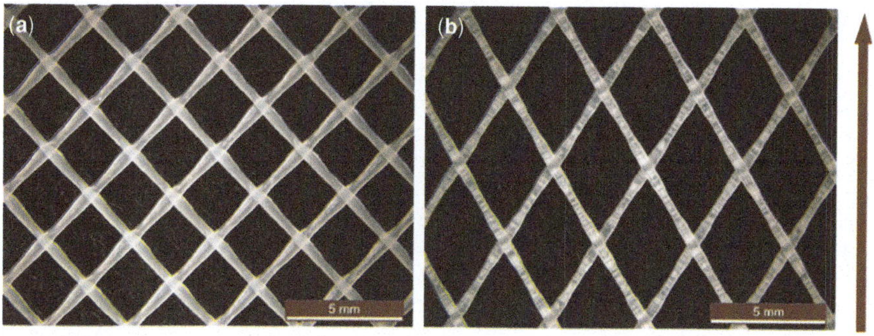

Figure 13.9 Microscopic top-view images of a 34 mil thick (a) SSP and (b) 3D printed modified geometry spacer (3DM). Arrow indicates flow direction of water in MFS.

in accordance with the simulated data (Figure 13.10). E.g. at a linear flow velocity of 0.18 m s^{-1} (20 L h^{-1}) 3DM had a 60% lower pressure drop than SSP.

Biofilm development was observed in MFSs containing 3DM and SSP by monitoring the pressure drop increase over a period of 7 days and analyzing the accumulated biomass at the end of the study (Figure 13.11). Compared to SSP, 3DM had (i) a 60% lower initial pressure drop and (ii) a 34% lower pressure drop

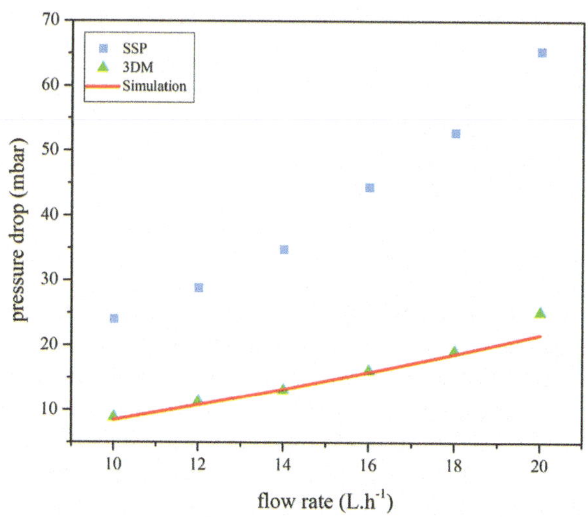

Figure 13.10 Feed channel pressure drop (mbar) as a function of the feed water flow (L · h^{-1}) simulated for feed spacer 3DM using numerical modeling and measured for the feed spacers SSP and 3DM using MFS.

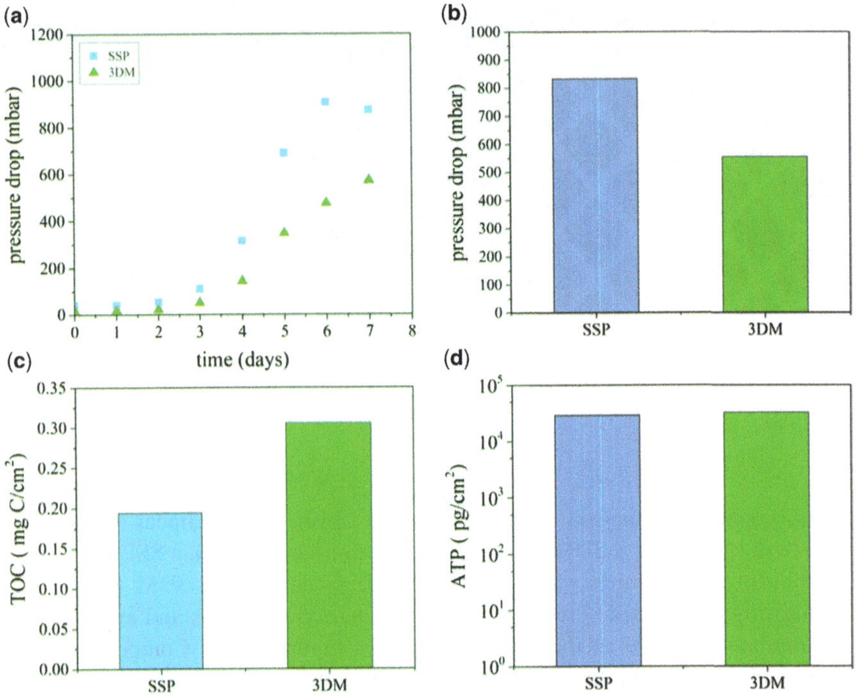

Figure 13.11 Pressure drop increase in time (a), pressure drop increase (b) and amount of accumulated biomass in the MFS (c and d) after 7 days of operation at constant feed flow.

increase after 7 days with about the same accumulated biomass amount. The results indicated that modifying the spacer geometry can reduce the impact of accumulated biomass on the pressure drop increase.

DISCUSSION

The objective of the study was to develop a strategy to design and evaluate advanced feed spacers with an adapted geometry to improve membrane performance, using the combination of numerical modeling and experimental studies with 3D printed spacers in MFSs. Results of numerical modeling and experimental studies on spacer hydrodynamics were in agreement (Figures 13.7 and 13.10). The feed spacer from practice and the 3D printed feed spacer with the same geometry showed similar hydrodynamics (Figure 13.7) and biofouling development (Figure 13.8). A modified 3D printed feed spacer showed improved hydrodynamics (Figure 13.10) and a lower impact of biofouling on membrane performance (Figure 13.11).

3D printing technique to design modified geometry feed spacers

Computer aided designing can help create modifications in feed spacer geometry. Detailed numerical modeling studies on feed spacer geometries have been done to optimize the feed spacer structure in relation to mass transfer and hydrodynamics in spiral-wound membrane modules. The studies addressed the impact of the spacer orientation, contact angle between the filaments and mesh size (Fimbres-Weihs & Wiley, 2007, 2010; Geraldes *et al.*, 2002; Koutsou *et al.*, 2007; Saeed *et al.*, 2012; Shakaib *et al.*, 2007). The spacer filament shape had a strong impact on hydrodynamics (Bucs *et al.*, 2014a; Picioreanu *et al.*, 2009). With the 3D printing technique, as addressed in this paper, the impact of the spacer filament shape can also be tested experimentally. In our study, the focus was on modifying the spacer geometry by increasing the mesh-size and reducing the contact angle. Numerical modeling showed that by these modifications of the spacers, the pressure drop is reduced significantly (Figure 13.5). The modified geometry spacer C showed a slight improvement in pressure drop performance over modified geometry spacer D (Figure 13.5). However, due to wider strand spacing, modified spacer C may have lower dimensional stability than modified spacer D. In spiral-wound elements lower dimensional stability of feed spacers is often associated with feed spacer extrusion or "telescoping" of the membrane element. For this reason modified spacer D was chosen over modified spacer C (Figure 13.4) for 3D printing and the experimental studies.

A 3D printing technique was selected to produce feed spacers for lab-scale studies. Emphasis of the study was the replication of the spacer geometry from practice applying polypropylene like material. The first 3D printing technique FDM proved to be unsuitable for producing feed spacers due to limitation in achieving the same thickness as the spacer from practice and the same material strength. Polyjet printing technique was then selected and it was shown that this technique can produce thin materials with a high resolution.

The Polyjet 3D printing technique was used to produce a spacer with a geometry and thickness identical to a feed spacer from practice. The experimental hydrodynamic characterization and biofouling study showed that the 3D printed spacer (3DS) was identical to the feed spacer from practice (SSP). Subsequently, the modified geometry spacer (3DM) was printed and evaluated experimentally using the MFS. At the end of the experimental period the same amount of biomass was accumulated using the SSP and 3DM spacer geometries in the MFS. The lower performance decline (lower pressure drop increase) for the monitor with the 3DM spacer geometry suggests that the accumulated biomass had less effect on the hydrodynamics of the 3DM spacer compared to the SSP spacer. Lower impact of the same amount of biomass can be explained by differences in the biofilm structure and/or compactness (volume) and/or spatial distribution compared to the biofilm formed in the flow channel with

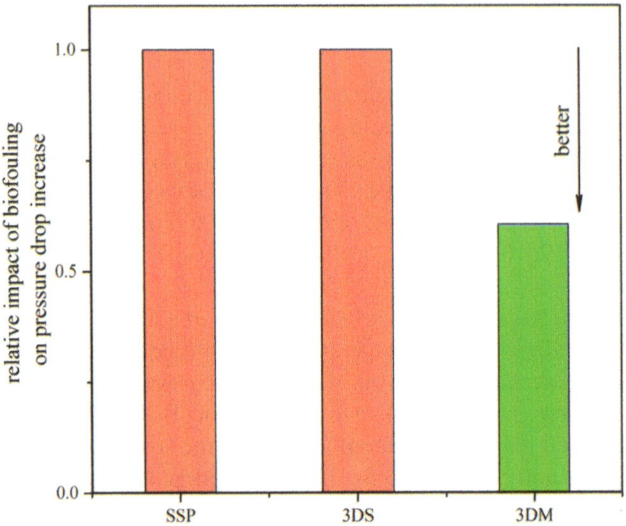

Figure 13.12 Impact of biofouling on pressure drop increase in MFSs containing the practice feed spacer (SSP), 3D printed same geometry spacer (3DS) and the 3D printed modified geometry spacer (3DM). With the same amount of accumulated biomass in the MFSs, the pressure drop increase was (i) similar for SSP and 3DS and was (ii) much lower for the 3DM compared to SSP.

the SSP spacer geometry. In Section 4.3 tools suitable for future studies are presented and discussed. In summary, results from the studies showed that the selected spacer with increased mesh-size and reduced contact angle reduced the impact of biofouling on pressure drop development thereby improving membrane performance (Figure 13.12).

A novel strategy to modify feed spacer geometry

3D printing has been used to manufacture bulky objects like flow cells, but it is not described in literature, that this technique is used to produce feed spacers for membrane based water treatment such as desalination and reuse. In this study, a new strategy is proposed for evaluating the performance of standard and modified spacers with a variety of geometries (Figure 13.13).

After designing and numerical modeling, the modified 3D printed spacers can be evaluated effectively on lab-scale. Once a modified spacer proves to perform better than current feed spacers in practice, the next stage will be to manufacture spacers in bulk using the conventional manufacturing technique or possibly using 3D printing techniques. The modified spacers can then be used in spiral-wound membrane modules in pilot plants and eventually industrial installations for evaluating the performance on large-scale.

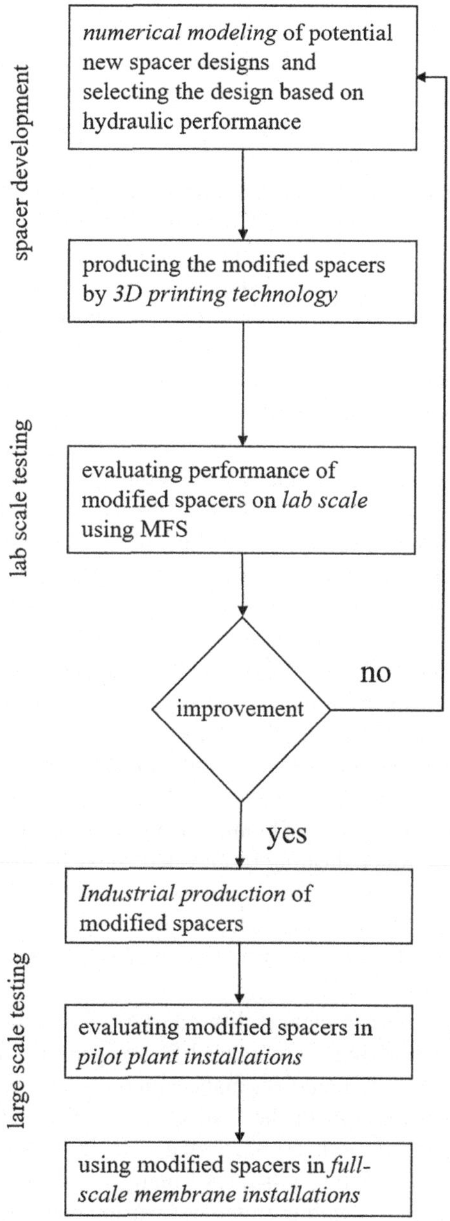

Figure 13.13 Scheme for cost effective development and evaluation of novel feed spacers for biofouling and scaling control.

Suggestions for future studies

3D printed spacers with a modified geometry can be used to study several factors affecting the performance of spiral-wound membrane systems, such as the permeate production in NF and RO membrane systems as described in numerical studies (Ma & Song, 2006). Modifying the spacer design will impact the permeation/product formation. Feed spacers enable proper mixing of feed water in the membrane module in order to avoid concentration polarization and consequently, scaling (Saeed *et al.*, 2012). Therefore, the effect of modified feed spacers on concentration polarization (scaling) should be studied as well.

Periodic cleanings for biofouling control of membrane systems are inevitable. Pre-treatment and dosage of chemicals such as anti-scalants with a low nutrient content will delay the build-up of biofouling, while use of modified spacers may reduce the impact of accumulated biomass on membrane performance. In the long run, despite the best organic nutrient removal by pre-treatment and thicker/modified spacer use, cleanings of membrane modules are unavoidable (Cornelissen *et al.*, 2007b; Vrouwenvelder *et al.*, 2009b). Current chemical cleaning strategies of membrane systems containing standard geometry feed spacers do not restore the original membrane performance (Creber *et al.*, 2010a, b). Chemical cleanings inactivate biomass but are not effective in a complete biomass removal from spiral-wound membrane modules (Bereschenko *et al.*, 2011; Creber *et al.*, 2010a, b; Vrouwenvelder & Van Der Kooij, 2001). Due to the presence of spacers, the biomass inactivated by chemicals remains trapped in the feed channel leading to a rapid regrowth of biofilm. In other words, the net-like structure of the feed spacer hampers biomass removal from the membrane module during conventional cleaning using chemicals. Designing a feed spacer with e.g. a larger mesh-size and reduced contact angle may lead to better removal of biomass from the modules. Advanced cleaning strategies such as air water flushing or elevated shear by using air (gas) bubbles (Cornelissen *et al.*, 2007b; Vrouwenvelder *et al.*, 2009b) may enhance biofilm detachment and removal. Therefore, effective biofilm removal from the membrane module may be possible using modified geometry feed spacers combined with advanced cleaning strategies.

The phenomenon of easy cleanability can be investigated in lab-scale MFS studies with in-situ imaging techniques like oxygen sensing optodes, optical coherence tomography, magnetic resonance imaging etc. by understanding the spatial distribution of biofilm in the feed channel in the presence of modified geometry spacers (Dreszer *et al.*, 2014b; Farhat *et al.*, 2015; Manz *et al.*, 2005; Prest *et al.*, 2012; Staal *et al.*, 2011a; Valladares Linares *et al.*, 2016; West *et al.*, 2016). There is a need to develop cleaning strategies focusing on the removal of biofouling from membrane systems. Use of modified feed spacers in membrane modules may facilitate reducing the impact of accumulated biomass on performance and removal of detached biofilm from spiral-wound membrane modules and the membrane installation, providing a robust solution for biofouling control.

The use of spacers is not limited to NF and RO spiral-wound membrane systems. Forward osmosis (FO), reverse electrodialysis (RED), membrane distillation (MD), and electrodeionisation (EDI) are examples of other membrane based processes where spacers play an important role. FO is a membrane separation process driven by a difference in osmotic pressure across the membrane, meaning FO operates at low or no hydrodynamic pressures compared to NF/RO (Cath *et al.*, 2006). RED is the process of obtaining energy from a salinity gradient i.e. energy is retrieved from the difference in the salt concentration between saline water and fresh water both let through a stack of alternating cation and anion exchange membranes (Lacey, 1980). MD is a hybrid process that uses membranes and operates on the basis of evaporation, where mass is transported by the difference in vapor pressures between feed and permeate (Phattaranawik *et al.*, 2003).

EDI is an ionic separation technology using electricity, ion exchange membranes and resins to produce high purity deionized water (Alvarado & Chen, 2014), involving non-conducting and various conducting spacers (Hou *et al.*, 2010; Nikonenko *et al.*, 2008).

Unlike NF/RO where biofouling has been reported on the feed side of the membrane, FO and RED systems face fouling problems on either side of the membrane. A recent biofouling study of FO systems with varying spacer thicknesses showed that spacer biofouling impacted FO membrane performance, emphasizing the importance of spacers (Valladares Linares *et al.*, 2014). Studies done in RED with and without spacers showed that fouling had less impact on performance in the absence of spacers (Post *et al.*, 2010; Vermaas *et al.*, 2013). MD studies showed that performance could be improved by using spacers and that a coarse spacer showed better performance than a fine spacer (Martínez-Díez *et al.*, 1998). Besides membrane based water filtration systems, spacers are also an essential part of membrane installations in food and dairy industry (Flint *et al.*, 2011; Mohammad *et al.*, 2015). EDI studies with ion conducting spacers showed to affect performance emphasizing the role of spacers (Nikonenko *et al.*, 2008).

In summary, the proposed strategy to design and evaluate advanced geometry spacers, using the combination of numerical modeling of spacer geometries and experimental testing of 3D printed spacers can help in characterizing and controlling operational problems in various membrane treatment processes from e.g. water desalination and reuse, energy production, food and dairy industry and can be used to develop new antifouling strategies.

CONCLUSIONS

Studies for spiral-wound NF and RO systems were done involving (i) the creation of several types of feed spacer geometries in CAD software and comparison of hydrodynamic behavior using numerical modeling, (ii) 3D printing of a practice and modified feed spacer, and (iii) experimental membrane fouling simulator (MFS) studies of a practice and 3D printed feed spacer with the same geometry and (iv) MFS studies of a practice and a 3D printed modified geometry spacer. The spacers were experimentally compared on geometry, hydrodynamic behavior and pressure drop development during biomass accumulation in MFSs. Based on the results, the following can be concluded:

- Numerical modeling and experimental studies showed the same hydrodynamic behavior for practice and 3D printed feed spacers with the same geometry.
- 3D printing technology is a suitable tool to produce spacers with a thin and complex geometry: feed spacer from practice and 3D printed spacer with same geometry showed similar hydrodynamics and biofouling development.
- Compared to a feed spacer from practice, a 3D printed feed spacer with a modified geometry showed a low feed channel pressure drop and low biofouling impact on performance.
- A strategy is proposed to design and evaluate spacers with advanced geometry, using the combination of numerical modeling of spacer geometries and experimental MFS testing of 3D printed spacers.

The proposed strategy may also be suitable for developing spacers in e.g. forward osmosis (FO), reverse electrodialysis (RED), membrane distillation (MD), and electrodeionisation (EDI) membrane systems.

Chapter 14

Strategies for biofouling mitigation in spiral wound membrane systems

INTRODUCTION

Currently, more than two billion people live in highly water-stressed areas (Vörösmarty *et al.*, 2000). Because of the uneven distribution of fresh water in time and space, the situation is likely to worsen in the future as a large number of regions are expected to experience more extreme climate conditions and rapidly growing demands in water-use sectors: agriculture (crop production, livestock), domestic (municipal), and industry (energy, manufacturing) (Vörösmarty *et al.*, 2000).

Since more than 97 percent of the water in the world is seawater, desalination technologies have the potential to solve the fresh water crisis. Seawater desalination is already used in many countries mainly in water scarce regions such as the Middle East, as well as in countries with adequate freshwater resources.

Desalination technologies can be divided into two major groups: thermal and membrane desalination. While thermal desalination was the main technology in the past, membrane-based desalination technologies gained importance in the last decade, reaching 60 percent of the global desalination capacity in 2015 with a continuously increasing trend (Pankratz, 2012). This is caused by the improved efficiency and lower energy demand of the membrane-based desalination processes, lowering thus the cost of water production. Reverse osmosis (RO) and nanofiltration (NF) membrane systems currently hold the largest desalination capacity globally (Lattemann *et al.*, 2010). Besides RO and NF there are also alternative emerging membrane-based desalination processes including electrodialysis (ED), membrane distillation (MD) and forward osmosis (FO).

Author(s): Szilárd Bucs, Joop Kruithof, Mark van Loosdrecht and Johannes Vrouwenvelder
doi: 10.2166/9781780409597_0277

Osmosis is the naturally occurring process where the water from solution passes through a semipermeable membrane to dilute a more concentrated solution. Reverse osmosis (RO) applies hydraulic pressure on the concentrated solution so that the water transport through the membrane is reversed and fresh water can be separated from saline water. RO membranes are able to reject colloidal and dissolved matter from aqueous solutions, resulting in a more concentrated solution called "brine" and fresh water, usually referred as "permeate".

Commercially available membrane modules include spiral-wound, hollow fibre, tubular and modules (Mulder, 1996). Amongst these, spiral wound modules are most commonly applied, due to their high membrane area to volume ratio. The major components of a spiral-wound module are the membrane, the feed and permeate channels, spacers keeping the membrane leaves apart, the permeate tube and the membrane housing (Dickson *et al.*, 1994). The feed channel spacer may enhance mass transfer near the membrane, but inevitably increases pressure loss along the membrane leaf (Da Costa *et al.*, 1994; Radu *et al.*, 2010). Membrane sheets with spacers in between are glued together on three sides to form an envelope and multiple envelopes are attached to and rolled up around the permeate tube to create the feed and permeate channels. A pressurized membrane module housing holds the membrane leaves in place. Usually, three or more modules are connected in series in a pressure vessel (Schwinge *et al.*, 2004a).

The performance of the modules is affected by many factors: (i) spacers geometry, which greatly affects local mixing, mass transfer (concentration polarization) and pressure loss, (ii) fouling propensity and cleaning ability, (iii) plant design and operating conditions, such as feed pre-treatment, feed concentration, feed pressure and permeate recovery.

Four major types of fouling can occur in spiral wound membrane systems: colloidal (suspended particles such as silica), inorganic (salt precipitates such as metal hydroxides and carbonates causing scale formation), organic (natural organic matters such as humic acids), and biological (such as bacteria and fungi). Because the reverse osmosis membranes are nonporous, the formation of a fouling layer on the membrane surface is the dominant fouling mechanism (Kang & Cao, 2012). RO membrane fouling is closely related to the interaction between the membrane surface and the foulant. Previous studies indicated that the physicochemical properties of the RO membrane surface, such as hydrophilicity, roughness, and surface charge, and the feed spacer geometry are major factors influencing membrane fouling (Louie *et al.*, 2011; Vrouwenvelder *et al.*, 2009a).

Biofouling is considered the major fouling type of the membrane process because microorganisms can multiply over time. Even if 99.9% of microorganisms are removed with pre-treatment of the feed water, there are still enough microbial cells remaining to grow by utilizing biodegradable substances in the membrane installation feed water (Flemming, 1997). Biofouling can be considered as a biotic form of organic fouling while fouling caused by organic matter derived from microbial cellular debris can be considered as an abiotic form of biofouling (Flemming, 2002). Biofouling has been known as a contributing factor to more than 45 percent of all membrane fouling (Nguyen *et al.*, 2012) and has been reported as a major problem in nanofiltration (NF) and reverse osmosis (RO) membrane filtration (Flemming, 1997; Nguyen *et al.*, 2012).

Biofouling of the RO membrane results in a decline in permeate water flux and a decrease of salt rejection. The decline in membrane performance it is due to the increase in the hydraulic resistance and the trans-membrane osmotic pressure of the fouled membrane. The increase in the trans-membrane osmotic pressure it is the result of bacterial cells deposition, which enhance the concentration polarization of salt near the membrane surface (Herzberg & Elimelech, 2007; Radu *et al.*, 2015). The greatest effect of biofilms on membrane systems may be attributed to the physical properties of the extracellular polymeric substance (EPS) matrix produced by the embedded microorganisms by increasing the hydraulic resistance and thus reducing permeate production.

Several fouling control strategies have been developed and tested in full-scale membrane installations. Colloidal, inorganic and organic fouling can generally be controlled by pre-treatment or by dosage of chemicals (e.g. antiscalants). However, biofouling can only be restricted and delayed by pre-treatment, but not eliminated (Bucs *et al.*, 2014b). Direct dosage of oxidizing biocides such as free chlorine is not possible due to damage of the membrane structure causing reduced membrane performance. Several non-oxidizing biocides would be used as nutrient by the microorganisms, thus enhancing biofilm growth (Creber *et al.*, 2010b; Vrouwenvelder *et al.*, 2000). Current research is focused on membrane surface modification, non-oxidizing biocides application and modification of the feed channel geometry and operating condition in order to reduce the biofouling in spiral wound membrane systems (Bertheas *et al.*, 2009; Bucs *et al.*, 2015; Cornelissen *et al.*, 2007a; Geraldes *et al.*, 2003; Siddiqui *et al.*, 2016; Tang *et al.*, 2009; Wang *et al.*, 2012). Quorum sensing is another approach for biofouling control, aiming at biofilm dispersal in response to certain biochemical compounds; however, no application has been implemented in practice (Dobretsov *et al.*, 2009). Despite the efforts on controlling fouling in spiral wound membrane systems, biofouling remains the major problem in membrane filtration processes, causing increased energy demand and unreliable water production. It is therefore crucial to gain more fundamental understanding of the biofilm formation in spiral wound membrane systems, in order to develop strategies to control and keep biofouling at an acceptable level. If biofouling cannot be avoided in the long run, biofouling control strategies should focus on delaying biofilm formation, reducing its impact on membrane performance and removing biofilms by advanced cleaning strategies.

The objective of this review is to provide an overview on the state of the art strategies to control biofouling in spiral wound reverse osmosis membrane systems and point to possible future research directions.

Biofouling control strategies

Biofouling is considered the major fouling problem in membrane systems for water treatment. In spite of extensive research to prevent and eliminate biofouling, no successful control strategy has been developed yet. Most common biofouling control strategies are: (i) feed water pre-treatment, (ii) membrane surface modification, (iii) feed spacer design and (iv) chemical/mechanical membrane cleaning.

Pre-treatment by water filtration and bacterial inactivation

For a constant and reliable operation of reverse osmosis membrane systems good quality feed water is essential. Good feed water quality is defined by the membrane manufacturers as water with a turbidity lower than one Nephelometric Turbidity Unit, NTU, silt density index SDI < 3 (or SDI < 4), oil and grease <1 mg $\cdot$ L^{-1} (Alawadhi, 1997). When the source water does not meet these criteria, the feed water has to be pre-treated before entering the reverse osmosis membrane system. Most commonly applied pre-treatment technologies are based on water filtration (e.g., filtration over granular media and low pressure membrane filtration) and disinfection.

Filtration over granular materials can be categorized into single, dual and mixed media filtration (Prihasto *et al.*, 2009), meaning that one or more materials can form the filter bed. The most important filtration mechanism here is deposition of the suspended particles to filter media grains, as the raw water passes through the filter bed. The most commonly used filter media are sand and anthracite (Chua *et al.*, 2003). Furthermore, to protect the reverse osmosis systems from particle fouling, cartridge filters with a pore size range between 1 and 20 μm are applied after the media filter (Alawadhi, 1997).

Low pressure membrane filtration such as microfiltration (MF) and ultrafiltration (UF) has gained importance in the past years as pre-treatment for reverse osmosis systems. Although MF and UF pre-treatment removes very well the bacterial cells from the feed water, biodegradable nutrients can pass UF membranes enabling eventual microbial growth in the subsequent RO installation.

In some cases, activated carbon or biofiltration is used to remove dissolved organic matter from the feed water. Activated carbon adsorption, either in granular or powder form, has also been considered as a feasible mean for reducing membrane fouling, either alone or in combination with other pre-treatment processes (Gur-Reznik *et al.*, 2008; Meier & Melin, 2005; Zhao & Yu, 2015). Chinu *et al.* (Chinu *et al.*, 2009) delayed fouling development in a lab scale setup by using biofiltration as pre-treatment using real seawater.

To protect the reverse osmosis membranes from biological fouling the raw water is usually disinfected by addition strong oxidant, such as chlorine, monochloramine, hypochlorite, chlorine dioxide, ozone, or UV irradiation.

In case of addition of chemical oxidants such as chlorine a further step is required. Since the reverse osmosis membranes are not resistant to oxidants, residual chlorine has to be removed from the raw water prior to entering the RO membrane system, commonly achieved by dosage of sodium bisulphite or by activated carbon filtration (Prihasto *et al.*, 2009).

Oxidation of organic compounds by addition of ozone has been used as pre-treatment for spiral wound membrane systems. Ozone reacts with the hydrophobic parts of organic foulants and transforms into more hydrophilic groups (Van Geluwe *et al.*, 2011; Yu *et al.*, 2016).

Even with extensive pre-treatment, a very low amount of biodegradable nutrients can be found in the feed water of the RO installation. With the large amount of water provided per membrane surface per day, even minimal amounts of substrate (microgram per litre level) in the feed water lead to a significant organic substrate supply for biofilm growth, which will occur over weeks or months of operation.

Non-oxidising biocides can be used as an effective control biofouling control strategy. An alternative non-oxidising biocide, monochloramine 2,2-dibromo-3-nitrilopropionamide (DBNPA) has been applied in limited number of water treatment plants. DBNPA is rapidly degrading in alkaline aqueous solutions and is compatible with polyamide based membranes and shows high rejection rates for RO membranes (Bertheas *et al.*, 2009). The antimicrobial effect is due to the fast reaction between DBNPA and sulfur-containing organic molecules in microorganisms such as glutathione or cysteine. Siddiqui *et al.* (Siddiqui *et al.*, 2017b), found that DBNPA can be successfully applied to prevent biofouling when it is continuously dosed, however it was not effective as cleaning agent on fouled membranes. Dosage of DBNPA to a biofouled membrane system inactivated the accumulated biomass but did not restore the original membrane performance (Siddiqui *et al.*, 2017b).

Membrane modification

Physical properties (e.g., hydrophilicity, surface charge, roughness) of the membrane surface impact membrane biofouling (Louie *et al.*, 2006, 2011). It is generally accepted that hydrophilic membranes are more resistant to fouling (Kang & Cao, 2012). Deposition of foulant is less likely on neutrally or close to neutrally charged membrane surfaces (Kang & Cao, 2012; Kim & Lee, 2006). Also surface morphology plays a significant role on membrane biofouling, because foulant are more likely to be trapped by rougher topologies than by the smoother ones (Louie *et al.*, 2006; Sagle *et al.*, 2009). Many studies showed that, compared to uncoated membranes, less fouling developed when membranes surfaces had a hydrophilic coating (Araújo *et al.*, 2012b; Hadidi & Zydney, 2014; Matin *et al.*, 2014; Ostuni *et al.*, 2001; Tang *et al.*, 2009; Venault *et al.*, 2014; Zhao *et al.*, 2007, 2010; Zhu *et al.*, 2009).

In many investigations, only short-term (2–6 h) static protein or bacterial cell (pure strain) adhesion tests were performed (Adout *et al.*, 2010; Miller *et al.*, 2012;

Park *et al.*, 2006; Yi *et al.*, 2010). In some other cases, short-term (2–24 h) cross-flow or dead-end filtration tests were used to evaluate the impact of the coating on fouling layer accumulation (Asatekin *et al.*, 2009; Madaeni & Ghaemi, 2007; Sagle *et al.*, 2009; Wang *et al.*, 2012; Zhao *et al.*, 2013). Short-term studies provide insight into initial protein or bacterial cell attachment to the membrane, but do not predict long term biofilm development (Asatekin *et al.*, 2009; Matin *et al.*, 2014; Miller *et al.*, 2012; van Loosdrecht *et al.*, 1989). Therefore, long-term studies (i.e., several months) are needed to investigate the effective impact of membrane surface modification on biofouling development. It has also been shown experimentally that long-term biofouling studies are more representative for practice than short-term protein and bacterial adhesion tests (Bucs *et al.*, 2017; Louie *et al.*, 2011; Miller *et al.*, 2012).

Few studies investigating coated membranes to control fouling reported on the coating stability (Brzozowska *et al.*, 2011; Bucs *et al.*, 2017; Louie *et al.*, 2006; Ronen *et al.*, 2015). Brzozowska *et al.* (Brzozowska *et al.*, 2011) reported a weak attachment of polymer brushes to the membrane surface, with the coating layer easily removed by the water flow. Experiments with silver nanoparticle-coated feed spacers showed silver leaching during a fouling study using a flow cell (Yang *et al.*, 2009). Louie *et al.* (Louie *et al.*, 2006) investigated the antifouling performance of a polyether-polyamide block copolymer coating on the membrane surface. At the end of a 106 days experimental period with oil/surfactant/water emulsion the presence of the polyether-polyamide coating was shown, meaning that the coating was not removed from the membrane surface during operation. In spite of many studies reporting anti-biofouling effects of various coatings, currently no coating has been developed that can prevent biofouling in membrane systems. Coated membranes can contribute to a delay in biofilm formation, but cannot prevent biofouling.

Feed spacer and hydrodynamics

Feed spacers are used in reverse osmosis membrane modules to keep the membrane sheets apart and to enhance mixing of the feed water thus reducing concentration polarization in the vicinity of the membrane. Extensive research was conducted to optimize the feed spacer geometries to reduce concentration polarization. Koutsou *et al.* (2007, 2009) and Koutsou and Karabelas (2010, 2015) determined optimal ratios between the feed spacer filament distance and filament diameter, mean flow angle and angle between filaments. It was shown feed spacers with a mesh angle of 60°, has the highest water flux, however, the associated pressure drops are slightly higher compared to nonwoven geometries. Middle layer geometries with a mesh angle of 30° produce the lowest water flux, while feed spacers with a mesh angle of 90° show the lowest pressure drop among all the filament arrangements (Amokrane *et al.*, 2015; Gu *et al.*, 2017; Koutsou & Karabelas, 2010).

Besides the impact on the hydrodynamics, it was shown that feed spacers provide a surface for initial deposition of biofouling that accumulates and eventually spreads to the free membrane area (Baker *et al.*, 1995a). Tran *et al.* (Tran *et al.*, 2007) reported that the membrane area in the vicinity of the feed spacer filaments is mostly affected by fouling. It was also shown that the impact of biofouling on the pressure drop was higher in the presence of a feed spacer (Bucs *et al.*, 2014a; Vrouwenvelder *et al.*, 2009b). Therefore, feed spacers are important for membrane performance by affecting the concentration polarization and biofouling. Siddiqui *et al.* (Siddiqui *et al.*, 2016) tested novel 3D printed feed spacer and showed that by changing the geometry of the feed spacer the impact of biofouling on membrane performance can be reduced. Further optimization of the feed spacer geometry, not only for enhanced mass transfer but also to reduce the impact of biofouling on the process performance, may be possible. Another important aspect of spacer design would be to enable the easy removal of biomass from membrane modules during cleaning (Vrouwenvelder *et al.*, 2011b). In this respect, experiments should be carried out on applying different cleaning steps and analysing the effectiveness of biomass removal for different feed spacer geometries. Again, as for the coating strategies, the feed spacer modification cannot prevent biofouling, but it can only reduce the impact of the accumulated biomass and enhance membrane module cleanability.

In conclusion, if biofouling cannot be avoided in the long run (Bucs *et al.*, 2015), then biofouling control strategies should focus on delaying the biofilm formation, reducing its impact on membrane performance and allowing biofilm removal by advanced cleaning strategies.

Spiral wound membrane modules cleaning strategies

In order to maintain a constant clean water recovery without compromising energy input, it is important to control fouling in the system. When performance declines membrane modules need to be chemically cleaned to maintain the plant operation. Curative cleaning is the conventional approach practiced for quite a few decades where chemicals are flushed into the system when there is a performance decline. Chemical cleaning broadly includes weakening of the biofilm matrix mostly by chemicals such chelating agents, detergents, bases, acids and enzymes and removal of the biofilm by mechanical forces such as rinsing with water or air (Cornelissen *et al.*, 2007a; Wibisono *et al.*, 2015a).

Most commonly a combination of acidic and/or basic (alkaline) chemicals are used to clean the membrane modules. Common acidic solutions (pH 2) include hydrochloric acid, phosphoric acid, sodium hydrosulphate and sulfamic acid, while alkaline (pH 12) chemicals include sodium lauryl sulphate, sodium hydroxide, sodium ethylene diamine tetra acetic acid. Commercial cleaning agents are found to be more effective than conventional cleaning agents recommended by the membrane manufacturers in some cases (DOW, 2014).

Hydraulic cleaning is the most commonly used physical cleaning method wherein water is flushed through the system in forward or backward direction to remove the accumulated biomass and organic foulants. Forward flushing in fact can cause further biofouling as the biomass accumulated in the lead membrane is pushed to the ones downstream, where they form new biofilms almost instantly. Due to this reason, some plants perform a backwash by reversing the module, thereby reducing the chance of spreading the biomass to all the adjacent membrane modules (reference). Pneumatic cleaning refers to the use of air or gas mixed with water for flushing (air-water flushing). A series of experiments done by Cornelissen *et al.*, shows promising results on pilot scale for the use of air/water flushing (Cornelissen *et al.*, 2007a; Wibisono *et al.*, 2014, 2015a). CO_2 dissolved in water has proved to restore initial hydraulic resistance as well as visible reduction in biofouling (Ngene *et al.*, 2010a).

FUTURE RESEARCH DIRECTIONS

Experimental and modelling results show that the same amount of biofilm impacts differently the membrane process performance (i.e., the feed channel pressure drop and flux) (Bucs *et al.*, 2014a, 2016b). It was hypothesized that the biofilm location (membrane and spacer), biofilm geometrical properties (porosity, thickness, and roughness), biofilm mechanical characteristics (rigidity, viscoelasticity, and density) and its hydraulic properties (permeability) all contribute to the membrane process performance decline. Therefore, future studies should focus on: (i) biofilm structural characterization; (ii) understanding to what extent biofilm properties affect membrane filtration performance, (iii) developing methods to engineer biofilm properties such that biofouling would have only a low or delayed impact on the filtration process and biomass should be easily removable.

Biofilm structural characterization

Traditionally biofilm characterization is carried out by microscopy methods. However, sample preparation for Scanning Electron Microscopy (SEM) may affect the biofilm structure and the method is off-line. Therefore, *in-situ* imaging methods should be pursued. Confocal laser scanning microscopy (CLSM) can be applied on-line or off-line, maintaining the biofilm structure. However, the sample must be fluorescent (either auto-fluorescent or stained), which may influence the biofilm behaviour and structure. In addition, the size of the observed area with CLSM is rather small for this application. Computed Tomography (CT) can acquire three-dimensional images of the biofilm surface attached to support materials with highly complex geometry, but contrast agents must be added (Davit *et al.*, 2011). There is a clear need for further development of *in-situ*, online characterization methods of the biofouling layer under representative conditions for practice.

Recent progress in the three-dimensional *in-situ* non-destructive biofilm imaging has been achieved by Magnetic Resonance Imaging (MRI) (Fridjonsson *et al.*, 2015b; Graf von der Schulenburg *et al.*, 2008) and Optical Coherence Tomography (OCT) (Blauert *et al.*, 2015; Wagner *et al.*, 2010b; West *et al.*, 2016). MRI imaging can cover large areas (a few cm^2) of actual spiral wound membrane modules, and

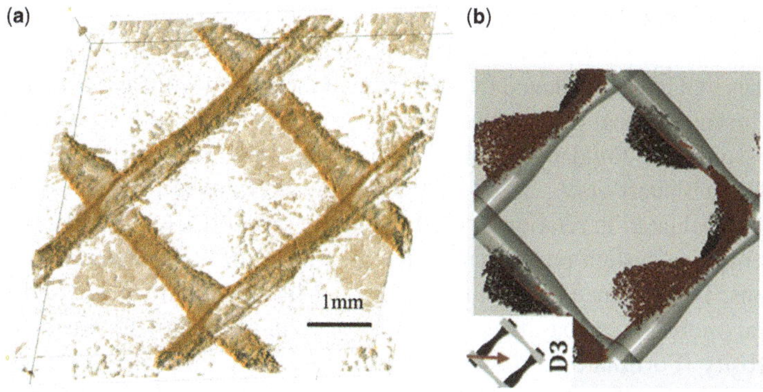

Figure 14.1 (a) Three-dimensional OCT image with biomass (brown colour), feed spacer, membrane and cover glass. Adapted from Fortunato *et al.* (2017). (b) Three-dimensional simulation of particle deposition on top (red) and bottom (black) membranes, in a spacer-filled feed channel, showing fouling patterns similar to those experimentally observed.

evaluate the biofilm distribution and preferential flow channel formation over the module length (Vrouwenvelder *et al.*, 2010b).

Optical coherence tomography can be used to investigate biofilm formation *in-situ* and without staining. The rationale to focus on OCT in this paper is because this the only technique able for direct, in situ, non-destructive and high resolution imaging of biofilms in membrane fouling simulators. OCT has been used to study the change in performance parameters (e.g., feed channel pressure drop, flux) using different types of membranes (RO, UF, MF) (Derlon *et al.*, 2013; Dreszer *et al.*, 2014b; Fortunato *et al.*, 2017, n.d.; West *et al.*, 2016).

The main advantages of the OCT technique are: (i) it enables three-dimensional observation and quantification of the biofilm over a large area (millimetres); (ii) it is totally non-invasive, requires no staining and can be performed during the operation of a lab-scale membrane setup and (iii) biofilm formation can be observed in time. A recent OCT study (Fortunato *et al.*, 2017) clearly showing specific and reproducible locations of biofilm growth in a spacer-filled channel with permeation is represented in Figure 14.1. Moreover, this technique can supply valuable data to be used in conjunction with numerical modelling of fluid dynamics and biofilm formation, as also shown in Figure 14.1.

Biofilm mechanical characterization

Mechanical biofilm properties should also be determined, as these can be strongly connected with the rate of biofilm removal (e.g., biofilm detachment) and thus with the efficiency of several chemical or mechanical treatments.

A suite of (classical) methods, such as rheology, nuclear magnetic resonance (NMR) imaging, and atomic force microscopy (AFM) techniques, and advances in instrumentation for determining biofilm mechanical properties has broadened the technological advances in instrumentation and interactions between multiple disciplines that have broadened the spectrum of methods available to conduct studies on biofilm mechanical properties (Billings *et al.*, 2015). Biofilm elasticity modulus and Poisson ratio could be estimated from real-time cross sectional OCT scans of biofilms (Blauert *et al.*, 2015). However, the mechanical properties of biofilms should be evaluated in relation to their hydraulic properties (e.g., permeability) and with membrane system performance decline, all in different membrane operating conditions. Three-dimensional biofilm scans can be numerically processed to extract structural (morphology) parameters such as: biovolume, biofilm thickness, roughness and porosity (Fortunato *et al.*, n.d.). Hardly accessible mechanical properties can be evaluated with time-dependent OCT scans, which is a clear advantage of this technique. For example, biofilm rigidity, viscoelasticity and density should be measured, as these strongly correlate with biofilm detachment and therefore with cleanability of the membrane system (Billings *et al.*, 2015; Dreszer *et al.*, 2014b; Hohne *et al.*, 2009; Houari *et al.*, 2008; Towler *et al.*, 2003; Valladares Linares *et al.*, 2016). Biofilm response under operational changes can be immediately observed, such as compaction with increased water flux (Figure 14.2). In this way,

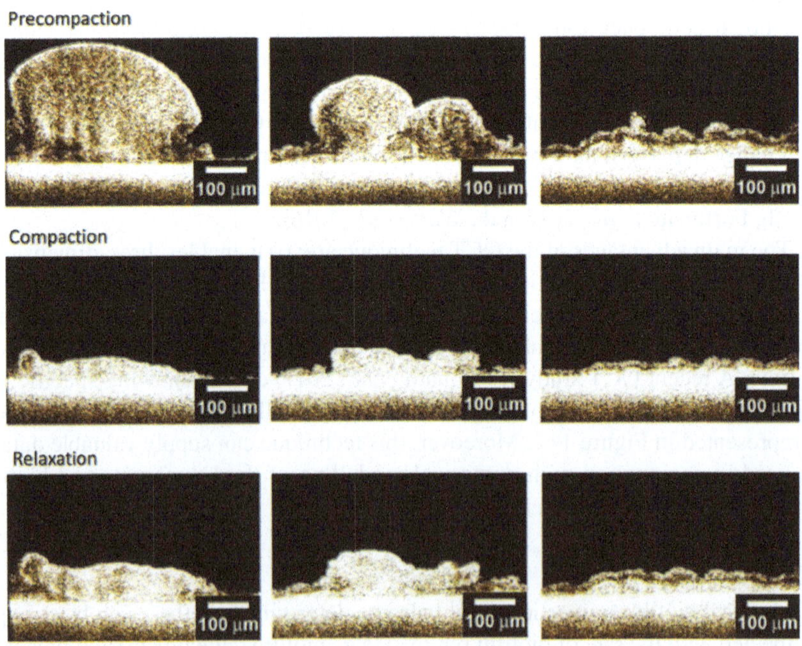

Figure 14.2 Optical coherence tomography observations of the biofilm at different locations in the flow channel: (top) initially; (middle) compacted by flux increase; (bottom) relaxed to the initial permeate flux. Figure from: Valladares Linares *et al.* (2016).

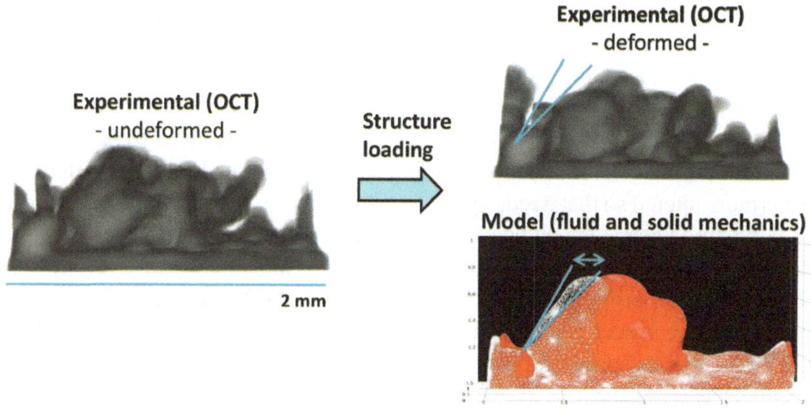

Figure 14.3 Determination of the elastic modulus by using deformed and undeformed biofilm geometry from optical coherence tomography in a fluid-structure interaction numerical model.

the behaviour of biofilms formed in different membrane filtration processes (i.e., reversed osmosis versus forward osmosis) can be evaluated. Knowledge on biofilms formed in different membrane systems varying in driving force may lead to the development of more effective biofouling control strategies.

Furthermore, the determined biofilm properties can provide input values in numerical models, which can be used to increase the understanding and to predict biofouling effects on membrane process performance. Based on the OCT scans, the biofilm mechanical properties can be individually evaluated by computational models, which would be advantageous when decoupling effects is not possible experimentally (Figure 14.3).

Biofouling mitigation strategies

Membrane and spacer surface modification

It is believed that understanding the early stages of biofilm formation could lead to the development of "antifouling" coatings for membrane and spacer surfaces. However, until now, no reported coating could prevent biofilm formation on long term. We believe therefore that the focus of research should change from biofouling prevention, to find ways to restrict biofilm formation and to facilitate biofilm removal.

New spacer designs

Based on the current literature, the spacer geometry appear to play an important role on biofilm formation and, most probably, also on the efficient biofilm removal from the membrane system (Siddiqui *et al.*, 2016, 2017a).

A possible optimization loop leading to the design of better spacers is proposed in Figure 14.4. Several commercial feed spacers can be screened for their hydrodynamics, mass transport and fouling properties. An accurate geometry of a selected spacer can be obtained by X-ray computed tomography (CT scanning step (Haaksman *et al.*, 2017)). This geometry is imported in a computer aided design (CAD) package and a virtual three-dimensional (3D) spacer model can be generated. This virtual model is furthermore altered so that a redesigned spacer geometry can be proposed, according to pre-existing knowledge on how different spacer elements contribute to membrane performance. The new spacer design is first evaluated numerically for hydrodynamics and mass transfer properties (e.g., (Haaksman *et al.*, 2017)). Based on the numerical simulations, the spacer geometry could be further improved. Once the simulations indicate a superior spacer, a prototype will be manufactured by three-dimensional

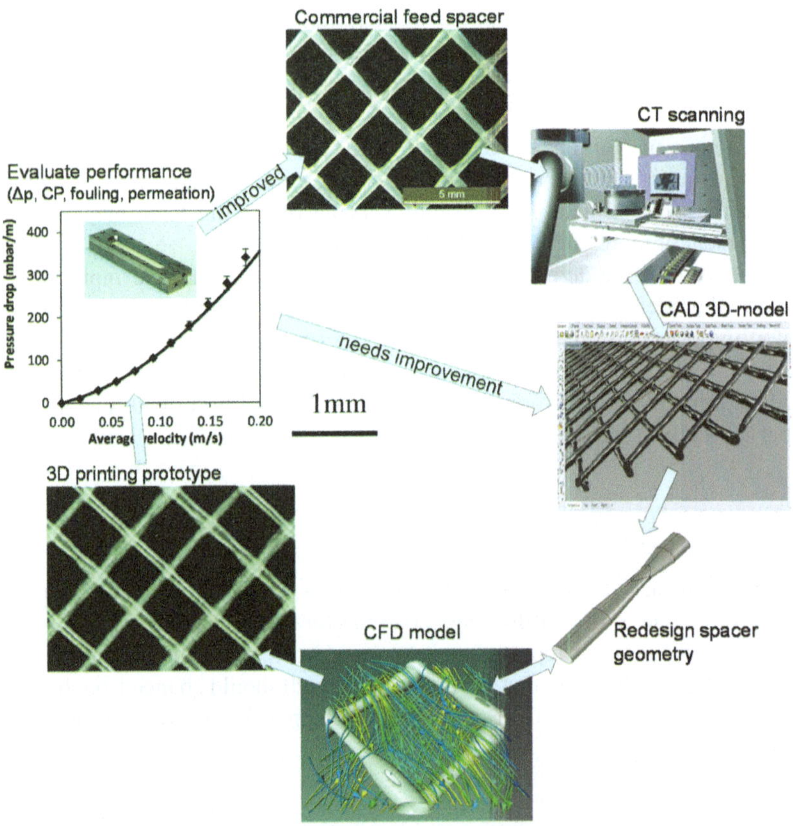

Figure 14.4 Proposed steps in the development of improved spacers for membrane filtration systems involving, 3D printing, CT scanning and numerical modelling in conjunction with experimental testing.

printing (Siddiqui *et al.*, 2016). The prototype is further evaluated in lab-scale experimental setups (e.g., in membrane fouling simulators) for hydrodynamics (pressure drop and permeation), mass transfer (concentration polarization), fouling behaviour (biofouling, scaling) and associated cleanability. If the lab-scale tests indicate a successful spacer design, pilot scale experiments (spiral wound membrane module) should be carried out and eventually the spacer could be commercialized. If the tests reveal that the spacer needs further improvement, a new virtual design should be proposed and the testing cycle continues (Figure 14.4).

Numerical evaluation

For an accurate assessment of the virtual design of the feed spacer, progress in numerical modelling has to be achieved in several areas. The computational fluid dynamics would need to include, beside the laminar steady flow (e.g., Bucs *et al.*, 2014a; Picioreanu *et al.*, 2009) also unsteady and time-dependent flow. As shown by Bucs *et al.* (Bucs *et al.*, 2015) at velocities used in practice in RO systems the flow begins to develop unsteady behaviour. Especially under fouling conditions the flow channel porosity changes which may lead to flow instability. This requires the use of more demanding computational fluid dynamics methods, which can represent both steady and unsteady flows (Fortunato *et al.*, 2017; Koutsou *et al.*, 2007). Another important aspect to be considered in numerical simulations is the representativeness of the spacer geometry. Picioreanu *et al.* (Picioreanu *et al.*, 2009) showed that the simplification of the spacer geometry in numerical models (e.g., cylindrical spacer filaments with uniform diameters) leads to non-representative hydrodynamics, therefore not desirable when simulating spacers used in practice. Moreover, since the feed spacer geometries used in practice are not completely symmetrical, the orientation of the feed spacer versus the main flow direction has to be considered as well in numerical models (Radu *et al.*, 2014b).

As the spacer fibres in practice do not have a perfectly circular cross-section and their diameter is variable along the fibres (Haaksman *et al.*, 2017), the simplified spacer geometries (e.g., cylinders) may not be sufficient to characterize the hydraulic impact of the virtual spacer (Figure 14.5). Therefore, spacers geometries derived from X-ray computed tomography scans should be used in numerical simulations.

In summary, details of spacer geometry and flow conditions are important for accurate numerical simulations, and proper results can only be obtained if actual feed spacer geometries are used. We believe that the proposed development cycle (Figure 14.4) can improve the rational design of spacers with a lower fouling potential, higher cleanability and thereby a better operational effectiveness of the whole membrane process.

Membrane cleaning strategies

Evaluations of new cleaning strategies for biofilm removal are still scarcely reported. By understanding how the biofilm forms and reacts to operation

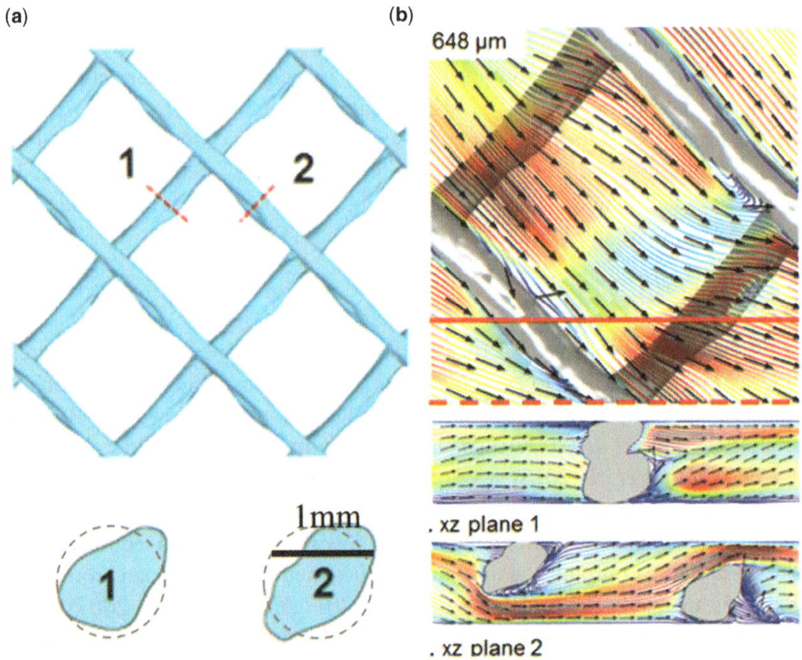

Figure 14.5 (a) Top view and cross sections of a spacer geometry scanned by X-ray computed tomography (CT); (b) Simulated water flow in laminar steady conditions, with streamlines showing flow velocity in a colour scale (red: high and blue: low velocity). The dashed-line circles in (a) represent cross-sections as they would appear from top-view microscopic observation assuming perfectly cylindrical spacer fibres.

conditions, more effective cleaning methods can be proposed. With advanced imaging techniques like optical coherence tomography the immediate impact of various physical and chemical agents on biofilms could be assessed. Examples of physical cleaning methods include: variation in shear by pulsating flow (Radu *et al.*, 2012b), air sparging (Cornelissen *et al.*, 2007a), back-washing and osmotic shock (Mi & Elimelech, 2010). Although chemical cleaning is widely used in industrial applications, current cleaning strategies inactivate most biofilm bacterial cells without removing the biofilm from the membrane module. Rather than inactivation of the bacterial cells, the focus for chemical cleaning should be on lysing the natural polymeric matrix, followed by removal of the biofilm (Bereschenko *et al.*, 2011).

CONCLUSIONS

- Biofouling cannot be avoided, thus control strategies should focus on: (i) delayed biofilm formation, (ii) reduced or delayed impact of accumulated biofilm on performance and (iii) biofilm removal by advanced cleaning strategies.

- Controlling biofouling and biofilm formation requires better understanding of biofilm development under hydraulic conditions and over time periods representative in practice.
- Novel imaging and analytical methods allow to study biofilm formation on-line and in-situ, and are able to deliver information on the biofilm morphology and mechanical properties.
- By understanding biofilm mechanics and its interactions with the water flow, novel control and engineering strategies can be developed to mitigate biofouling in spiral wound membrane systems.
- Cleaning strategies should focus on lysing and removal of biofilms, rather than on mechanical destruction or bacterial cell inactivation.
- Computational methods should be improved to: (i) represent time-dependent and unsteady flows; (ii) represent the fluid-biofilm interaction effects on permeation, pressure drop and biofilm removal; (iii) describe more accurately the spacer geometry.

References

Adamczyk Z., Warszyński P., Szyk-Warszyńska L. and Weroński P. X. (2000). Role of convection in particle deposition at solid surfaces. *Colloids and Surfaces A: Physicochemical and Engineering Aspects*, **165**, 157–187.

Adout A., Kang S., Asatekin A., Mayes A. M. and Elimelech M. (2010). Ultrafiltration membranes incorporating amphiphilic comb copolymer additives prevent irreversible adhesion of bacteria. *Environmental Science and Technology*, **44**, 2406–11.

Ahmad A. L., Lau K. K. and Abu Bakar M. Z. (2005). Impact of different spacer filament geometries on concentration polarization control in narrow membrane channel. *Journal of Membrane Science*, **262**, 138–152.

Al-Amoudi A. and Lovitt R. W. (2007). Fouling strategies and the cleaning system of NF membranes and factors affecting cleaning efficiency. *Journal of Membrane Science*, **303**, 4–28.

Al Ashhab A., Gillor O. and Herzberg M. (2014). Biofouling of reverse-osmosis membranes under different shear rates during tertiary wastewater desalination: Microbial community composition. *Water Research*, **67**, 86–95.

Alawadhi A. A. (1997). Pretreatment plant design – key to a successful reverse osmosis desalination plant. *Desalination*, **110**, 1–10.

Al-Juboori R. A. and Yusaf T. (2012). Biofouling in RO system: Mechanisms, monitoring and controlling. *Desalination*, **302**, 1–23.

Altena F. W. and Belfort G. (1984). Lateral migration of spherical particles in porous flow channels: Application to membrane filtration. *Chemical Engineering Science*, **39**, 343–355.

Altmann J. and Ripperger S. (1997). Particle deposition and layer formation at the crossflow microfiltration. *Journal of Membrane Science*, **124**, 119–128.

Alvarado L. and Chen A. (2014). Electrodeionization: Principles, strategies and applications. *Electrochimica Acta*, **132**, 583–597.

Amokrane M., Sadaoui D., Koutsou C. P., Karabelas A. J. and Dudeck M. (2015). A study of flow field and concentration polarization evolution in membrane channels with two-dimensional spacers during water desalination. *Journal of Membrane Science*, **477**, 139–150.

An G., Lin J., Li J., Li X. and Jian X. (2011). Non-invasive measurement of membrane scaling and cleaning in spiral-wound reverse osmosis modules by ultrasonic time-domain reflectometry with sound intensity calculation. *Desalination*, **283**, 3–9.

Ang W. S., Lee S. and Elimelech M. (2006). Chemical and physical aspects of cleaning of organic-fouled reverse osmosis membranes. *Journal of Membrane Science*, **272**, 198–210.

Ang W. S., Tiraferri A., Chen K. L. and Elimelech M. (2011). Fouling and cleaning of RO membranes fouled by mixtures of organic foulants simulating wastewater effluent. *Journal of Membrane Science*, **376**, 196–206.

Antony A., Low J. H., Gray S., Childress A. E., Le-Clech P. and Leslie G. (2011). Scale formation and control in high pressure membrane water treatment systems: A review. *Journal of Membrane Science*, **383**, 1–16.

Araújo P. A., Kruithof J. C., Van Loosdrecht M. C. M. and Vrouwenvelder J. S. (2012a). The potential of standard and modified feed spacers for biofouling control. *Journal of Membrane Science*, **403–404**, 58–70.

Araújo P. A., Miller D. J., Correia P. B., Van Loosdrecht M. C. M., Kruithof J. C., Freeman B. D., Paul D. R. and Vrouwenvelder J. S. (2012b). Impact of feed spacer and membrane modification by hydrophilic, bactericidal and biocidal coating on biofouling control. *Desalination*, **295**, 1–10.

Arvin E. and Kristensen G. H. (1983). Phosphate precipitation in biofilms and flocs. *Water Science and Technology*, **15**, 65–85.

Asatekin A., Olivetti E. A. and Mayes A. M. (2009). Fouling resistant, high flux nanofiltration membranes from polyacrylonitrile-graft-poly(ethylene oxide). *Journal of Membrane Science*, **332**, 6–12.

Bacchin P., Aimar P. and Sanchez V. (1995). Model for colloidal fouling of membranes. *AIChE Journal*, **41**, 368–376.

Bacchin P., Aimar P. and Field R. W. (2006). Critical and sustainable fluxes: Theory, experiments and applications. *Journal of Membrane Science*, **281**, 42–69.

Bacon J. R. and Demas J. N. (1987). Determination of oxygen concentrations by luminescence quenching of a polymer-immobilized transition-metal complex. *Analytical Chemistry*, **59**, 2780–2785.

Baker J. S. and Dudley L. Y. (1998a). Biofouling in membrane systems – a review. *Desalination*, **118**, 81–90.

Baker J. S. and Dudley L. Y. (1998b). Biofouling in membrane systems – a review. *Desalination*, **118**, 81–89.

Baker J., Stephenson T., Dard S. and Côté P. (1995). Characterisation of fouling of nanofiltration membranes used to treatsurface waters. *Environmental Technology*, **16**, 977–985.

Bakke R., Kommedal R. and Kalvenes S. (2001). Quantification of biofilm accumulation by an optical approach. *Journal of Microbiological Methods*, **44**, 13–26.

Bartels C., Hirose M. and Fujioka H. (2008). Performance advancement in the spiral wound RO/NF element design. *Desalination*, **221**, 207–214.

Bar-Zeev E., Berman-Frank I., Girshevitz O. and Berman T. (2012). Revised paradigm of aquatic biofilm formation facilitated by microgel transparent exopolymer particles. *Proceedings of the National Academy of Sciences*, **109**, 9119–9124.

Beaufort S., Alfenore S. and Lafforgue C. (2011). Use of fluorescent microorganisms to perform in vivo and in situ local characterization of microbial deposits. *Journal of Membrane Science*, **369**, 30–39.

Beech I. B. and Sunner J. (2004). Biocorrosion: Towards understanding interactions between biofilms and metals. *Current Opinion in Biotechnology*, **15**, 181–186.

Behrens J. W., Stahl H. J., Steffensen J. F. and Glud R. N. (2007). Oxygen dynamics around buried lesser sandeels Ammodytes tobianus (Linnaeus 1785): Mode of ventilation and oxygen requirements. *Journal of Experimental Biology*, **210**, 1006–1014.

Belfort G., Davis R. H. and Zydney A. L. (1994). The behavior of suspensions and macromolecular solutions in crossflow microfiltration. *Journal of Membrane Science*, **96**, 1–58.

Béné G. J. (1980). Nuclear magnetism of liquid systems in the earth field range. *Physics Reports*, **58**, 213–267.

Ben-Sasson M., Zodrow K. R., Genggeng Q., Kang Y., Giannelis E. P. and Elimelech M. (2014). Surface functionalization of thin-film composite membranes with copper nanoparticles for antimicrobial surface properties. *Environmental Science & Technology*, **48**, 384–393.

Bereschenko L. A., Stams A. J. M., Heilig G. H. J., Euverink G. J. W., Nederlof M. M. and Van Loosdrecht M. C. M. (2007). Investigation of microbial communities on reverse osmosis membranes used for process water production. *Water Science and Technology* **55**, 181–190.

Bereschenko L. A., Stams A. J. M., Euverink G. J. W. and Van Loosdrecht M. C. M. (2010). Biofilm formation on reverse osmosis membranes is initiated and dominated by Sphingomonas spp. *Applied and Environmental Microbiology*, **76**, 2623–2632.

Bereschenko L. A., Prummel H., Euverink G. J. W., Stams A. J. M. and van Loosdrecht M. C. M. (2011). Effect of conventional chemical treatment on the microbial population in a biofouling layer of reverse osmosis systems. *Water Research*, **45**, 405–416.

Bertheas U., Majamaa K., Arzu A. and Pahnke R. (2009). Use of DBNPA to control biofouling in RO systems. *Desalination and Water Treatment*, **3**, 175–178.

Beuling E. E., Van Dusschoten D., Lens P., Van Den Heuvel J. C., Van As H. and Ottengraf S. P. P. (1998). Characterization of the diffusive properties of biofilms using pulsed field gradient-nuclear magnetic resonance. *Biotechnology and Bioengineering*, **60**, 283–291.

Beyenal H., Donovan C., Lewandowski Z. and Harkin G. (2004a). Three-dimensional biofilm structure quantification. *Journal of Microbiological Methods*, **59**, 395–413.

Beyenal H., Lewandowski Z. and Harkin G. (2004b). Quantifying biofilm structure: Facts and fiction. *Biofouling*, **20**, 1–23.

Billings N., Birjiniuk A., Samad T. S., Doyle P. S. and Ribbeck K. (2015). Material properties of biofilms-a review of methods for understanding permeability and mechanics. *Reports on Progress in Physics*, **78**, 0.3601.

Blauert F., Horn H. and Wagner M. (2015). Time-resolved biofilm deformation measurements using optical coherence tomography. *Biotechnology and Bioengineering*, **112**, 1893–1905.

Blümich B., Casanova F. and Appelt S. (2009). NMR at low magnetic fields. *Chemical Physics Letters*, **477**, 231–240.

Borisov S. M., Nuss G. and Klimant I. (2008). Red light-excitable oxygen sensing materials based on platinum(II) and palladium(II) benzoporphyrins. *Analytical Chemistry*, **80**, 9435–9442.

Bowen B. D., Levine S. and Epstein N. (1976). Fine particle deposition in laminar flow through parallel-plate and cylindrical channels. *Journal of Colloid and Interface Science*, **54**, 375–390.

Brinkman H. C. (1949). A calculation of the viscous force exerted by a flowing fluid on a dense swarm of particles. *Applied Scientific Research*, **1**, 27–34.

Bryers J. D. and Characklis W. G. (1982). Processes governing primary biofilm formation. *Biotechnology and Bioengineering*, **24**, 2451–2476.

Brzozowska A. M., Spruijt E., de Keizer A., Cohen Stuart M. A. and Norde W. (2011). On the stability of the polymer brushes formed by adsorption of Ionomer Complexes on hydrophilic and hydrophobic surfaces. *Journal of Colloid and Interface Science*, **353**, 380–391.

Bucs S. S., Radu A. I., Lavric V., Vrouwenvelder J. S. and Picioreanu C. (2014a). Effect of different commercial feed spacers on biofouling of reverse osmosis membrane systems: A numerical study. *Desalination*, **343**, 26–37.

Bucs S. S., Valladares Linares R., van Loosdrecht M. C. M., Kruithof J. C. and Vrouwenvelder J. S. (2014b). Impact of organic nutrient load on biomass accumulation, feed channel pressure drop increase and permeate flux decline in membrane systems. *Water Research*, **67**, 227–242.

Bucs S. S., Valladares Linares R., Marston J. O., Radu A. I., Vrouwenvelder J. S. and Picioreanu C. (2015). Experimental and numerical characterization of the water flow in spacer-filled channels of spiral-wound membranes. *Water Research*, **87**, 299–310.

Bucs S. S., Farhat N., Siddiqui A., Linares R. V., Radu A., Kruithof J. C., Vrouwenvelder J. S., Valladares Linares R., Radu A., Kruithof J. C. and Vrouwenvelder J. S. (2016a). Development of a setup to enable stable and accurate flow conditions for membrane biofouling studies. *Desalination and Water Treatment*, **57**, 12893–12901.

Bucs S. S., Valladares Linares R., Vrouwenvelder J. S. and Picioreanu C. (2016b). Biofouling in forward osmosis systems: An experimental and numerical study. *Water Research*, **106**, 86–97.

Bucs S. S., Linares R. V., Farhat N., Matin A., Khan Z., Van Loosdrecht M. C. M., Yang R., Wang M., Gleason K. K., Kruithof J. C. and Vrouwenvelder J. S. (2017). Coating of reverse osmosis membranes with amphiphilic copolymers for biofouling control. *Desalination and Water Treatment*, **68**, 1–11.

Busscher H. J. and Van Der Mei H. C. (2006). Microbial adhesion in flow displacement systems. *Clinical Microbiology Reviews*, **19**, 127–141.

Callaghan P. T. and Xia Y. (1991). Velocity and diffusion imaging in dynamic NMR microscopy. *Journal of Magnetic Resonance*, **91**, 326–352.

Cao Z., Wiley D. E. and Fane A. G. (2001). CFD simulations of net-type turbulence promoters in a narrow channel. *Journal of Membrane Science*, **185**, 157–176.

Carraway E. R., Demas J. N., DeGraff B. A. and Bacon J. R. (1991). Photophysics and photochemistry of oxygen sensors based on luminescent transition-metal complexes. *Analytical Chemistry*, **63**, 337–342.

Casey T. A., Sakakibara J. and Thoroddsen S. T. (2013). Scanning tomographic particle image velocimetry applied to a turbulent jet. *Physics of Fluids*, **25**, 025102.

Cath T. Y., Childress A. E. and Elimelech M. (2006). Forward osmosis: Principles, applications, and recent developments. *Journal of Membrane Science*, **281**, 70–87.

Chambless J. D. and Stewart P. S. (2007). A three-dimensional computer model analysis of three hypothetical biofilm detachment mechanisms. *Biotechnology and Bioengineering*, **97**, 1573–1584.

Chaumeil F. and Crapper M. (2013). DEM simulations of initial deposition of colloidal particles around non-woven membrane spacers. *Journal of Membrane Science*, **442**, 254–263.

Chellam S., Wiesner M. R., Shankararaman C. and Wlesner M. R. (1992). Particle transport in clean membrane filters in laminar flow. *Environmental Science & Technology*, **26**, 1611–1621.

Chen X. and Stewart P. S. (2002). Role of electrostatic interactions in cohesion of bacterial biofilms. *Applied Microbiology and Biotechnology*, **59**, 718–720.

Cheng H.-C. and Liu Y.-C. (2010). Simultaneous measurement of group refractive index and thickness of optical samples using optical coherence tomography. *Applied Optics*, **49**, 790–797.

Chesters S. P. (2009). Innovations in the inhibition and cleaning of reverse osmosis membrane scaling and fouling. *Desalination*, **238**, 22–29.

Chiang P. and Burrows L. L. (2003). Biofilm formation by hyperpiliated mutants of Pseudomonas aeruginosa. *Journal of Bacteriology*, **185**, 2374–2378.

Chicote E., Moreno D. A., Garcia A. M., Sarro M. I., Lorenzo P. I. and Montero F. (2004). Biofouling on the walls of a spent nuclear fuel pool with radioactive ultrapure water. *Biofouling*, **20**, 35–42.

Chilcott T., Antony A., Coster H. and Leslie G. (2013). In situ characterization of fouling in reverse osmosis membranes using electrical impedance spectroscopy. *Journal of Physics: Conference Series*, **434**, 89–97.

Chinu K. J., Johir A. H., Vigneswaran S., Shon H. K. and Kandasamy J. (2009). Biofilter as pretreatment to membrane based desalination: Evaluation in terms of fouling index. *Desalination*, **247**, 77–84.

Chong T. H., Wong F. S. and Fane A. G. (2008a). The effect of imposed flux on biofouling in reverse osmosis: Role of concentration polarisation and biofilm enhanced osmotic pressure phenomena. *Journal of Membrane Science*, **325**, 840–850.

Chong T. H., Wong F. S. and Fane A. G. (2008b). Implications of critical flux and cake enhanced osmotic pressure (CEOP) on colloidal fouling in reverse osmosis: Experimental observations. *Journal of Membrane Science*, **314**, 101–111.

Chua K. T., Hawlader M. N. A. and Malek A. (2003). Pretreatment of seawater: Results of pilot trials in Singapore. *Desalination*, **159**, 225–243.

Codd S. L., Vogt S. J., Hornemann J. A., Phillips A. J., Maneval J. E., Romanenko K. R., Hansen L., Cunningham A. B. and Seymour J. D. (2011). NMR relaxation measurements of biofouling in model and geological porous media. *Organic Geochemistry*, **42**, 965–971.

Cooke A. J., Rowe R. K., Rittmann B. E. and Fleming I. R. (1999). Modeling biochemically driven mineral precipitation in anaerobic biofilms. *Water Science and Technology*, pp. 57–64.

Cornelissen E. R., Vrouwenvelder J. S., Heijman S. G. J., Viallefont X. D. and van der Kooij D. (2007a). Periodic air/water cleaning for control of biofouling in spiral wound membrane elements. *Journal of Membrane Science*, **287**, 94–101.

Cornelissen E. R., Vrouwenvelder J. S., Heijman S. G. J., Viallefont X. D., van der Kooij D. and Wessels L. P. (2007b). Air/water cleaning for biofouling control in spiral wound membrane elements. *Desalination*, **204**, 145–147.

Costerton J. W., Lewandowski Z., Caldwell D. E., Korber D. R. and Lappin-Scott H. M. (1995). Microbial Biofilms. *Annual Review of Microbiology*, **49**(1), 711–45.

Creber S. A., Pintelon T. R. R., Graf Von Der Schulenburg D. A. W., Vrouwenvelder J. S., Van Loosdrecht M. C. M. and Johns M. L. (2010a). Magnetic resonance imaging and 3D simulation studies of biofilm accumulation and cleaning on reverse osmosis membranes. *Food and Bioproducts Processing*, **88**, 401–408.

Creber S. A., Vrouwenvelder J. S., van Loosdrecht M. C. M. and Johns M. L. (2010b). Chemical cleaning of biofouling in reverse osmosis membranes evaluated using magnetic resonance imaging. *Journal of Membrane Science*, **362**, 202–210.

Da Costa A. R., Fane A. G., Fell C. J. D. and Franken A. C. M. (1991). Optimal channel spacer design for ultrafiltration. *Journal of Membrane Science*, **62**, 275–291.

Da Costa A. R., Fane A. G. and Wiley D. E. (1994). Spacer characterization and pressure drop modelling in spacer-filled channels for ultrafiltration. *Journal of Membrane Science*, **87**, 79–98.

Darton T., Annunziata U., del Vigo Pisano F. and Gallego S. (2004). Membrane autopsy helps to provide solutions to operational problems. *Desalination*, **167**, 239–245.

Davit Y., Iltis G., Debenest G., Veran-Tissoires S., Wildenschild D., Gerino M. and Quintard M. (2011). Imaging biofilm in porous media using X-ray computed microtomography. *Journal of Microscopy*, **242**, 15–25.

de Beer D. and Stoodley P. (1995). Relation between the structure of an aerobic biofilm and transport phenomena. *Water Science and Technology*, **32**, 11–18.

de Beer D., Stoodley P., Roe F. and Lewandowski Z. (1994). Effects of biofilm structures on oxygen distribution and mass transport. *Biotechnology and Bioengineering*, **43**, 1131–1138.

De Bivar Xavier J., Picioreanu C. and Van Loosdrecht M. C. M. (2005). A general description of detachment for multidimensional modelling of biofilms. *Biotechnology and Bioengineering*, **91**, 651–669.

De Boer J. F., Milner T. E. and Stuart Nelson J. (1999). Determination of the depth-resolved Stokes parameters of light backscattered from turbid media by use of polarization-sensitive optical coherence tomography. *Optics Letters*, **24**, 300–302.

Dendukuri D., Karode S. K. and Kumar A. (2005). Flow visualization through spacer filled channels by computational fluid dynamics-II: Improved feed spacer designs. *Journal of Membrane Science*, **249**, 41–49.

Derlon N., Peter-Varbanets M., Scheidegger A., Pronk W. and Morgenroth E. (2012). Predation influences the structure of biofilm developed on ultrafiltration membranes. *Water Research*, **46**, 3323–3333.

Derlon N., Koch N., Eugster B., Posch T., Pernthaler J., Pronk W. and Morgenroth E. (2013). Activity of metazoa governs biofilm structure formation and enhances permeate flux during Gravity-Driven Membrane (GDM) filtration. *Water Research*, **47**, 2085–2095.

Dickson J. M., Whitacker G., DeLeeuw J. and Spencer J. (1994). Dilute single and mixed solute systems in a spiral wound reverse osmosis module. Part II. Experimental data and application of the model. *Desalination*, **99**, 1–18.

Dobretsov S., Teplitski M. and Paul V. (2009). Mini-review: Quorum sensing in the marine environment and its relationship to biofouling. *Biofouling*, **25**, 413–427.

DOW (2014). DOW FILMTEC™ Membranes, Cleaning Procedures for DOW FILMTEC FT30 Elements.

Dreszer C., Vrouwenvelder J. S., Paulitsch-Fuchs A. H., Zwijnenburg A., Kruithof J. C. and Flemming H.-C. (2013). Hydraulic resistance of biofilms. *Journal of Membrane Science*, **429**, 436–447.

Dreszer C., Flemming H.-C., Wexler A. D., Zwijnenburg A., Kruithof J. C. and Vrouwenvelder J. S. (2014a). Development and testing of a transparent membrane biofouling monitor. *Desalination and Water Treatment*, **52**, 1807–1819.

Dreszer C., Wexler A. D., Drusová S., Overdijk T., Zwijnenburg A., Flemming H.-C., Kruithof J. C. and Vrouwenvelder J. S. (2014b). In-situ biofilm characterization in membrane systems using Optical Coherence Tomography: Formation, structure, detachment and impact of flux change. *Water Research*, **67**, 243–254.

Duddridge J. E., Kent C. A. and Laws J. F. (1982). Effect of surface shear stress on the attachment of Pseudomonas fluorescens to stainless steel under defined flow conditions. *Biotechnology and Bioengineering*, **24**, 153–164.

Duddu R., Chopp D. L. and Moran B. (2009). A two-dimensional continuum model of biofilm growth incorporating fluid flow and shear stress based detachment. *Biotechnology and Bioengineering*, **103**, 92–104.

Dupraz C., Reid R. P., Braissant O., Decho A. W., Norman R. S. and Visscher P. T. (2009). Processes of carbonate precipitation in modern microbial mats. *Earth-Science Reviews*, **96**, 141–162.

Dydo P., Turek M., Ciba J., Wandachowicz K. and Misztal J. (2004). The nucleation kinetic aspects of gypsum nanofiltration membrane scaling. *Desalination*, **164**, 41–52.

Eberl H. J., Picioreanu C., Heijnen J. J. and Van Loosdrecht M. C. M. (2000). Three-dimensional numerical study on the correlation of spatial structure, hydrodynamic conditions, and mass transfer and conversion in biofilms. *Chemical Engineering Science*, **55**, 6209–6222.

Elimelech M. and O'Melia C. R. (1990). Kinetics of deposition of colloidal particles in porous media. *Environmental Science & Technology*, **24**, 1528–1536.

Elsinga G. E., Scarano F., Wieneke B. and Van Oudheusden B. W. (2006). Tomographic particle image velocimetry. *Experiments in Fluids*, **41**, 933–947.

Eshed L., Yaron S. and Dosoretz C. G. (2008). Effect of permeate drag force on the development of a biofouling layer in a pressure-driven membrane separation system. *Applied and Environmental Microbiology*, **74**, 7338–7347.

Farhat N. M., Staal M., Siddiqui A., Borisov S. M., Bucs S. S. and Vrouwenvelder J. S. (2015). Early non-destructive biofouling detection and spatial distribution: Application of oxygen sensing optodes. *Water Research*, **83**, 10–20.

Fejerskov O. (2004). Changing paradigms in concepts on dental caries: Consequences for oral health care. *Caries Research*, **38**, 182–191.

Fenchel T. and Finlay B. (2008). Oxygen and the spatial structure of microbial communities. *Biology Reviews*, **83**, 553–569.

Fercher A. F., Drexler W., Hitzenberger C. K. and Lasser T. (2003). Optical coherence tomography – principles and applications. *Reports on Progress in Physics*, **66**, 239–303.

Fernández-Sempere J., Ruiz-Beviá F., García-Algado P. and Salcedo-Díaz R. (2010). Experimental study of concentration polarization in a crossflow reverse osmosis system using Digital Holographic Interferometry. *Desalination*, **257**, 36–45.

Field R. W., Wu D., Howell J. A. and Gupta B. B. (1995). Critical flux concept for microfiltration fouling. *Journal of Membrane Science*, **100**, 259–272.

Fimbres-Weihs G. A. and Wiley D. E. (2007). Numerical study of mass transfer in three-dimensional spacer-filled narrow channels with steady flow. *Journal of Membrane Science*, **306**, 228–243.

Fimbres-Weihs G. A. and Wiley D. E. (2010). Review of 3D CFD modeling of flow and mass transfer in narrow spacer-filled channels in membrane modules. *Chemical Engineering and Processing: Process Intensification*, **49**, 759–781.

Fimbres-Weihs G. A., Wiley D. E. and Fletcher D. F. (2006). Unsteady flows with mass transfer in narrow zigzag spacer-filled channels: A numerical study. *Industrial & Engineering Chemistry Research*, **45**, 6594–6603.

Flemming H.-C. (1997). Reverse osmosis membrane biofouling. *Experimental Thermal and Fluid Science*, **14**, 382–391.

Flemming H.-C. (2002). Biofouling in water systems–cases, causes and countermeasures. *Applied Microbiology and Biotechnology*, **59**, 629–640.

Flemming H. C. (2003). Role and levels of real-time monitoring for successful anti-fouling strategies – An overview. *Water Science and Technology*, pp. 1–8.

Flemming H.-C., Schaule G., Griebe T., Schmitt J. and Tamachkiarowa A. (1997). Biofouling – the Achilles heel of membrane processes. *Desalination*, **113**, 215–225.

Fletcher D. F. and Wiley D. E. (2004). A computational fluids dynamics study of buoyancy effects in reverse osmosis. *Journal of Membrane Science*, **245**, 175–181.

Flint S., Tang X., Bennett R. and Brooks J. (2011). Controlling the microbiology of whey and whey products. In: *Whey: Types, Composition and Health Implications*. Nova Science Publishers, Inc. Hauppauge NY, United States of America, pp. 115–135.

Flory P. J. (1977). Theory of elasticity of polymer networks. The effect of local constraints on junctions. *Journal of Chemical Physics*, **66**, 5720–5729.

Fortunato L., Bucs S., Linares R. V., Cali C., Vrouwenvelder J. S. and Leiknes T. (n.d.). Spatially-resolved in-situ quantification of biofouling using optical coherence tomography (OCT) and 3D image analysis in a spacer filled channel. *Journal of Membrane Science*, 673–681.

Fortunato L., Qamar A., Wang Y., Jeong S. and Leiknes T. (2017). In-situ assessment of biofilm formation in submerged membrane system using optical coherence tomography and computational fluid dynamics. *Journal of Membrane Science*, **521**, 84–94.

Fowler J. D. and Robertson C. R. (1991). Hydraulic permeability of immobilized bacterial cell aggregates. *Applied and Environmental Microbiology*, **57**, 102–113.

Fridjonsson E. O., Flux L. S. and Johns M. L. (2012). Determination of mean droplet sizes of water-in-oil emulsions using an Earth's field NMR instrument. *Journal of Magnetic Resonance*, **221**, 97–102.

Fridjonsson E. O., Stanwix P. L. and Johns M. L. (2014). Earth's field NMR flow meter: Preliminary quantitative measurements. *Journal of Magnetic Resonance*, **245**, 110–115.

Fridjonsson E. O., Creber S. A., Vrouwenvelder J. S. and Johns M. L. (2015a). Magnetic resonance signal moment determination using the Earth's magnetic field. *Journal of Magnetic Resonance*, **252**, 145–150.

Fridjonsson E. O., Vogt S. J., Vrouwenvelder J. S. and Johns M. L. (2015b). Early non-destructive biofouling detection in spiral wound RO membranes using a mobile earth's field NMR. *Journal of Membrane Science*, **489**, 227–236.

Fritzmann C., Löwenberg J., Wintgens T. and Melin T. (2007). State-of-the-art of reverse osmosis desalination. *Desalination*, **216**, 1–76.

Fukushima E. (1999). Nuclear magnetic resonance as a tool to study flow. *Annual Review of Fluid Mechanics*, **31**, 95–123.

Gao Y., Haavisto S., Tang C. Y., Salmela J. and Li W. (2013). Characterization of fluid dynamics in spacer-filled channels for membrane filtration using Doppler optical coherence tomography. *Journal of Membrane Science*, **448**, 198–208.

Geraldes V., Semião V. and Pinho M. N. (2000). Numerical modelling of mass transfer in slits with semi-permeable membrane walls. *Engineering Computations*, **17**, 192–217.

Geraldes V., Semião V. and De Pinho M. N. (2002). The effect of the ladder-type spacers configuration in NF spiral-wound modules on the concentration boundary layers disruption. *Desalination*, **146**, 187–194.

Geraldes V., Semião V., Norberta Pinho M., Semiao V. and Pinho M. N. (2003). Hydrodynamics and concentration polarization in NF/RO spiral-wound modules with ladder-type spacers. *Desalination*, **157**, 395–402.

Ghidossi R., Veyret D. and Moulin P. (2006). Computational fluid dynamics applied to membranes: State of the art and opportunities. *Chemical Engineering and Processing: Process Intensification*, **45**, 437–454.

Gimmelshtein M. and Semiat R. (2005). Investigation of flow next to membrane walls. *Journal of Membrane Science*, **264**, 137–150.

Gjaltema A., Van Der Marel N., Van Loosdrecht M. C. M. and Heijnen J. J. (1997). Adhesion and biofilm development on suspended carriers in airlift reactors: Hydrodynamic conditions versus surface characteristics. *Biotechnology and Bioengineering*, **55**, 880–889.

Glud R. N., Ramsing N. B., Gundersen J. K. and Klimant I. (1996). Planar optrodes: A new tool for fine scale measurements of two-dimensional O_2 distribution in benthic communities. *Marine Ecology Progress Series*, **140**, 217–226.

Glud R. N., Santegoeds C. M., De Beer D., Kohls O. and Ramsing N. B. (1998). Oxygen dynamics at the base of a biofilm studied with planar optodes. *Aquatic Microbial Ecology*, **14**, 223–233.

Glud R. N., Kühl M., Kohls O. and Ramsing N. B. (1999). Heterogeneity of oxygen production and consumption in a photosynthetic microbial mat as studied by planar optodes. *Journal of Phycology*, **35**, 270–279.

Goosen M. F. A., Sablani S. S., Al-Hinai H., Al-Obeidani S., Al-Belushi R. and Jackson D. (2004). Fouling of reverse osmosis and ultrafiltration membranes: A critical review. *Separation Science and Technology*, **39**, 2261–2297.

Graf von der Schulenburg D. A., Vrouwenvelder J. S., Creber S. A., van Loosdrecht M. C. M. and Johns M. L. (2008). Nuclear magnetic resonance microscopy studies of membrane biofouling. *Journal of Membrane Science*, **323**, 37–44.

Greenlee L. F., Lawler D. F., Freeman B. D., Marrot B. and Moulin P. (2009). Reverse osmosis desalination: Water sources, technology, and today's challenges. *Water Research*, **43**, 2317–2348.

Gu B., Adjiman C. S. and Xu X. Y. (2017). The effect of feed spacer geometry on membrane performance and concentration polarisation based on 3D CFD simulations. *Journal of Membrane Science*, **527**, 78–91.

Guillen G. and Hoek E. M. V. (2009). Modeling the impacts of feed spacer geometry on reverse osmosis and nanofiltration processes. *Chemical Engineering Journal*, **149**, 221–231.

Guo W., Ngo H.-H. and Li J. (2012). A mini-review on membrane fouling. *Bioresource Technology*, **122**, 27–34.

Gur-Reznik S., Katz I. and Dosoretz C. G. (2008). Removal of dissolved organic matter by granular-activated carbon adsorption as a pretreatment to reverse osmosis of membrane bioreactor effluents. *Water Research*, **42**, 1595–1605.

Haaksman V. A., Siddiqui A., Schellenberg C., Kidwell J., Vrouwenvelder J. S. and Picioreanu C. (2017). Characterization of feed channel spacer performance using geometries obtained by X-ray computed tomography. *Journal of Membrane Science*, **522**, 124–139.

Habimana O., Semião A. J. C. and Casey E. (2014). The role of cell-surface interactions in bacterial initial adhesion and consequent biofilm formation on nanofiltration/reverse osmosis membranes. *Journal of Membrane Science*, **454**, 82–96.

Hadidi M. and Zydney A. L. (2014). Fouling behavior of zwitterionic membranes: Impact of electrostatic and hydrophobic interactions. *Journal of Membrane Science*, **452**, 97–103.

Hahn M. W. and O'Melia C. R. (2004). Deposition and reentrainment of brownian particles in porous media under unfavorable chemical conditions: Some concepts and applications. *Environmental Science & Technology*, **38**, 210–220.

Haisch C. and Niessner R. (2007). Visualisation of transient processes in biofilms by optical coherence tomography. *Water Research*, **41**, 2467–2472.

Hamachi M. and Mietton-Peuchot M. (2001). Cake thickness measurement with an optical laser sensor. *Chemical Engineering Research and Design*, **79**, 151–155.

Hassan A. N., Anand S. and Avadhanula M. (2010). Microscopic observation of multispecies biofilm of various structures on whey concentration membranes. *Journal of Dairy Science*, **93**, 2321–2329.

Hennig J., Nauerth A. and Friedburg H. (1986). RARE imaging: A fast imaging method for clinical MR. *Magnetic Resonance in Medicine*, **3**, 823–833.

Henry C., Minier J.-P. and Lefèvre G. (2012). Towards a description of particulate fouling: From single particle deposition to clogging. *Advances in Colloid and Interface Science*, **185–186**, 34–76.

Herman G. T. (2009). Fundamentals of Computerized Tomography, Advances in Pattern Recognition. Springer, London.

Hermanowicz S. W. (2001). A simple 2D biofilm model yields a variety of morphological features. *Mathematical Biosciences*, **169**, 1–14.

Herzberg M. and Elimelech M. (2007). Biofouling of reverse osmosis membranes: Role of biofilm-enhanced osmotic pressure. *Journal of Membrane Science*, **295**, 11–20.

Herzberg M., Kang S. and Elimelech M. (2009). Role of extracellular polymeric substances (EPS) in biofouling of reverse osmosis membranes. *Environmental Science & Technology*, **43**, 4393–4398.

Herzberg M., Berry D. and Raskin L. (2010). Impact of microfiltration treatment of secondary wastewater effluent on biofouling of reverse osmosis membranes. *Water Research*, **44**, 167–176.

Hohne D. N., Younger J. G. and Solomon M. J. (2009). Flexible microfluidic device for mechanical property characterization of soft viscoelastic solids such as bacterial biofilms. *Langmuir*, **25**, 7743–7751.

Horn H. and Morgenroth E. (2006). Transport of oxygen, sodium chloride, and sodium nitrate in biofilms. *Chemical Engineering Science*, **61**, 1347–1356.

Horn H., Reiff H. and Morgenroth E. (2003). Simulation of growth and detachment in biofilm systems under defined hydrodynamic conditions. *Biotechnology and Bioengineering*, **81**, 607–617.

Hoskins B. C., Fevang L., Majors P. D., Sharma M. M. and Georgiou G. (1999). Selective imaging of biofilms in porous media by NMR relaxation. *Journal of Magnetic Resonance*, **139**, 67–73.

Hou Y. L., Li J. X., Gao Y. N., Xu X. C., Cai Y. and Yang N. (2010). Monitoring of CaSO(4) deposition behaviors on biofilm during nanofiltration via ultrasonic time-domain reflectometry (UTDR). *Water Science & Technology*, **61**, 2853–2861.

Houari A., Picard J., Habarou H., Galas L., Vaudry H., Heim V. and Di Martino P. (2008). Rheology of biofilms formed at the surface of NF membranes in a drinking water production unit. *Biofouling*, **24**, 235–240.

Howell J. A. (1995). Sub-critical flux operation of microfiltration. *Journal of Membrane Science*, **107**, 165–171.

Huang D., Swanson E. A., Lin C. P., Schuman J. S., Stinson W. G., Chang W., Hee M. R., Flotte T., Gregory K., Puliafito C. A. and Fujimoto J. G. (1991). Optical coherence tomography. *Science*, **254**, 1178–1181.

Hunter J. D. (2007). MatplotlibL A 2D graphics environment. *Computing in Science & Engineering*, **9**, 90–95.

Ilie O., van Loosdrecht M. C. M. and Picioreanu C. (2012). Mathematical modelling of tooth demineralisation and pH profiles in dental plaque. *Journal of Theoretical Biology*, **309**, 159–175.

Iltis G. C., Armstrong R. T., Jansik D. P., Wood B. D. and Wildenschild D. (2011). Imaging biofilm architecture within porous media using synchrotron-based X-ray computed microtomography. *Water Resources Research*, **47**, W02601.

Ishigami T., Amano K., Fujii A., Ohmukai Y., Kamio E., Maruyama T. and Matsuyama H. (2012). Fouling reduction of reverse osmosis membrane by surface modification via layer-by-layer assembly. *Separation and Purification Technology*, **99**, 1–7.

Janjaroen D., Ling F., Monroy G., Derlon N., Mogenroth E., Boppart S. A., Liu W.-T. and Nguyen T. H. (2013). Roles of ionic strength and biofilm roughness on adhesion kinetics of Escherichia coli onto groundwater biofilm grown on PVC surfaces. *Water Research*, **47**, 2531–2542.

Jensen S. I., Kühl M., Glud R. N., Jørgensen L. B. and Priemé A. (2005). Oxic microzones and radial oxygen loss from roots of Zostera marina. *Marine Ecology Progress Series*, **293**, 49–58.

Johnson E. M. and Deen W. M. (1996). Hydraulic permeability of agarose gels. *AIChE Journal*, **42**, 1220–1224.

Kaghazchi T., Mehri M., Ravanchi M. T. and Kargari A. (2010). A mathematical modeling of two industrial seawater desalination plants in the Persian Gulf region. *Desalination*, **252**, 135–142.

Kang I. S. and Chang H. N. (1982). The effect of turbulence promoters on mass transfer-numerical analysis and flow visualization. *International Journal of Heat and Mass Transfer*, **25**, 1167–1181.

Kang G.-D. G. and Cao Y. (2012). Development of antifouling reverse osmosis membranes for water treatment: A review. *Water Research*, **46**, 584–600.

Kang S. and Choi H. (2005). Effect of surface hydrophobicity on the adhesion of S. cerevisiae onto modified surfaces by poly(styrene-ran-sulfonic acid) random copolymers. *Colloids and Surfaces B: Biointerfaces*, **46**, 70–77.

Kang S.-T., Subramani A., Hoek E. M. V., Deshusses M. A. and Matsumoto M. R. (2004). Direct observation of biofouling in cross-flow microfiltration: Mechanisms of deposition and release. *Journal of Membrane Science*, **244**, 151–165.

Kapellos G. E., Alexiou T. S. and Payatakes A. C. (2007). Hierarchical simulator of biofilm growth and dynamics in granular porous materials. *Advances in Water Resources*, **30**, 1648–1667.

Kappelhof J. W. N. M., Vrouwenvelder H. S., Schaap M., Kruithof J. C., Van Der Kooij D. and Schippers J. C. (2003). An in situ biofouling monitor for membrane systems. *Water Science and Technology: Water Supply*, **3**, 205–210.

Karode S. K. and Kumar A. (2001). Flow visualization through spacer filled channels by computational fluid dynamics I.: pressure drop and shear rate calculations for flat sheet geometry. *Journal of Membrane Science*, **193**, 69–84.

Kim D.-S. and Fogler H. S. (2000). Biomass evolution in porous media and its effects on permeability under starvation conditions. *Biotechnology and Bioengineering*, **69**, 47–56.

Kim S. and Hoek E. M. V. (2005). Modeling concentration polarization in reverse osmosis processes. *Desalination*, **186**, 111–128.

Kim I.-C. and Lee K.-H. (2006). Dyeing process wastewater treatment using fouling resistant nanofiltration and reverse osmosis membranes. *Desalination*, **192**, 246–251.

Kim M.-m. and Zydney A. L. (2006). Theoretical analysis of particle trajectories and sieving in a two-dimensional cross-flow filtration system. *Journal of Membrane Science*, **281**, 666–675.

Kim A. S., Chen H. and Yuan R. (2006). EPS biofouling in membrane filtration: An analytic modeling study. *Journal of Colloid and Interface Science*, **303**, 243–249.

Kim D., Jung S., Sohn J., Kim H. and Lee S. (2009). Biocide application for controlling biofouling of SWRO membranes – an overview. *Desalination*, **238**, 43–52.

Klapper I., Rupp C. J., Cargo R., Purvedorj B. and Stoodley P. (2002). Viscoelastic fluid description of bacterial biofilm material properties. *Biotechnology and Bioengineering*, **80**, 289–296.

Knutsen J. S. and Davis R. H. (2006). Deposition of foulant particles during tangential flow filtration. *Journal of Membrane Science*, **271**, 101–113.

Körstgens V., Flemming H.-C., Wingender J. and Borchard W. (2001). Uniaxial compression measurement device for investigation of the mechanical stability of biofilms. *Journal of Microbiological Methods*, **46**, 9–17.

Koutsou C. P. and Karabelas A. J. (2010). Towards optimization of spacer geometrical characteristics for spiral wound membrane modules. *Desalination and Water Treatment*, **18**, 139–150.

Koutsou C. P. and Karabelas A. J. (2015). A novel retentate spacer geometry for improved spiral wound membrane (SWM) module performance. *Journal of Membrane Science*, **488**, 129–142.

Koutsou C. P., Yiantsios S. G. and Karabelas A. J. (2004). Numerical simulation of the flow in a plane-channel containing a periodic array of cylindrical turbulence promoters. *Journal of Membrane Science*, **231**, 81–90.

Koutsou C. P., Yiantsios S. G. and Karabelas A. J. (2007). Direct numerical simulation of flow in spacer-filled channels: Effect of spacer geometrical characteristics. *Journal of Membrane Science*, **291**, 53–69.

Koutsou C. P., Yiantsios S. G. and Karabelas A. J. (2009). A numerical and experimental study of mass transfer in spacer-filled channels: Effects of spacer geometrical characteristics and Schmidt number. *Journal of Membrane Science*, **326**, 234–251.

Krebs C. J. (1999). *Ecological Methodology*, 2nd ed. Benjamin/Cummings, California.

Kreft J.-U., Picioreanu C., Wimpenny J. W. T. and Van Loosdrecht M. C. M. (2001). Individual-based modelling of biofilms. *Microbiology*, **147**, 2897–2912.

Kwok W. K., Picioreanu C., Ong S. L., Van Loosdrecht M. C. M., Ng W. J. and Heijnen J. J. (1998). Influence of biomass production and detachment forces on biofilm structures in a biofilm airlift suspension reactor. *Biotechnology and Bioengineering*, **58**, 400–407.

Lacey R. E. (1980). Energy by reverse electrodialysis. *Ocean Engineering*, **7**, 1–47.

Larsen M., Borisov S. M., Grunwald B., Klimant I. and Glud R. N. (2011). A simple and inexpensive high resolution color ratiometric planar optode imaging approach: Application to oxygen and pH sensing. *Limnology and Oceanography: Methods*, **9**, 348–360.

Lattemann S., Kennedy M. D., Schippers J. C. and Amy G. (2010). Chapter 2 global desalination situation. *Sustainability Science and Engineering*, **2**, 7–39.

Lau K. K., Abu Bakar M. Z., Ahmad A. L. and Murugesan T. (2009). Feed spacer mesh angle: 3D modeling, simulation and optimization based on unsteady hydrodynamic in spiral wound membrane channel. *Journal of Membrane Science*, **343**, 16–33.

Lawler W., Bradford-Hartke Z., Cran M. J., Duke M., Leslie G., Ladewig B. P. and Le-Clech P. (2012). Towards new opportunities for reuse, recycling and disposal of used reverse osmosis membranes. *Desalination*, **299**, 103–112.

Lecuyer S., Rusconi R., Shen Y., Forsyth A., Vlamakis H., Kolter R. and Stone H. A. (2011). Shear stress increases the residence time of adhesion of Pseudomonas aeruginosa. *Biophysical Journal*, **100**, 341–350.

Lee K.-C. and Rittmann B. E. (2003). Effects of pH and precipitation on autohydrogenotrophic denitrification using the hollow-fiber membrane-biofilm reactor. *Water Research*, **37**, 1551–1556.

Lee K. P., Arnot T. C. and Mattia D. (2011). A review of reverse osmosis membrane materials for desalination-development to date and future potential. *Journal of Membrane Science*, **370**, 1–22.

Lewandowski Z., Altobelli S. A., Majors P. D. and Fukushima E. (1992). NMR imaging of hydrodynamics near microbially colonized surfaces. *Water Science and Technology*, **26**, 577–584.

Lewandowski Z., Altobelli S. A. and Fukushima E. (1993). NMR and microelectrode studies of hydrodynamics and kinetics in biofilms. *Biotechnology Progress*, **9**, 40–45.

Lewandowski Z., Beyenal H. and Stookey D. (2004). Reproducibility of biofilm processess and the meaning of steady state in biofilm reactors. *Water Science and Technology*, **49**, 359–364.

Li Q. and Elimelech M. (2006). Synergistic effects in combined fouling of a loose nanofiltration membrane by colloidal materials and natural organic matter. *Journal of Membrane Science*, **278**, 72–82.

Li H., Fane A. G., Coster H. G. L. and Vigneswaran S. (1998). Direct observation of particle deposition on the membrane surface during crossflow microfiltration. *Journal of Membrane Science*, **149**, 83–97.

Li F., Meindersma G. W., De Haan A. B. and Reith T. (2002a). Optimization of non-woven spacers by CFD and validation by experiments. *Desalination*, **146**, 209–212.

Li F., Meindersma W., De Haan A. B. and Reith T. (2002b). Optimization of commercial net spacers in spiral wound membrane modules. *Journal of Membrane Science*, **208**, 289–302.

Li H., Fane A. G., Coster H. G. L. and Vigneswaran S. (2003). Observation of deposition and removal behaviour of submicron bacteria on the membrane surface during crossflow microfiltration. *Journal of Membrane Science*, **217**, 29–41.

Li F., Meindersma W., De Haan A. B. and Reith T. (2005). Novel spacers for mass transfer enhancement in membrane separations. *Journal of Membrane Science*, **253**, 1–12.

Li Y.-L., Chang T.-H., Wu C.-Y., Chuang C.-J. and Tung K.-L. (2006). CFD analysis of particle deposition in the spacer-filled membrane module. *Journal of Water Supply: Research and Technology–AQUA*, **55**, 589–601.

Li Y.-L., Tung K.-L., Chen Y.-S. and Hwang K.-J. (2012). CFD analysis of the initial stages of particle deposition in spiral-wound membrane modules. *Desalination, Special Issue in Honour of Professor Takeshi Matsuura on His 75th Birthday*, **287**, 200–208.

Liikanen R., Yli-Kuivila J. and Laukkanen R. (2002). Efficiency of various chemical cleanings for nanofiltration membrane fouled by conventionally-treated surface water. *Journal of Membrane Science*, **195**, 265–276.

Lin Y. M., Lotti T., Sharma P. K. and van Loosdrecht M. C. M. (2013). Apatite accumulation enhances the mechanical property of anammox granules. *Water Research*, **47**, 4556–4566.

Liu Y. and Mi B. (2012). Combined fouling of forward osmosis membranes: Synergistic foulant interaction and direct observation of fouling layer formation. *Journal of Membrane Science*, **407–408**, 136–144.

Liu J., Liu Z., Xu X. and Liu F. (2015). Saw-tooth spacer for membrane filtration: Hydrodynamic investigation by PIV and filtration experiment validation. *Chemical Engineering and Processing: Process Intensification*, **91**, 23–34.

Louie J. S., Pinnau I., Ciobanu I., Ishida K. P., Ng A. and Reinhard M. (2006). Effects of polyether–polyamide block copolymer coating on performance and fouling of reverse osmosis membranes. *Journal of Membrane Science*, **280**, 762–770.

Louie J. S., Pinnau I. and Reinhard M. (2011). Effects of surface coating process conditions on the water permeation and salt rejection properties of composite polyamide reverse osmosis membranes. *Journal of Membrane Science*, **367**, 249–255.

Lyster E. and Cohen Y. (2007). Numerical study of concentration polarization in a rectangular reverse osmosis membrane channel: Permeate flux variation and hydrodynamic end effects. *Journal of Membrane Science*, **303**, 140–153.

Lyster E., Au J., Rallo R., Giralt F. and Cohen Y. (2009). Coupled 3-D hydrodynamics and mass transfer analysis of mineral scaling-induced flux decline in a laboratory plate-and-frame reverse osmosis membrane module. *Journal of Membrane Science*, **339**, 39–48.

Lyster E., Kim M.-M., Au J. and Cohen Y. (2010). A method for evaluating antiscalant retardation of crystal nucleation and growth on RO membranes. *Journal of Membrane Science*, **364**, 122–131.

Ma S. and Song L. (2006). Numerical study on permeate flux enhancement by spacers in a crossflow reverse osmosis channel. *Journal of Membrane Science*, **284**, 102–109.

Madireddi K., Babcock R. B., Levine B., Kim J. H. and Stenstrom M. K. (1999). An unsteady-state model to predict concentration polarization in commercial spiral wound membranes. *Journal of Membrane Science*, **157**, 13–34.

Madaeni S. S. and Ghaemi N. (2007). Characterization of self-cleaning RO membranes coated with TiO2 particles under UV irradiation. *Journal of Membrane Science*, **303**, 221–233.

Majamaa K., Johnson J. E. and Aerts P. E. M. (2009). Improved reverse osmosis element construction reducing differential pressure in high fouling conditions. In: *Proceedings IDA World Congress*, Dubai.

Majamaa K., Aerts P. E. M., Groot C., Paping L. L. M. J., van den Broek W. and van Agtmaal S. (2010). Industrial water reuse with integrated membrane system increases the sustainability of the chemical manufacturing. *Desalination and Water Treatment*, **18**, 17–23.

Malek A., Hawlader M. N. A. and Ho J. C. (1996). Design and economics of RO seawater desalination. *Desalination*, **105**, 245–261.

Manz B., Volke F., Goll D. and Horn H. (2003). Measuring local flow velocities and biofilm structure in biofilm systems with Magnetic Resonance Imaging (MRI). *Biotechnology and Bioengineering*, **84**, 424–432.

Manz B., Volke F., Goll D. and Horn H. (2005). Investigation of biofilm structure, flow patterns and detachment with magnetic resonance imaging. *Water Science and Technology*, **52**, 1–6.

Margalit E., Leshansky A. and Freger V. (2013). Modeling and analysis of hydrodynamic and physico-chemical effects in bacterial deposition on surfaces. *Biofouling*, **29**, 977–989.

Marmur A. (2006). Super-hydrophobicity fundamentals: Implications to biofouling prevention. *Biofouling*, **22**, 107–115.

Marselina Y., Lifia, Le-Clech P., Stuetz R. M. and Chen V. (2009). Characterisation of membrane fouling deposition and removal by direct observation technique. *Journal of Membrane Science*, **341**, 163–171.

Martin K. J., Picioreanu C. and Nerenberg R. (2013). Multidimensional modeling of biofilm development and fluid dynamics in a hydrogen-based, membrane biofilm reactor (MBfR). *Water Research*, **47**, 4739–4751.

Martínez-Díez L., Vázquez-González M. I. and Florido-Díaz F. J. (1998). Study of membrane distillation using channel spacers. *Journal of Membrane Science*, **144**, 45–56.

Matin A., Khan Z., Zaidi S. M. J. and Boyce M. C. (2011). Biofouling in reverse osmosis membranes for seawater desalination: Phenomena and prevention. *Desalination*, **281**, 1–16.

Matin A., Shafi H. Z., Khan Z., Khaled M., Yang R., Gleason K. and Rehman F. (2014). Surface modification of seawater desalination reverse osmosis membranes: Characterization studies & performance evaluation. *Desalination*, **343**, 128–139.

Mattila K., Carpen L., Hakkarainen T. and Salkinoja-Salonen M. S. (1997). Biofilm development during ennoblement of stainless steel in Baltic Sea water: A microscopic study. *International Biodeterioration & Biodegradation*, **40**, 1–10.

McDonogh R., Schaule G. and Flemming H.-C. (1994). The permeability of biofouling layers on membranes. *Journal of Membrane Science*, **87**, 199–217.

Meier J. and Melin T. (2005). Wastewater reclamation by the PAC-NF process. *Desalination*, **178**, 27–40.

Meinders J. M., van der Mei H. C. and Busscher H. J. (1995). Deposition efficiency and reversibility of bacterial adhesion under flow. *Journal of Colloid and Interface Science*, **176**, 329–341.

Mendret J., Guigui C., Schmitz P., Cabassud C. and Duru P. (2007). An optical method for in situ characterization of fouling during filtration. *AIChE Journal*, **53**, 2265–2274.

Mendret J., Guigui C., Schmitz P. and Cabassud C. (2009). In situ dynamic characterisation of fouling under different pressure conditions during dead-end filtration: Compressibility properties of particle cakes. *Journal of Membrane Science*, **333**, 20–29.

Mi B. and Elimelech M. (2010). Organic fouling of forward osmosis membranes: Fouling reversibility and cleaning without chemical reagents. *Journal of Membrane Science*, **348**, 337–345.

Miller D. J., Araújo P. A., Correia P. B., Ramsey M. M., Kruithof J. C., van Loosdrecht M. C., Freeman B. D., Paul D. R., Whiteley M. and Vrouwenvelder J. S. (2012). Short-term adhesion and long-term biofouling testing of polydopamine and poly(ethylene glycol) surface modifications of membranes and feed spacers for biofouling control. *Water Research*, **46**, 3737–3753.

Mitchell A. C., Dideriksen K., Spangler L. H., Cunningham A. B. and Gerlach R. (2010). Microbially enhanced carbon capture and storage by mineral-trapping and solubility-trapping. *Environmental Science & Technology*, **44**, 5270–5276.

Mo H. and Ng H. Y. (2010). An experimental study on the effect of spacer on concentration polarization in a long channel reverse osmosis membrane cell. *Water Science & Technology*, **61**, 2035–2041.

Mohammad A. W., Teow Y. H., Ang W. L., Chung Y. T., Oatley-Radcliffe D. L. and Hilal N. (2015). Nanofiltration membranes review: Recent advances and future prospects. *Desalination*, **356**, 226–254.

Möhle R. B., Langemann T., Haesner M., Augustin W., Scholl S., Neu T. R., Hempel D. C. and Horn H. (2007). Structure and shear strength of microbial biofilms as determined with confocal laser scanning microscopy and fluid dynamic gauging using a novel rotating disc biofilm reactor. *Biotechnology and Bioengineering*, **98**, 747–755.

Mojab S. M., Hanff E. S., Beale S. B., Pollard A. and Pharoah J. G. (2010). Particle image velocimetry investigation of flows within spacer-filled channels. In: *15th Int Symp on Applications of Laser Techniques to Fluid Mechanics*, Lisbon.

Mueller E., Deimling M. and Reinhardt E. R. (1986). Quantification of pulsatile flow in MRI by an analysis of T_2 changes in ECG-gated multiecho experiments. *Magnetic Resonance in Medicine*, **3**, 331–335.

Mulder M. (1996). *Basic Principles of Membrane Technology*, 2nd edn. *Journal of Membrane Science*. Springer Netherlands.

Neal P. R., Li H., Fane A. G. and Wiley D. E. (2003). The effect of filament orientation on critical flux and particle deposition in spacer-filled channels. *Journal of Membrane Science*, **214**, 165–178.

Neu T. R. and Lawrence J. R. (1997). Development and structure of microbial biofilms in river water studied by confocal laser scanning microscopy. *FEMS Microbiology Ecology*, **24**, 11–25.

Ngene I. S., Lammertink R. G. H., Kemperman A. J. B., Van De Ven W. J. C., Wessels L. P., Wessling M. and Van Der Meer W. G. J. (2010a). CO_2 nucleation in membrane spacer channels remove biofilms and fouling deposits. *Industrial & Engineering Chemistry Research*, **49**, 10034–10039.

Ngene I. S., Lammertink R. G. H., Wessling M. and Van der Meer W. G. J. (2010b). Particle deposition and biofilm formation on microstructured membranes. *Journal of Membrane Science*, **364**, 43–51.

Nguyen T., Roddick F. A. and Fan L. (2012). Biofouling of water treatment membranes: A review of the underlying causes, monitoring techniques and control measures. *Membranes*, **2**, 804–840.

Nikonenko V. V., Pismenskaya N. D., Istoshin A. G., Zabolotsky V. I. and Shudrenko A. A. (2008). Description of mass transfer characteristics of ED and EDI apparatuses by using the similarity theory and compartmentation method. *Chemical Engineering and Processing: Process Intensification*, **47**, 1118–1127.

Ohl A. L., Horn H. and Hempel D. C. (2004). Behaviour of biofilm systems under varying hydrodynamic conditions. *Water Science and Technology*, **49**, 345–351.

Olivera-Nappa A., Picioreanu C. and Asenjo J. A. (2010). Non-homogeneous biofilm modeling applied to bioleaching processes. *Biotechnology and Bioengineering*, **106**, 660–676.

Ostuni E., Chapman R. G., Holmlin R. E., Takayama S. and Whitesides G. M. (2001). A survey of structure-property relationships of surfaces that resist the adsorption of protein. *Langmuir*, **17**, 5605–5620.

Pang C. M., Hong P., Guo H. and Liu W.-T. (2005). Biofilm formation characteristics of bacterial isolates retrieved from a reverse osmosis membrane. *Environmental Science & Technology*, **39**, 7541–7550.

Pankratz T. (2000). Dissecting desalination. *Water Environmental Internationl*, **9**, 8–9.

Pankratz T. (2012). Water desalination report. *Desalination*, **48**, 1–4.

Paramonova E., Zerfoss E. L. and Logan B. E. (2006). Measurement of biocolloid collision efficiencies for granular activated carbon by use of a two-layer filtration model. *Applied and Environmental Microbiology*, **72**, 5190–5196.

Paramonova E., De Jong E. D., Krom B. P., Van Der Mei H. C., Busscher H. J. and Sharma P. K. (2007). Low-load compression testing: A novel way of measuring biofilm thickness. *Applied and Environmental Microbiology*, **73**, 7023–7028.

Paramonova E., Krom B. P., van der Mei H. C., Busscher H. J. and Sharma P. K. (2009). Hyphal content determines the compression strength of Candida albicans biofilms. *Microbiology*, **155**, 1997–2003.

Park J. Y., Acar M. H., Akthakul A., Kuhlman W. and Mayes A. M. (2006). Polysulfone-graft-poly(ethylene glycol) graft copolymers for surface modification of polysulfone membranes. *Biomaterials*, **27**, 856–65.

Paris T., Skali-Lami S. and Block J.-C. (2007). Effect of wall shear rate on biofilm deposition and grazing in drinking water flow chambers. *Biotechnology and Bioengineering*, **97**, 1550–1561.

Passmore J. M., Rudolph D. L., Mesquita M. M. F., Cey E. E. and Emelko M. B. (2010). The utility of microspheres as surrogates for the transport of E. coli RS2g in partially saturated agricultural soil. *Water Research*, **44**, 1235–1245.

Pendlebury J. M., Smith K., Unsworth P., Greene G. L. and Mampe W. (1979). Precision field averaging NMR magnetometer for low and high fields, using flowing water. *Review of Scientific Instruments*, **50**, 535–540.

Peppas N. A. and Barr-Howell B. D. (1986). Characterization of the cross-linked structure of hydrogels, Hydrogels in Medicine and Pharmacy, Volume I: Fundamentals. CRC Press Inc.

Pervov A. G. (1991). Scale formation prognosis and cleaning procedure schedules in reverse osmosis systems operation. *Desalination*, **83**, 77–118.

Phattaranawik J., Jiraratananon R. and Fane A. G. (2003). Heat transport and membrane distillation coefficients in direct contact membrane distillation. *Journal of Membrane Science*, **212**, 177–193.

Picioreanu C., Van Loosdrecht M. C. M. and Heijnen J. J. (1998a). A new combined differential-discrete cellular automaton approach for biofilm modeling: Application for growth in gel beads. *Biotechnology and Bioengineering*, **57**, 718–731.

Picioreanu C., Van Loosdrecht M. C. M. and Heijnen J. J. J. (1998b). Mathematical modeling of biofilm structure with a hybrid differential-discrete cellular automaton approach. *Biotechnology and Bioengineering*, **58**, 101–116.

Picioreanu C., Van Loosdrecht M. C. M. and Heijnen J. J. J. (2000). Effect of diffusive and convective substrate transport on biofilm structure formation: A two-dimensional modeling study. *Biotechnology and Bioengineering*, **69**, 504–515.

Picioreanu C., Van Loosdrecht M. C. M. and Heijnen J. J. (2001). Two-dimensional model of biofilm detachment caused by internal stress from liquid flow. *Biotechnology & Bioengineering*, **72**, 205–218.

Picioreanu C., Kreft J.-U. and Van Loosdrecht M. C. M. (2004). Particle-based multidimensional multispecies biofilm model. *Applied and Environmental Microbiology*, **70**, 3024–3040.

Picioreanu C., Kreft J.-U., Klausen M., Haagensen J. A. J., Tolker-Nielsen T. and Molin S. (2007). Microbial motility involvement in biofilm structure formation – A 3D modelling study. *Water Science and Technology*, pp. 337–343.

Picioreanu C., Vrouwenvelder J. S. and van Loosdrecht M. C. M. (2009). Three-dimensional modeling of biofouling and fluid dynamics in feed spacer channels of membrane devices. *Journal of Membrane Science*, **345**, 340–354.

Pintelon T. R. R., Creber S. A., Graf Von Der Schulenburg D. A. and Johns M. L. (2010). Validation of 3D simulations of reverse osmosis membrane biofouling. *Biotechnology and Bioengineering*, **106**, 677–689.

Pintelon T. R., Picioreanu C., van Loosdrecht M. C. and Johns M. L. (2012). The effect of biofilm permeability on bio-clogging of porous media. *Biotechnology and Bioengineering*, **109**, 1031–1042.

Post J. W., Goeting C. H., Valk J., Goinga S., Veerman J., Hamelers H. V. M. and Hack P. J. F. M. (2010). Towards implementation of reverse electrodialysis for power generation from salinity gradients. *Desalination and Water Treatment*, **16**, 182–193.

Potts D. E., Ahlert R. C. and Wang S. S. (1981). A critical review of fouling of reverse osmosis membranes. *Desalination*, **36**, 235–264.

Prest E. I., Staal M., Kühl M., van Loosdrecht M. C. M. and Vrouwenvelder J. S. (2012). Quantitative measurement and visualization of biofilm O_2 consumption rates in membrane filtration systems. *Journal of Membrane Science*, **392–393**, 66–75.

Prihasto N., Liu Q.-F. and Kim S.-H. (2009). Pre-treatment strategies for seawater desalination by reverse osmosis system. *Desalination*, **249**, 308–316.

Radu A. I., Vrouwenvelder J. S., van Loosdrecht M. C. M. and Picioreanu C. (2010). Modeling the effect of biofilm formation on reverse osmosis performance: Flux, feed channel pressure drop and solute passage. *Journal of Membrane Science*, **365**, 1–15.

Radu A. I., Vrouwenvelder J. S., van Loosdrecht M. C. M. and Picioreanu C. (2012a). Effect of flow velocity, substrate concentration and hydraulic cleaning on biofouling of reverse osmosis feed channels. *Chemical Engineering Journal*, **188**, 30–39.

Radu A. I., Vrouwenvelder J. S., van Loosdrecht M. C. M. and Picioreanu C. (2012b). Modeling biofouling, scaling and combined fouling in reverse osmosis membrane devices. *Procedia Engineering*, **44**, 341–342.

Radu A. I., Bergwerff L., van Loosdrecht M. C. M. and Picioreanu C. (2014a). A two-dimensional mechanistic model for scaling in spiral wound membrane systems. *Chemical Engineering Journal*, **241**, 77–91.

Radu A. I., van Steen M. S. H., Vrouwenvelder J. S., van Loosdrecht M. C. M. and Picioreanu C. (2014b). Spacer geometry and particle deposition in spiral wound membrane feed channels. *Water Research*, **64**, 160–176.

Radu A. I., Bergwerff L., van Loosdrecht M. C. M. and Picioreanu C. (2015). Combined biofouling and scaling in membrane feed channels: A new modeling approach. *Biofouling*, **31**, 83–100.

Rahardianto A., Shih W.-Y., Lee R.-W. and Cohen Y. (2006). Diagnostic characterization of gypsum scale formation and control in RO membrane desalination of brackish water. *Journal of Membrane Science*, **279**, 655–668.

Ramon G. Z. and Hoek E. M. V. (2012). On the enhanced drag force induced by permeation through a filtration membrane. *Journal of Membrane Science*, **392–393**, 1–8.

Rana D. and Matsuura T. (2010). Surface modifications for antifouling membranes. *Chemical Reviews*, **110**, 2448–2471.

Ranade V. V. and Kumar A. (2006). Fluid dynamics of spacer filled rectangular and curvilinear channels. *Journal of Membrane Science*, **271**, 1–15.

Rand J. L., Hofmann R., Alam M. Z. B., Chauret C., Cantwell R., Andrews R. C. and Gagnon G. A. (2007). A field study evaluation for mitigating biofouling with chlorine dioxide or chlorine integrated with UV disinfection. *Water Research*, **41**, 1939–1948.

Renslow R. S., Majors P. D., McLean J. S., Fredrickson J. K., Ahmed B. and Beyenal H. (2010). In situ effective diffusion coefficient profiles in live biofilms using pulsed-field gradient nuclear magnetic resonance. *Biotechnology and Bioengineering*, **106**, 928–937.

Ridgway H. F., Kelly A., Justice C. and Olson B. H. (1983). Microbial fouling of reverse-osmosis membranes used in advanced wastewater treatment technology: Chemical, bacteriological, and ultrastructural analyses. *Applied and Environmental Microbiology*, **45**, 1066–1084.

Riding R. (2000). Microbial carbonates: The geological record of calcified bacterial-algal mats and biofilms. *Sedimentology*, **47**, 179–214.

Roache P. J. (1994). Perspective: A method for uniform reporting of grid refinement studies. *Journal of Fluids Engineering*, **116**, 405–413.

Romo R. F. V. and Pitts M. M. (1999). Application of electrotechnology for removal and prevention of reverse osmosis biofouling. *Environmental Progress*, **18**, 107–112.

Ronen A., Lerman S., Ramon G. Z. and Dosoretz C. G. (2015). Experimental characterization and numerical simulation of the anti-biofuling activity of nanosilver-modified feed spacers in membrane filtration. *Journal of Membrane Science*, **475**, 320–329.

Ronen A., Resnick A., Lerman S., Eisen M. S. and Dosoretz C. G. (2016). Biofouling suppression of modified feed spacers: Localized and long-distance antibacterial activity. *Desalination*, **393**, 159–165.

Rukapan W., Khananthai B., Chiemchaisri C., Chiemchaisri W. and Srisukphun T. (2012). Short- and long-term fouling characteristics of reverse osmosis membrane at full scale leachate treatment plant. *Water Science & Technology*, **65**, 127–134.

Sablani S., Goosen M., Al-Belushi R. and Wilf M. (2001). Concentration polarization in ultrafiltration and reverse osmosis: A critical review. *Desalination*, **141**, 269–289.

Sablani S. S., Goosen M. F. A., Al-Belushi R. and Gerardos V. (2002). Influence of spacer thickness on permeate flux in spiral-wound seawater reverse osmosis systems. *Desalination*, **146**, 225–230.

Saeed M. O., Jamaluddin A. T., Tisan I. A., Lawrence D. A., Al-Amri M. M. and Chida K. (2000). Biofouling in a seawater reverse osmosis plant on the Red Sea coast, Saudi Arabia. *Desalination*, **128**, 177–190.

Saeed A., Vuthaluru R., Yang Y. and Vuthaluru H. B. (2012). Effect of feed spacer arrangement on flow dynamics through spacer filled membranes. *Desalination*, **285**, 163–169.

Sagle A. C., Van Wagner E. M., Ju H., McCloskey B. D., Freeman B. D. and Sharma M. M. (2009). PEG-coated reverse osmosis membranes: Desalination properties and fouling resistance. *Journal of Membrane Science*, **340**, 92–108.

Salvador Cob S., Beaupin C., Hofs B., Nederlof M. M., Harmsen D. J. H., Cornelissen E. R., Zwijnenburg A., Genceli Güner F. E. and Witkamp, G. J. (2012). Silica and silicate precipitation as limiting factors in high-recovery reverse osmosis operations. *Journal of Membrane Science*, **423–424**, 1–10.

Sanderlin A. B., Vogt S. J., Grunewald E., Bergin B. A. and Codd S. L. (2013). Biofilm detection in natural unconsolidated porous media using a low-field magnetic resonance system. *Environmental Science & Technology*, **47**, 987–992.

Santos J. L. C., Geraldes V., Velizarov S. and Crespo J. G. (2007). Investigation of flow patterns and mass transfer in membrane module channels filled with flow-aligned spacers using computational fluid dynamics (CFD). *Journal of Membrane Science*, **305**, 103–117.

Schneider R. P., Ferreira L. M., Binder P. and Ramos J. R. (2005). Analysis of foulant layer in all elements of an RO train. *Journal of Membrane Science*, **261**, 152–162.

Schock G. and Miquel A. (1987). Mass transfer and pressure loss in spiral wound modules. *Desalination*, **64**, 339–352.

Schramm A., De Beer D., Gieseke A. and Amann R. (2000). Microenvironments and distribution of nitrifying bacteria in a membrane-bound biofilm. *Environmental Microbiology*, **2**, 680–686.

Schwinge J., Wiley D. E. and Fletcher D. F. (2002a). Simulation of the flow around spacer filaments between channel walls. 2. Mass-transfer enhancement. *Industrial & Engineering Chemistry Research*, **41**, 4879–4888.

Schwinge J., Wiley D. E. and Fletcher D. F. (2002b). Simulation of the flow around spacer filaments between narrow channel walls. 1. Hydrodynamics. *Industrial & Engineering Chemistry Research*, **41**, 2977–2987.

Schwinge J., Wiley D. E. and Fletcher D. F. (2003). Simulation of unsteady flow and vortex shedding for narrow spacer-filled channels. *Industrial & Engineering Chemistry Research*, **42**, 4962–4977.

Schwinge J., Neal P. R., Wiley D. E., Fletcher D. F. and Fane A. G. (2004a). Spiral wound modules and spacers: Review and analysis. *Journal of Membrane Science*, **242**, 129–153.

Schwinge J., Wiley D. E. and Fane A. G. (2004b). Novel spacer design improves observed flux. *Journal of Membrane Science*, **229**, 53–61.

Seymour J. D., Gage J. P., Codd S. L. and Gerlach R. (2004). Anomalous fluid transport in porous media induced by biofilm growth. *Physical Review Letters*, **93**, 198103.

Seymour J. D., Gage J. P., Codd S. L. and Gerlach R. (2007). Magnetic resonance microscopy of biofouling induced scale dependent transport in porous media. *Advances in Water Resources*, **30**, 1408–1420.

Shahan D. (1990). Reverse osmosis membranes for seawater desalination state-of-the-art. *Separation and Purification Methods*, **19**, 121–131.

Shakaib M., Hasani S. M. F. and Mahmood M. (2007). Study on the effects of spacer geometry in membrane feed channels using three-dimensional computational flow modeling. *Journal of Membrane Science*, **297**, 74–89.

Shakaib M., Hasani S. M. F. and Mahmood M. (2009). CFD modeling for flow and mass transfer in spacer-obstructed membrane feed channels. *Journal of Membrane Science*, **326**, 270–284.

Shannon M. A., Bohn P. W., Elimelech M., Georgiadis J. G., Mariñas B. J. and Mayes A. M. (2008). Science and technology for water purification in the coming decades. *Nature*, **452**, 301–310.

Sharp R. R., Stoodley P., Adgie M., Gerlach R. and Cunningham A. (2005). Visualization and characterization of dynamic patterns of flow, growth and activity of biofilms growing in porous media. *Water Science and Technology*, **52**, 85–90.

Siddiqui A., Farhat N., Bucs S. S., Linares R. V., Picioreanu C., Kruithof J. C., van Loosdrecht M. C. M., Kidwell J. and Vrouwenvelder J. S. (2016). Development and characterization of 3D-printed feed spacers for spiral wound membrane systems. *Water Research*, **91**, 55–67.

Siddiqui A., Lehmann S., Bucs S. S., Fresquet M., Fel L., Prest E. I. E. C., Ogier J., Schellenberg C., van Loosdrecht M. C. M., Kruithof J. C. and Vrouwenvelder J. S. (2017a). Predicting the impact of feed spacer modification on biofouling by hydraulic characterization and biofouling studies in membrane fouling simulators. *Water Research*, **110**, 281–287.

Siddiqui A., Pinel I., Prest E. I., Bucs S. S., van Loosdrecht M. C. M., Kruithof J. C. and Vrouwenvelder J. S. (2017b). Application of DBNPA dosage for biofouling control in spiral wound membrane systems. *Desalination and Water Treatment*, **68**, 12–22.

Sim S. T. V., Suwarno S. R., Chong T. H., Krantz W. B. and Fane A. G. (2013). Monitoring membrane biofouling via ultrasonic time-domain reflectometry enhanced by silica dosing. *Journal of Membrane Science*, **428**, 24–37.

Simões M., Pereira M. O. and Vieira M. J. (2005). Effect of mechanical stress on biofilms challenged by different chemicals. *Water Research*, **39**, 5142–5152.

Simon A., Nghiem L. D., Le-Clech P., Khan S. J. and Drewes J. E. (2009). Effects of membrane degradation on the removal of pharmaceutically active compounds (PhACs) by NF/RO filtration processes. *Journal of Membrane Science*, **340**, 16–25.

Sivan A., Szanto M. and Pavlov V. (2006). Biofilm development of the polyethylene-degrading bacterium Rhodococcus ruber. *Applied Microbiology and Biotechnology*, **72**, 346–352.

Song L. and Elimelech M. (1995). Particle deposition onto a permeable surface in laminar flow. *Journal of Colloid and Interface Science*, **173**, 165–180.

Song L. and Ma S. (2005). Numerical studies of the impact of spacer geometry on concentration polarization in spiral wound membrane modules. *Industrial & Engineering Chemistry Research*, **44**, 7638–7645.

Sonune A. and Ghate R. (2004). Developments in wastewater treatment methods. *Desalination*, **167**, 55–63.

Sousa P., Soares A., Monteiro E. and Rouboa A. (2014). A CFD study of the hydrodynamics in a desalination membrane filled with spacers. *Desalination*, **349**, 22–30.

Staal M., Borisov S. M., Rickelt L. F., Klimant I. and Kühl M. (2011a). Ultrabright planar optodes for luminescence life-time based microscopic imaging of O_2 dynamics in biofilms. *Journal of Microbiological Methods*, **85**, 67–74.

Staal M., Prest E. I., Vrouwenvelder J. S., Rickelt L. F. and Kühl M. (2011b). A simple optode based method for imaging O_2 distribution and dynamics in tap water biofilms. *Water Research*, **45**, 5027–5037.

Steiner Z., Rapaport H., Oren Y. and Kasher R. (2010). Effect of surface-exposed chemical groups on calcium-phosphate mineralization in water-treatment systems. *Environmental Science & Technology*, **44**, 7937–7943.

Stewart P. S. (1998). A review of experimental measurements of effective diffusive permeabilities and effective diffusion coefficients in biofilms. *Biotechnology and Bioengineering*, **59**, 261–272.

Stoodley P., Lewandowski Z., Boyle J. D. and Lappin-Scott H. M. (1998a). Oscillation characteristics of biofilm streamers in turbulent flowing water as related to drag and pressure drop. *Biotechnology and Bioengineering*, **57**, 536–544.

Stoodley P., Dodds I., Boyle J. D. and Lappin-Scott H. M. (1998b). Influence of hydrodynamics and nutrients on biofilm structure. *Journal of Applied Microbiology*, **85**, 19S–28S.

Stoodley P., Boyle J. D., Debeer D. and Lappin-Scott H. M. (1999a). Evolving perspectives of biofilm structure. *Biofouling*, **14**, 75–90.

Stoodley P., Lewandowski Z., Boyle J. D. and Lappin-Scott H. M. (1999b). Structural deformation of bacterial biofilms caused by short-term fluctuations in fluid shear: An in situ investigation of biofilm rheology. *Biotechnology and Bioengineering*, **65**, 83–92.

Subramani A. and Hoek E. M. V. (2008). Direct observation of initial microbial deposition onto reverse osmosis and nanofiltration membranes. *Journal of Membrane Science*, **319**, 111–125.

Subramani A., Kim S. and Hoek E. M. V. (2006). Pressure, flow, and concentration profiles in open and spacer-filled membrane channels. *Journal of Membrane Science*, **277**, 7–17.

Suidan M. T., Rittmann B. E. and Traegner U. K. (1987). Criteria establishing biofilm-kinetic types. *Water Research*, **21**, 491–498.

Suwarno S. R., Chen X., Chong T. H., Puspitasari V. L., McDougald D., Cohen Y., Rice S. A. and Fane A. G. (2012). The impact of flux and spacers on biofilm development on reverse osmosis membranes. *Journal of Membrane Science*, **405–406**, 219–232.

Sweity A., Oren Y., Ronen Z. and Herzberg M. (2013). The influence of antiscalants on biofouling of RO membranes in seawater desalination. *Water Research*, **47**, 3389–3398.

Taherzadeh D., Picioreanu C., Küttler U., Simone A., Wall W. A. and Horn H. (2010). Computational study of the drag and oscillatory movement of biofilm streamers in fast flows. *Biotechnology and Bioengineering*, **105**, 600–610.

Tamachkiarow A. and Flemming H.-C. (2003). On-line monitoring of biofilm formation in a brewery water pipeline system with a fibre optical device. *Water Science and Technology*, pp. 19–24.

Tang C. Y., Kwon Y.-N. and Leckie J. O. (2009). Effect of membrane chemistry and coating layer on physiochemical properties of thin film composite polyamide RO and NF membranes. *Desalination*, **242**, 168–182.

Tang C. Y., Chong T. H. and Fane A. G. (2011). Colloidal interactions and fouling of NF and RO membranes: A review. *Advances in Colloid and Interface Science*, **164**, 126–143.

Tasaka K., Katsura T., Iwahori H. and Kamiyama Y. (1994). Analysis of RO elements operated at more than 80 plants in Japan. *Desalination*, **96**, 259–272.

Thompson J., Lin N., Lyster E., Arbel R., Knoell T., Gilron J. and Cohen Y. (2012). RO membrane mineral scaling in the presence of a biofilm. *Journal of Membrane Science*, **415–416**, 181–191.

Thullner M., Mauclaire L., Schroth M. H., Kinzelbach W. and Zeyer J. (2002). Interaction between water flow and spatial distribution of microbial growth in a two-dimensional flow field in saturated porous media. *Journal of Contaminant Hydrology*, **58**, 169–189.

Towler B. W., Rupp C. J., Cunningham A. L. B. and Stoodley P. (2003). Viscoelastic properties of a mixed culture biofilm from rheometer creep analysis. *Biofouling*, **19**, 279–285.

Tran T., Bolto B., Gray S., Hoang M. and Ostarcevic E. (2007). An autopsy study of a fouled reverse osmosis membrane element used in a brackish water treatment plant. *Water Research*, **41**, 3915–3923.

Tung K.-L., Wang S., Lu W.-M. and Pan C.-H. (2001). In situ measurement of cake thickness distribution by a photointerrupt sensor. *Journal of Membrane Science*, **190**, 57–67.

Valladares Linares R., Bucs S. S. S., Li Z., AbuGhdeeb M., Amy G. and Vrouwenvelder J. S. S. (2014). Impact of spacer thickness on biofouling in forward osmosis. *Water Research*, **57**, 223–233.

Valladares Linares R., Wexler A. D., Bucs S. S., Dreszer C., Zwijnenburg A., Flemming H. C., Kruithof J. C. and Vrouwenvelder J. S. (2016). Compaction and relaxation of biofilms. *Desalination and Water Treatment*, **57**, 12902–12914.

van der Bruggen B. (2003). Desalination by distillation and by reverse osmosis – trends towards the future. *Membrane Technology*, **2003**, 6–9.

Van der Bruggen B. and Vandecasteele C. (2002). Distillation vs. membrane filtration: Overview of process evolutions in seawater desalination. *Desalination*, **143**, 207–218.

Van der Hoek J. P., Hofman J. A. M. H., Bonné P. A. C., Nederlof M. M. and Vrouwenvelder H. S. (2000). RO treatment: Selection of a pretreatment scheme based on fouling characteristics and operating conditions based on environmental impact. *Desalination*, **127**, 89–101.

Van Geluwe S., Vinckier C., Braeken L. and Van der Bruggen B. (2011). Ozone oxidation of nanofiltration concentrates alleviates membrane fouling in drinking water industry. *Journal of Membrane Science*, **378**, 128–137.

van Loosdrecht M. C. M., Lyklema J., Norde W. and Zehnder A. J. B. (1989). Bacterial adhesion: A physicochemical approach. *Microbial Ecology*, **17**, 1–15.

Van Loosdrecht M. C. M., Lyklema J., Norde W. and Zehnder A. J. B. (1990). Influence of interfaces on microbial activity. *Microbiology Reviews*, **54**, 75–87.

van Loosdrecht M. C. M., Eikelboom D., Gjaltema A., Mulder A., Tijhuis L. and Heijnen J. J. (1995). Biofilm structures. *Water Science and Technology*, **32**, 35–43.

Van Loosdrecht M. C. M., Bereschenko L., Radu A., Kruithof J. C., Picioreanu C., Johns M. L. and Vrouwenvelder H. S. (2012). New approaches to characterizing and understanding biofouling of spiral wound membrane systems. *Water Science & Technology*, **66**, 88–94.

Van Paassen J. A. M., Kruithof J. C., Bakker S. M. and Kegel F. S. (1998). Integrated multi-objective membrane systems for surface water treatment: Pre-treatment of nanofiltration by riverbank filtration and conventional ground water treatment. *Desalination*, **118**, 239–248.

van Paassen J. A. M., van der Meer W. G. J. and Post J. (2005). Optiflux®: From innovation to realisation. *Desalination*, **178**, 325–331.

Venault A., Liu Y.-H., Wu J.-R., Yang H.-S., Chang Y., Lai J.-Y. and Aimar P. (2014). Low-biofouling membranes prepared by liquid-induced phase separation of the PVDF/polystyrene-b-poly (ethylene glycol) methacrylate blend. *Journal of Membrane Science*, **450**, 340–350.

Verliefde A. R. D., Cornelissen E. R., Heijman S. G. J., Petrinic I., Luxbacher T., Amy G. L., Van der Bruggen B. and van Dijk J. C. (2009). Influence of membrane fouling by (pretreated) surface water on rejection of pharmaceutically active compounds (PhACs) by nanofiltration membranes. *Journal of Membrane Science*, **330**, 90–103.

Vermaas D. A., Kunteng D., Saakes M. and Nijmeijer K. (2013). Fouling in reverse electrodialysis under natural conditions. *Water Research*, **47**, 1289–1298.

Viljoen F. C. (2010). The World Health Organization's water safety plan is much more than just an integrated drinking water quality management plan. *Water Science and Technology*, **61**, 173–179.

Vogt S. J., Sanderlin A. B., Seymour J. D. and Codd S. L. (2013). Permeability of a growing biofilm in a porous media fluid flow analyzed by magnetic resonance displacement-relaxation correlations. *Biotechnology and Bioengineering*, **110**, 1366–1375.

Vörösmarty C. J., Green P., Salisbury J. and Lammers R. B. (2000). Global water resources: Vulnerability from climate change and population growth. *Science*, **289**, 284–288.

Vrijenhoek E. M., Hong S. and Elimelech M. (2001). Influence of membrane surface properties on initial rate of colloidal fouling of reverse osmosis and nanofiltration membranes. *Journal of Membrane Science*, **188**, 115–128.

Vrouwenvelder J. S. and Van Der Kooij D. (2001). Diagnosis, prediction and prevention of biofouling of NF and RO membranes. *Desalination*, **139**, 65–71.

Vrouwenvelder J. S. and Van Der Kooij D. (2003). Integral diagnosis of fouling problems by analysing biomass and inorganic compounds in membrane elements used in water treatment. *Water Science and Technology: Water Supply*, **3**, 211–215.

Vrouwenvelder H. S., Van Paassen J. A. M., Folmer H. C., Hofman J. A. M. H., Nederlof M. M. and Van Der Kooij D. (1998). Biofouling of membranes for drinking water production. *Desalination*, **118**, 157–166.

Vrouwenvelder J. S., Manolarakis S. A., Veenendaal H. R. and Van Der Kooij D. (2000). Biofouling potential of chemicals used for scale control in RO and NF membranes. *Desalination*, **132**, 1–10.

Vrouwenvelder J. S., Kappelhof J. W. N. M., Heijman S. G. J., Schippers J. C. and van der Kooij D. (2003). Tools for fouling diagnosis of NF and RO membranes and assessment of the fouling potential of feed water. *Desalination*, **157**, 361–365.

Vrouwenvelder J. S., van Paassen J. A. M., Wessels L. P., van Dam A. F. and Bakker S. M. (2006). The Membrane Fouling Simulator: A practical tool for fouling prediction and control. *Journal of Membrane Science*, **281**, 316–324.

Vrouwenvelder J. S., Bakker S. M., Cauchard M., Le Grand R., Apacandié M., Idrissi M., Lagrave S., Wessels L. P., van Paassen J. A. M., Kruithof J. C. and van Loosdrecht M. C. M. (2007a). The membrane fouling simulator: A suitable tool for prediction and characterisation of membrane fouling. *Water Science and Technology*, **55**, 197–205.

Vrouwenvelder J. S., Bakker S. M., Wessels L. P. and van Paassen J. A. M. (2007b). The Membrane Fouling Simulator as a new tool for biofouling control of spiral-wound membranes. *Desalination*, **204**, 170–174.

Vrouwenvelder J. S., Hinrichs C., Sun A. R., Royer F., van Paassen J. A. M., Bakker S. M., van der Meer W. G. J., Kruithof J. C. and van Loosdrecht M. C. M. (2008a). Monitoring and control of biofouling in nanofiltration and reverse osmosis membranes. *Water Science and Technology: Water Supply*, **8**, 449–458.

Vrouwenvelder J. S., Manolarakis S. A., van der Hoek J. P., van Paassen J. A. M., van der Meer W. G. J., van Agtmaal J. M. C., Prummel H. D. M., Kruithof J. C. and van Loosdrecht M. C. M. (2008b). Quantitative biofouling diagnosis in full scale nanofiltration and reverse osmosis installations. *Water Research*, **42**, 4856–4868.

Vrouwenvelder J. S., Graf von der Schulenburg D. A., Kruithof J. C., Johns M. L. and van Loosdrecht M. C. M. (2009a). Biofouling of spiral-wound nanofiltration and reverse osmosis membranes: A feed spacer problem. *Water Research*, **43**, 583–594.

Vrouwenvelder J. S., Hinrichs C., van der Meer W. G. J., van Loosdrecht M. C. M. and Kruithof J. C. (2009b). Pressure drop increase by biofilm accumulation in spiral wound RO and NF membrane systems: Role of substrate concentration, flow velocity, substrate load and flow direction. *Biofouling*, **25**, 543–555.

Vrouwenvelder J. S., van Paassen J. A. M., Kruithof J. C. and van Loosdrecht M. C. M. (2009c). Sensitive pressure drop measurements of individual lead membrane elements for accurate early biofouling detection. *Journal of Membrane Science*, **338**, 92–99.

Vrouwenvelder J. S., van Paassen J. A. M., van Agtmaal J. M. C., van Loosdrecht M. C. M. and Kruithof J. C. (2009d). A critical flux to avoid biofouling of spiral wound nanofiltration and reverse osmosis membranes: Fact or fiction? *Journal of Membrane Science*, **326**, 36–44.

Vrouwenvelder J. S., Beyer F., Dahmani K., Hasan N., Galjaard G., Kruithof J. C. and Van Loosdrecht M. C. M. (2010a). Phosphate limitation to control biofouling. *Water Research*, **44**, 3454–3466.

Vrouwenvelder J. S., Picioreanu C., Kruithof J. C. and van Loosdrecht M. C. M. (2010b). Biofouling in spiral wound membrane systems: Three-dimensional CFD model based evaluation of experimental data. *Journal of Membrane Science*, **346**, 71–85.

Vrouwenvelder J. S., Buiter J., Riviere M., van der Meer W. G., van Loosdrecht M. C. M. and Kruithof J. C. (2010c). Impact of flow regime on pressure drop increase and biomass accumulation and morphology in membrane systems. *Water Research*, **44**, 689–702.

Vrouwenvelder J. S., van Loosdrecht M. C. M. and Kruithof J. C. (2011a). Early warning of biofouling in spiral wound nanofiltration and reverse osmosis membranes. *Desalination*, **265**, 206–212.

Vrouwenvelder J. S., Van Loosdrecht M. C. M. and Kruithof J. C. (2011b). A novel scenario for biofouling control of spiral wound membrane systems. *Water Research*, **45**, 3890–3898.

Wagner M., Ivleva N. P., Haisch C., Niessner R. and Horn H. (2009). Combined use of confocal laser scanning microscopy (CLSM) and Raman microscopy (RM): Investigations on EPS – Matrix. *Water Research*, **43**, 63–76.

Wagner M., Manz B., Volke F., Neu T. R. and Horn H. (2010a). Online assessment of biofilm development, sloughing and forced detachment in tube reactor by means of magnetic resonance microscopy. *Biotechnology and Bioengineering*, **107**, 172–181.

Wagner M., Taherzadeh D., Haisch C. and Horn H. (2010b). Investigation of the mesoscale structure and volumetric features of biofilms using optical coherence tomography. *Biotechnology and Bioengineering*, **107**, 844–853.

Wang F.-C., Marchetti L. and Wright P. K. (1999). Rapid prototyping using machining. *Technical Paper – Society of Manufacturing Engineers*, 99–118.

Wang C., Such G. K., Widjaya A., Lomas H., Stevens G., Caruso F. and Kentish S. E. (2012). Click poly(ethylene glycol) multilayers on RO membranes: Fouling reduction and membrane characterization. *Journal of Membrane Science*, **409–410**, 9–15.

Wanner O. and Gujer W. (1986). A multispecies biofilm model. *Biotechnology and Bioengineering*, **28**, 314–328.

Wanner O., Eberl H., Morgenroth E., Noguera D., Picioreanu C., Rittmann B. and van L. M. C. M. (2006). *Mathematical Modeling of Biofilms*. Scientific and Technical Report Series. IWA Publishing.

Wardeh S. and Morvan H. P. (2009). Detailed numerical simulations of flow mechanics and membrane performance in spacer-filled channels, flat and curved. *Desalination and Water Treatment*, **1**, 277–288.

West S., Wagner M., Engelke C. and Horn H. (2016). Optical coherence tomography for the in situ three-dimensional visualization and quantification of feed spacer channel fouling in reverse osmosis membrane modules. *Journal of Membrane Science*, **498**, 345–352.

Wibisono Y., Cornelissen E. R., Kemperman A. J. B., Van Der Meer W. G. J. and Nijmeijer K. (2014). Two-phase flow in membrane processes: A technology with a future. *Journal of Membrane Science*, **453**, 566–602.

Wibisono Y., El Obied K. E., Cornelissen E. R., Kemperman A. J. B. and Nijmeijer K. (2015a). Biofouling removal in spiral-wound nanofiltration elements using two-phase flow cleaning. *Journal of Membrane Science*, **475**, 131–146.

Wibisono Y., Yandi W., Golabi M., Nugraha R., Cornelissen E., Kemperman A. J. B., Ederth T. and Nijmeijer K. (2015b). Hydrogel-coated feed spacers in two-phase flow cleaning in spiral wound membrane elements: Anovel platform for eco-friendly biofouling mitigation. *Water Research*, **71**, 171–186.

Wildenschild D. and Sheppard A. P. (2013). X-ray imaging and analysis techniques for quantifying pore-scale structure and processes in subsurface porous medium systems. *Advances in Water Resources*, **51**, 217–246.

Wiley D. E. and Fletcher D. F. (2002). Computational fluid dynamics modelling of flow and permeation for pressure-driven membrane processes. *Desalination*, **145**, 183–186.

Wiley D. E. and Fletcher D. F. (2003). Techniques for computational fluid dynamics modelling of flow in membrane channels. *Journal of Membrane Science*, **211**, 127–137.

Willems P., Deen N. G., Kemperman A. J. B., Lammertink R. G. H., Wessling M., van Sint Annaland M., Kuipers J. A. M. and van der Meer W. G. J. (2010). Use of Particle Imaging Velocimetry to measure liquid velocity profiles in liquid and liquid/gas flows through spacer filled channels. *Journal of Membrane Science*, **362**, 143–153.

Windt C. W., Soltner H., Dusschoten D. V. and Blümler P. (2011). A portable Halbach magnet that can be opened and closed without force: The NMR-CUFF. *Journal of Magnetic Resonance*, **208**, 27–33.

Winters H. and Isquith I. R. (1979). In-plant microfouling in desalination. *Desalination*, **30**, 387–399.

Xi C., Marks D., Schlachter S., Luo W. and Boppart S. A. (2006). High-resolution three-dimensional imaging of biofilm development using optical coherence tomography. *Journal of Biomedical Optics*, **11**, 034001.

Xu P., Bellona C. and Drewes J. E. (2010). Fouling of nanofiltration and reverse osmosis membranes during municipal wastewater reclamation: Membrane autopsy results from pilot-scale investigations. *Journal of Membrane Science*, **353**, 111–121.

Yang X., Beyenal H., Harkin G. and Lewandowski Z. (2000). Quantifying biofilm structure using image analysis. *Journal of Microbiological Methods*, **39**, 109–119.

Yang H.-L., Lin J. C.-T. and Huang C. (2009). Application of nanosilver surface modification to RO membrane and spacer for mitigating biofouling in seawater desalination. *Water Res.*, **43**, 3777–86.

Yang X., Fridjonsson E. O., Johns M. L., Wang R. and Fane A. G. (2014). A non-invasive study of flow dynamics in membrane distillation hollow fiber modules using low-field nuclear magnetic resonance imaging (MRI). *Journal of Membrane Science*, **451**, 46–54.

Yi Z., Zhu L.-P., Xu Y.-Y., Zhao Y.-F., Ma X.-T. and Zhu B.-K. (2010). Polysulfone-based amphiphilic polymer for hydrophilicity and fouling-resistant modification of polyethersulfone membranes. *Journal of Membrane Science*, **365**, 25–33.

Yiantsios S. G. and Karabelas A. J. (2003). Deposition of micron-sized particles on flat surfaces: Effects of hydrodynamic and physicochemical conditions on particle attachment efficiency. *Chemical Engineering Science*, **58**, 3105–3113.

Yiantsios S. G., Sioutopoulos D. and Karabelas A. J. (2005). Colloidal fouling of RO membranes: An overview of key issues and efforts to develop improved prediction techniques. *Desalination*, **183**, 257–272.

Ying W., Gitis V., Lee J. and Herzberg M. (2013). Effects of shear rate on biofouling of reverse osmosis membrane during tertiary wastewater desalination. *Journal of Membrane Science*, **427**, 390–398.

Yu W., Brown M. and Graham N. J. D. (2016). Prevention of PVDF ultrafiltration membrane fouling by coating MnO_2 nanoparticles with ozonation. *Scientific Reports*, **6**, 30144.

Zhang T. C. and Bishop P. L. (1994). Density, porosity, and pore structure of biofilms. *Water Research*, **28**, 2267–2277.

Zhang T. and Klapper I. (2010). Mathematical model of biofilm induced calcite precipitation. *Water Science & Technology*, **61**, 2957–2964.

Zhao D. and Yu S. (2015). A review of recent advance in fouling mitigation of NF/RO membranes in water treatment: Pretreatment, membrane modification, and chemical cleaning. *Desalination and Water Treatment*, **55**, 870–891.

Zhao Y. F., Zhu L. P., Yi Z., Zhu B. K. and Xu Y. Y. (2013). Improving the hydrophilicity and fouling-resistance of polysulfone ultrafiltration membranes via surface zwitterionicalization mediated by polysulfone-based triblock copolymer additive. *Journal of Membrane Science*, **440**, 40–47.

Zhao Y.-H., Zhu B.-K., Kong L. and Xu Y.-Y. (2007). Improving Hydrophilicity and Protein Resistance of Poly(vinylidene fluoride) Membranes by Blending with Amphiphilic Hyperbranched-Star Polymer. *Langmuir*, **23**, 5779–5786.

Zhao Y. H., Wee K.-H. and Bai R. (2010). Highly hydrophilic and low-protein-fouling polypropylene membrane prepared by surface modification with sulfobetaine-based zwitterionic polymer through a combined surface polymerization method. *Journal of Membrane Science*, **362**, 326–333.

Zhu L.-P., Xu Y.-Y., Wei X.-Z. and Zhu B.-K. (2009). Hydrophilic modification of poly(phthalazine ether sulfone ketone) ultrafiltration membranes by the surface immobilization of poly(ethylene glycol) acrylates. *Desalination*, **242**, 96–109.

Zondervan E. and Roffel B. (2007). Evaluation of different cleaning agents used for cleaning ultra filtration membranes fouled by surface water. *Journal of Membrane Science*, **304**, 40–49.

Index